PROBABILITY AND RANDOM PROCESSES FOR ELECTRICAL AND COMPUTER ENGINEERS

SECOND EDITION

T0335502

PROBABILITY AND RANDOM PROCESSES FOR ELECTRICAL AND COMPUTER ENGINEERS

SECOND EDITION

CHARLES W. THERRIEN
MURALI TUMMALA

CRC Press
Taylor & Francis Group
Boca Raton London New York

CRC Press is an imprint of the
Taylor & Francis Group, an **informa** business

CRC Press
Taylor & Francis Group
6000 Broken Sound Parkway NW, Suite 300
Boca Raton, FL 33487-2742

© 2012 by Taylor & Francis Group, LLC
CRC Press is an imprint of Taylor & Francis Group, an Informa business

No claim to original U.S. Government works

Printed in the United States of America on acid-free paper
Version Date: 20111103

International Standard Book Number: 978-1-4398-2698-0 (Hardback)

Library of Congress Cataloging-in-Publication Data

Therrien, Charles W.
    Probability and stochastic processes for electrical and computer engineers / Charles Therrien, Murali Tummala. -- 2nd ed.
        p. cm.
    Previously published under title: Probability for electrical and computer engineers, 2004.
    Includes bibliographical references and index.
    ISBN 978-1-4398-2698-0 (hardback)
    1. Electrical engineering--Mathematics. 2. Computer engineering--Mathematics. 3. Probabilities. 4. Stochastic processes. I. Tummala, Murali. II. Title.

TK153.T44 2011
621.301'5192--dc23   2011034811

Visit the Taylor & Francis Web site at
http://www.taylorandfrancis.com

and the CRC Press Web site at
http://www.crcpress.com

To all the teachers of probability – especially Alvin Drake, who made it fun to learn and Athanasios Papoulis, who made it just rigorous enough for engineers.

Also, to our families for their enduring support.

Contents

Preface to the Second Edition

Several years ago we had the idea to offer a course in basic probability and random vectors for engineering students that would focus on the topics that they would encounter in later studies. As electrical engineers we find there is strong motivation for learning these topics if we can see immediate applications in such areas as binary and cellular communication, computer graphics, music, speech applications, multimedia, aerospace, control, and many more such topics.

The course offered was very successful; it was offered twice a year (in a quarter system) and was populated by students not only in electrical engineering but also in other areas of engineering and computer science. Instructors in higher level courses in communications, control, and signal processing were gratified by this new system because they did not need to spend long hours reviewing, or face blank stares when bringing up the topic of a random variable.

The course, called Probabilistic Analysis of Signals and Systems, was taught mainly from notes, and it was a few years before we came around to writing the first edition of this book. The first edition was successful, and it wasn't long before our publisher at CRC Press was asking for a second edition. True to form, and still recovering from our initial writing pains, it took some time before we actually agreed to sign a contract and even longer before we put down the first new words on paper. The good news is that our original intent has not changed; so we can use most of the earlier parts of the book with suitable enhancements. What's more, we have added some new topics, such as confidence intervals, and reorganized the chapter on random processes so that by itself it can serve as an introduction to this more advanced topic.

In line with these changes, we have retitled this edition to *Probability and Random Processes for Electrical and Computer Engineers* and divided the book into two parts, the second of which focuses on random processes. In the new effort, the book has grown from some 300 pages in the first edition to just over 400 pages in the second edition. We can attribute the increase in length to not only new text but also to added problems and computer assignments. In the second edition, there are almost as many problems as pages, and each chapter has at least two computer assignments for students to work using a very high-level language such as MATLAB®.

Let us highlight some of the changes that appear in the second edition.

- We have introduced a short chapter on random vectors following Chapter 6 on multiple random variables. This not only adds some advanced new material but also supports certain topics that relate to discrete random processes in particular.

- We have reorganized first edition Chapters 7 and 8 on random processes to make these topics more accessible. In particular, these chapters now consist of an introductory chapter on the theory and nature of random processes, a chapter focusing on random processes using first and second moments in the time domain, a chapter based on analysis in the frequency domain, and finally (a topic that appeared in the first edition) Markov and Poisson random processes.

- We have introduced a large number of new problems and computer projects, as mentioned earlier. We hope that instructors will not only welcome the new material but will also be inspired to assign new problems and computer-based investigations based on the material that now appears.

Throughout the text we have tried to maintain the style of the original edition and keep the explanations simple and intuitive. In places where this is more difficult, we have tried to provide adequate references to where a more advanced discussion can be found. We have also tried to make corrections and improvements based on feedback from our students, teachers, and other users. We are looking forward to using the new edition in our classes at the Naval Postgraduate School and hope that students and teachers will continue to express their enthusiasm for learning about a topic that is essential to modern methods of engineering and computer science.

MATLAB® is a registered trademark of The MathWorks, Inc. For product information, please contact: The MathWorks, Inc., 3 Apple Hill Drive, Natick, MA 01760-2098 USA; Tel: 508 647 7000, Fax: 508-647-7001, E-mail: info@mathworks.com, Web: www.mathworks.com.

Charles W. Therrien
Murali Tummala
Monterey, California

Preface

Beginnings

About ten years ago we had the idea to begin a course in probability for students of electrical engineering. Prior to that, electrical engineering graduate students at the Naval Postgraduate School specializing in communication, control, and signal processing were given a basic course in probability in another department and then began a course in random processes within the Electrical and Computer Engineering (ECE) department. ECE instructors consistently found that they were spending far too much time "reviewing" topics related to probability and random variables, and therefore could not devote the necessary time to teaching random processes.

The problem was not with the teachers; we had excellent instructors in all phases of the students' programs. We hypothesized (and it turned out to be true) that engineering students found it difficult to relate to the probability material because they could not see the immediate application to engineering problems that they cared about and would study in the future.

When we first offered the course Probabilistic Analysis of Signals and Systems in the ECE department, it became an immediate success. We found that students became interested and excited about probability and looked forward to (rather than dreading) the follow-on courses in stochastic signals and linear systems. We soon realized the need to include other topics relevant to computer engineering, such as basics of queueing theory. Today nearly every student in Electrical and Computer Engineering at the Naval Postgraduate School takes this course as a prerequisite to his or her graduate studies. Even students who have previously had some exposure to probability and random variables find that they leave with a much better understanding and the feeling of time well spent.

Intent

While we had planned to write this book a number a years ago, events overtook us and we did not begin serious writing until about three years ago; even then we did most of the work in quarters when we were teaching the course. In the meantime we had developed an extensive set of PowerPoint slides which we use for teaching and which served as an outline for the text. Although we benefited during this time from the experience of more course offerings, our intent never changed. That intent is to make the topic interesting (even fun) and *relevant* for students of electrical and computer engineering. In line with this, we have tried to make the text very readable for anyone with a background in first-year calculus, and have included a number of application topics and numerous examples.

As you leaf through this book, you may notice that topics such as the binary communication channel, which are often studied in an introductory course in communication theory, are included in Chapter 2, on the Probability Model. Elements of coding (Shannon-Fano and Huffman) are also introduced in Chapters 2 and 4. A simple introduction to detection and classification is also presented early in the text, in Chapter 3. These topics are provided with both homework problems and computer projects to be carried out in a language such as MATLAB®. (Some special MATLAB functions will be available on the CRC website.)

While we have intended this book to focus primarily on the topic of probability, some other topics have been included that are relevant to engineering. We have included a short chapter on random processes as an introduction to the topic (*not* meant to be complete!). For some students, who will not take a more advanced course in random processes, this may be all they need. The definition and meaning of a random process is also important, however, for Chapter 8, which develops the ideas leading up to queueing theory. These topics are important for students who will go on to study computer networks. Elements of parameter estimation and their application to communication theory are also presented in Chapter 6.

We invite you to peruse the text and note especially how the engineering applications are treated and how they appear as early as possible in the course of study. We have found that this keeps students motivated.

Suggested Use

We believe that the topics in this book should be taught at the earliest possible level. In most universities, this would represent the junior or senior undergraduate level. Although it is possible for students to study this material at an even earlier (e.g., sophomore) level, in this case students may have less appreciation of the applications and the instructor may want to focus more on the theory and only Chapters 1 through 6.

Ideally, we feel that a course based on this book should directly precede a more advanced course on communications or random processes for students studying those areas, as it does at the Naval Postgraduate School. While several engineering books devoted to the more advanced aspects of random processes and systems also include background material on probability, we feel that a book that focuses mainly on probability for engineers with engineering applications has been long overdue.

While we have written this book for students of electrical and computer engineering, we do not mean to exclude students studying other branches of engineering or physical sciences. Indeed, the theory is most relevant to these other disciplines, and the applications and examples, although drawn from our own discipline, are ubiquitous in other areas. We would welcome suggestions for possibly expanding the examples to these other areas in future editions.

Organization

After an introductory chapter, the text begins in Chapter 2 with the algebra of events and probability. Considerable emphasis is given to representation of the sample space for various types of problems. We then move on to random variables, discrete and continuous, and transformations of random variables in Chapter 3. Expectation is considered to be sufficiently important that a separate chapter (4) is devoted to the topic. This chapter also provides a treatment of moments and generating functions. Chapter 5 deals with two and more random variables and includes a short (more advanced) introduction to random vectors for instructors with students having sufficient background in linear algebra who may want to cover this topic. Chapter 6 groups together topics of convergence, limit theorems, and parameter estimation. Some of these topics are typically taught at a more advanced level, but we feel that introducing them at this more basic level and relating them to earlier topics in the text helps students appreciate the need for a strong theoretical basis to support practical engineering applications.

Chapter 7 provides a brief introduction to random processes and linear systems, while Chapter 8 deals with discrete and continuous Markov processes and an introduction to queueing theory. Either or both of these chapters may be skipped for a basic level course in probability; however, we have found that both are useful even for students who later plan to take a more advanced course in random processes.

As mentioned above, applications and examples are distributed throughout the text. Moreover, we have tried to introduce the applications at the earliest opportunity, as soon as the supporting probabilistic topics have been covered.

Acknowledgments

First and foremost, we want to acknowledge our students over many years for their inspiration and desire to learn. These graduate students at the Naval Postgraduate School, who come from many countries as well as the United States, have always kept us "on our toes" academically and close to reality in our teaching. We should also acknowledge our own teachers of these particular topics, who helped us learn and thus set directions for our careers. These teachers have included Alvin Drake (to whom this book is dedicated in part), Wilbur Davenport, and many others.

We also should acknowledge colleagues at the Naval Postgraduate School, both former and present, whose many discussions have helped in our own teaching and presentation of these topics. These colleagues include John M. ("Jack") Wozencraft, Clark Robertson, Tri Ha, Roberto Cristi, (the late) Richard Hamming, Bruno Shubert, Donald Gaver, and Peter A. W. Lewis.

We would also like to thank many people at CRC for their help in publishing this book, especially Nora Konopka, our acquisitions editor. Nora is an extraordinary person, who was enthusiastic about the project from the get-go and was never too busy to help with our many questions during the writing and production. In addition, we could never have gotten to this stage without the help of Jim Allen, our technical editor of many years. Jim is an expert in LaTeX and other publication software and has been associated with our teaching of Probabilistic Analysis of Signals and Systems long before the first words of this book were entered in the computer.

Finally, we acknowledge our families, who have always provided moral support and encouragement and put up with our many hours of work on this project. We love you!

Charles W. Therrien
Murali Tummala
Monterey, California

PART I

Probability and Random Variables

1 Introduction

In many situations in the real world, the outcome is uncertain. For example, consider the event that it rains in New York on a particular day in the spring. No one will argue with the premise that this event is uncertain, although some people may argue that the likelihood of the event is increased if one forgets to bring an umbrella.

In engineering problems, as in daily life, we are surrounded by events whose occurrence is either uncertain or whose outcome cannot be specified by a precise value or formula. The exact value of the power line voltage during high activity in the summer, the precise path that a space object may take upon reentering the atmosphere, and the turbulence that a ship may experience during an ocean storm are all examples of events that cannot be described in any deterministic way. In communications (which includes computer communications as well as personal communications over devices such as cellular phones), the events can frequently be reduced to a series of binary digits (0s and 1s). However, it is the *sequence* of these binary digits that is uncertain and carries the information. After all, if a particular message represented an event that occurs with complete certainty, why would we ever have to transmit the message?

The most successful method to date to deal with uncertainty is through the application of probability and its sister topics of statistics and random processes. These are the subjects of this book. In this chapter, we visit a few of the engineering applications for which methods of probability play a fundamental role.

1.1 The Analysis of Random Experiments

A model for the analysis of random experiments is depicted in Fig. 1.1.

This will be referred to as the basic probability model. The analysis begins with a "random experiment," i.e., an experiment in which the outcome is uncertain. All of the possible outcomes of the experiment can be listed (either explicitly or conceptually); this complete collection of outcomes comprises what is called the *sample space*. Events of interest are "user-defined," i.e., they can be defined in any way that supports analysis of the experiment. Events, however, must always have a representation in the sample space. Thus formally, an *event* is always a single outcome or a collection of outcomes from the sample space.

Probabilities are real numbers on a scale of 0 to 1 that represent the likelihood of events. In everyday speech we often use percentages to represent likelihood. For example, the morning weather report may state that "there is a 60% chance of rain." In a probability model, we would say that the probability of rain is 0.6. Numbers closer to 1 indicate more likely events while numbers closer to 0 indicate more unlikely events. (*There are no negative probabilities!*) The probabilities assigned to outcomes in the sample space should satisfy our intuition about the experiment and make good sense. (Frequently the outcomes are equally likely.) The probability of other events germane to the problem is then computed by rules which are studied in Chapter 2.

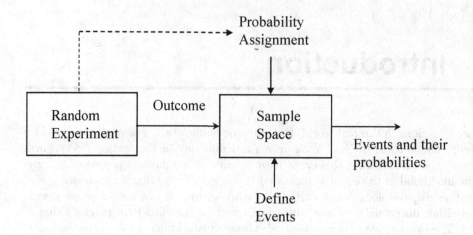

Figure 1.1 The probability model.

Some further elements of probabilistic analysis are random variables and random processes. In many cases there is an important numerical value resulting from the performance of a random experiment. That numerical value may be considered to be the outcome of the experiment, in which case the sample space could be considered to be the set of integers or the real line. Frequently the experiment is more basic, however, and the numerical value, although important, can be thought of as a derived result. If the sample space is chosen as the collection of these more basic events, then the number of interest is represented by a *mapping* from the sample space to the real line. Such a mapping is called a *random variable*. If the mapping is to n-dimensional Euclidean space the mapping is sometimes called a *random vector*. It is generally sufficient to deal with a scalar mapping, however, because a random vector can be considered to be an ordered set of random variables. A *random process* is like a random variable except each outcome in the sample space is mapped to a function of time (see Fig. 1.2). Random processes serve as appropriate models of many signals in control and communications as well as the interference that we refer to as noise. Random processes also are useful for representing the traffic and changes that occur in communications networks. Thus the study of probability, random variables, and random processes is an important component of engineering education.

The following example illustrates the concept of a random variable.

Example 1.1: Consider the rolling of a pair of dice at a game in Las Vegas. The sample
space could be represented as the list of pairs of numbers

$$(1,1), (1,2), \ldots, (6,6)$$

indicating what is shown on the dice. Each such pair represents an *outcome* of the
experiment. Notice that the outcomes are distinct; that is, if one of the outcomes occurs
the others do not. An important numerical value associated with this experiment is
the "number rolled," i.e., the total number of dots showing on the two dice. This can
be represented by a random variable N. It is a mapping from the sample space to the
real line since a value of N can be computed for each outcome in the sample space.

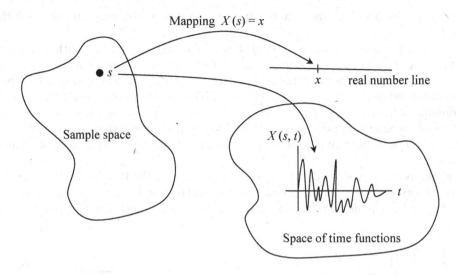

Figure 1.2 Random variables and random processes.

For example, we have the following mappings:

outcome		N
$(1, 2)$	$\longrightarrow$	3
$(2, 3)$	$\longrightarrow$	5
$(4, 1)$	$\longrightarrow$	5

The sample space chosen for this experiment contains more information than just the number N. For example, we can tell if "doubles" are rolled, which may be significant for some games.

☐

Another reason for modeling an experiment to include a random variable is that the necessary probabilities can be chosen more easily. In the game of dice, it is obvious that the outcomes (1,1), (1,2), ..., (6,6) should each have the same probability (unless someone is cheating), while the possible values of N are not equally likely because more than one outcome can map to the same value of N (see example above). By modeling N as a random variable, we can take advantage of well-developed procedures to compute its probabilities.

1.2 Probability in Electrical and Computer Engineering

In this section, a few examples related to the fields of electrical and computer engineering are presented to illustrate how a probabilistic model is appropriate. You can probably think of many more applications in these days of rapidly advancing electronic and information technology.

1.2.1 Signal detection and classification

A problem of considerable interest to electrical engineers is that of signal detection. Whether dealing with conversation on a cellular phone, data on a computer line, or scattering from a radar target, it is important to detect and properly interpret the signal.

In one of the simplest forms of this problem, we may be dealing with a message in the form of digital data sent to a radio or telephone where we want to decode the message bit by bit. In this case the transmitter sends a logical 0 or 1, and the receiver must decide whether a 0 or a 1 has been sent. The waveform representing the bits is modulated onto an analog radio frequency (RF) carrier signal for transmission. In the process of transmission, noise and perhaps other forms of distortion are applied to the signal so the waveform that arrives at the receiver is not an exact replica of what is transmitted.

The receiver consists of a demodulator, which processes the RF signal, and a decoder which observes the analog output of the demodulator and makes a decision: 0 or 1 (see Fig. 1.3). The demodulator as depicted does two things: First, it removes the

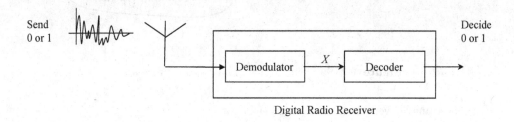

Digital Radio Receiver

Figure 1.3 Reception of a digital signal.

RF carrier from the received waveform; secondly, it compares the remaining signal to a prototype through a process known as "matched filtering." The result of this comparison is a single value X which is provided to the decoder.

Several elements of probability are evident in this problem. First of all, from the receiver's point of view, the transmission of a 0 or 1 is a random event. (If the receiver knew for certain what the transmission was going to be, there would be no need to transmit at all!) Secondly, the noisy waveform that arrives at the receiver is a random process because it represents a time function that corresponds to random events. The output of the demodulator is a real number whose value is determined by uncertain events. Thus it can be represented by a random variable. Finally, because of the uncertainty in the rest of the situation, the output decision (0 or 1) is itself a random event.

The performance of the receiver also needs to be measured probabilistically. In the best situation, the decision rule that the receiver uses will minimize the *probability* of making an error.

Elements of this problem have been known since the early part of the last century and are dealt with in Chapters 3 and 5 of the text. The core of results lies in the field that has come to be known as *statistical communication theory* [1].

1.2.2 Speech modeling and recognition

The engineering analysis of human speech has been going on for decades. Recently significant progress has been made thanks to the availability of small powerful computers and the development of advanced methods of signal processing. To grasp the complexity of the problem, consider the event of speaking a single English word, such as the word "hello," illustrated in Fig. 1.4.

Although the waveform has clear structure in some areas, it is obvious that there are significant variations in the details from one repetition of the word to another. Segments of the waveform representing distinct sounds or "phonemes" are typically modeled by a random process, while the concatenation of sounds and words to produce meaningful speech also follows a probabilistic model. Techniques have advanced through research such that synthetic speech *production* is quite good. Speech *recognition*, however, is still in a relatively embryonic stage although recognition in applications with a limited vocabulary such as telephone answering systems and vehicle navigation systems are commercially successful. In spite of difficulties of the problem, however, speech may become the primary interface for man/machine interaction, ultimately replacing the mouse and keyboard in present day computers [2]. In whatever form the future research takes, you can be certain that probabilistic methods will play an important if not central role.

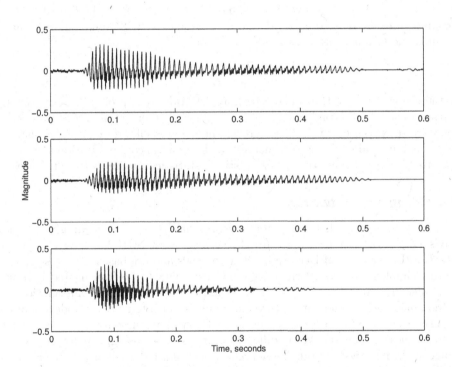

Figure 1.4 Waveform corresponding to the word "hello" (three repetitions, same speaker).

1.2.3 Coding and data transmission

The representation of information as digital data for storage and transmission has been rapidly replacing older analog methods. The technology is the same whether the data represents speech, music, video, telemetry, or a combination (multimedia). Simple fixed-length codes such as the ASCII standard are inefficient because of the considerable redundancy that exists in most information that is of practical interest.

In the late 1940s Claude Shannon [3, 4] studied fundamental methods related to the representation and transmission of information. This field, which later became known as *information theory*, is based on probabilistic methods. Shannon showed that given a set of symbols (such as letters of the alphabet) and their probabilities of occurrence in a message, one could compute a probabilistic measure of "information" for a source producing such messages. He showed that this measure of information, known as entropy H and measured in "bits," provides a lower bound on the average number of binary digits needed to code the message. Others, such as David Huffman, developed algorithms for coding the message into binary bits that approached the lower bound.[1] Some of these methods are introduced in Chapter 2 of the text.

Shannon also developed fundamental results pertaining to the communication channel over which a message is to be transmitted. A message to be transmitted is subject to errors made due to interference, noise, self interference between adjacent symbols, and many other effects. Putting all of these details aside, Shannon proposed that the channel could be characterized in a probabilistic sense by a number C, known as the *channel capacity*, and measured in bits. If the source produces a symbol every T_s seconds and that symbol is sent over the channel in T_c seconds, then the source rate and the channel rate are defined as H/T_s and C/T_c respectively. Shannon showed that if the source and channel rates satisfy the condition

$$\frac{H}{T_s} \leq \frac{C}{T_c}$$

then the message can be transmitted with an arbitrarily low probability of error. On the other hand, if this condition does not hold, then there is no way to achieve a low error rate on the channel. These ground-breaking results developed by Shannon became fundamental to the development of modern digital communication theory and emphasized the importance of a probabilistic or statistical framework for the area.

1.2.4 Computer networks

It goes without saying that computer networks have become exceedingly complex. Today's modern computer networks are descendents of the ARPAnet research project by the U.S. Department of Defense in the late 1960s and are based on the store-and-forward technology developed there. A set of user data to be transmitted is broken into smaller units or packets together with addressing and layering information and sent from one machine to another through a number or intermediate nodes (computers). The nodes provide routing to the final destination. In some cases the packets of a single message may arrive by different routes and are assembled to produce the message at the final destination. There are a variety of problems in computer networks including packet loss, packet delay, and packet buffer size that require the application of probabilistic and statistical concepts. In particular, the concepts of queueing systems, which preceded the development of computer networks, have been aptly applied to these systems by early researchers such a Kleinrock [5, 6]. Today, electrical and computer engineers study those principles routinely.

The analysis of computer networks deals with random processes that are *discrete* in nature. That is, the random processes take on finite discrete values as a function of time. Examples are the number of packets in the system (as a function of time), the length of a queue, or the number of packets arriving in some given period of time. Various random variables and statistical quantities also need to be dealt with such

---

[1] Huffman developed the method as a term paper in a class at M.I.T.

as the average length of a queue, the delay encountered by packets from arrival to departure, measures of throughput versus load on the system, and so on.

The topic of queuing brings together many probabilistic ideas that are presented in the book and so is dealt with in the last chapter. The topic seems to underscore the need for a knowledge of probability and statistics in almost all areas of the modern curriculum in electrical and computer engineering.

1.3 Outline of the Book

The overview of the probabilistic model in this chapter is meant to depict the general framework of ideas that are relevant for this area of study and to put some of these concepts in perspective. The set of examples is meant to show that this topic, which is rich in mathematics, is extremely important for engineers.

The chapters that follow attempt to present the study of probability in a context that continues to show its importance for engineering applications. Applications such as those cited above are discussed explicitly in each chapter. Our goal has been to bring in the applications as early as possible even if this requires considerable simplification of a real-world problem. We have also tried to keep the discussion friendly but not *totally* lacking in rigor, and to occasionally inject some light humor (especially in the problems).

From this point on, the book consists of two parts. The first part deals with probability and random variables while the second part deals with random processes.

Chapter 2, the first chapter in Part I, begins with a discussion of the probability model, events, and probability measure. This is followed in Chapter 3 by a discussion of random variables: continuous and discrete. Averages, known as statistical expectation, form an important part of the theory and are discussed in a fairly short Chapter 4. Chapter 5 then deals with multiple random variables, joint and conditional distributions, expectation, and sums of random variables. The topics of theorems, bounds, and estimation are left to fairly late in the book, in Chapter 6. While these topics are important, we have decided to present the more application-oriented material first. Chapter 7 presents some basic material on random vectors which is useful for more advanced topic in random processes and concludes Part I.

Chapter 8 begins Part II and provides an introduction to the concept of a random process. Both continuous time-processes and discrete-time processes are discussed with examples. Chapters 9 and 10 then discuss the analysis of random signals in the time domain and the frequency domain respectively. The discussion emphasizes use of correlation and power spectral density for many typical applications in signal processing and communications.

Chapter 11 deals with the special Markov and Poisson random processes that appear in models of network traffic, service and related topics. This chapter provides the mathematical tools for the discussion of queueing models which are used in these applications.

References

[1] David Middleton. *An Introduction to Statistical Communication Theory*. McGraw-Hill, New York, 1960.

[2] Michael L. Dertouzos. *What WILL be*. HarperCollins, San Francisco, 1997.

[3] Claude E. Shannon. A mathematical theory of communication. *Bell System Technical Journal*, 27(3):379–422, July 1948. (See also [7].).

[4] Claude E. Shannon. A mathematical theory of communication (concluded). *Bell System Technical Journal*, 27(4):623–656, October 1948. (See also [7].).

[5] Leonard Kleinrock. *Queueing Systems - Volume I: Theory.* John Wiley & Sons, New York, 1975.

[6] Leonard Kleinrock. *Queueing Systems - Volume II: Computer Applications.* John Wiley & Sons, New York, 1976.

[7] Claude E. Shannon and Warren Weaver. *The Mathematical Theory of Communication.* University of Illinois Press, Urbana, IL, 1963.

2 The Probability Model

This chapter describes a well-accepted model for the analysis of random experiments which is known as the *Probability Model*. The chapter also defines an algebra suitable for describing sets of events, and shows how measures of likelihood or *probabilities* are assigned to these events. Probabilities provide quantitative numerical values to the likelihood of occurrence of events.

Events do not always occur independently. In fact, it is the very *lack* of independence that allows inference of one fact from another. Here a mathematical meaning is given to the concept of independence and methods are developed to deal with probabilities when events are or are not independent.

Several illustrations and examples are provided throughout this chapter on basic probability. In addition, a number of applications of the theory to some basic electrical engineering problems are given to provide motivation for further study of this topic and those to come.

2.1 The Algebra of Events

It is seen in Chapter 1 that the collection of all possible outcomes of a random experiment comprise the *sample space*. Outcomes are members of the sample space and events of interest are represented as *sets* (see Fig. 2.1). In order to describe these events

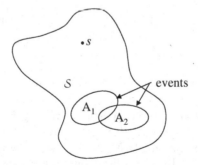

Figure 2.1 Abstract representation of the sample space S with element s and sets A_1 and A_2 representing events.

and compute their probabilities in a consistent manner it is necessary to have a formal representation for operations involving events. More will be said about representing the sample space in Section 2.1.2; for now, let us focus on the methods for describing relations among events.

2.1.1 Basic operations

In analyzing the outcome of a random experiment, it is usually necessary to deal with events that are derived from other events. For example, if A is an event, then A^c, known as the *complement* of A, represents the event that "A did not occur." The complement of the sample space is known as the *null event*, $\emptyset = S^c$. The operations of multiplication and addition are used to represent certain combinations of events (known as intersections and unions in set theory). The statement "$A_1 \cdot A_2$," or simply "$A_1 A_2$" represents the event that *both* event A_1 and event A_2 have occurred (intersection), while the statement "$A_1 + A_2$" represents the event that *either* A_1 or A_2 *or both* have occurred (union).[1]

Since complements and combinations of events are themselves events, a formal structure for representing events and derived events is needed. This formal structure is in fact a set of sets known in mathematics as an *algebra* or a *field* and referred to here as the *algebra of events*.[1] Table 2.1 lists the two postulates that define an algebra $\mathcal{A}$.

1. If $A \in \mathcal{A}$ *then* $A^c \in \mathcal{A}$.
2. If $A_1 \in \mathcal{A}$ *and* $A_2 \in \mathcal{A}$ *then* $A_1 + A_2 \in \mathcal{A}$.

Table 2.1   Postulates for an algebra of events.

Table 2.2 lists seven axioms that define the properties of the operations. Together these

$A_1 A_1{}^c = \emptyset$	Mutual exclusion
$A_1 S = A_1$	Inclusion
$(A_1{}^c)^c = A_1$	Double complement
$A_1 + A_2 = A_2 + A_1$	Commutative law
$A_1 + (A_2 + A_3) = (A_1 + A_2) + A_3$	Associative law
$A_1(A_2 + A_3) = A_1 A_2 + A_1 A_3$	Distributive law
$(A_1 A_2)^c = A_1{}^c + A_2{}^c$	DeMorgan's law

Table 2.2 Axioms of operations on events.

tables can be used to show all of the properties of the algebra of events. For example, the postulates state that the event $A_1 + A_2$ is included in the algebra. The postulates in conjunction with the last axiom (DeMorgan's law) show that the event "$A_1 A_2$" is also included in the algebra. Table 2.3 lists some other handy identities that can be derived from the axioms and the postulates. You will find that many of the results in Tables 2.2 and 2.3 are used either implicitly or explicitly in solving problems involving events and their probability. Notice especially the two distributive laws; addition is distributive over multiplication (Table 2.3) as well as *vice versa* (Table 2.2).

Since the events "$A_1 + A_2$" and "$A_1 A_2$" are included in the algebra, it is easy to

---

[1] The operations represented as multiplication and addition are commonly represented with the intersection ∩ and union ∪ symbols. Except for the case of multiple such operations, we will adhere to the notation introduced above.

$$S^c = \emptyset$$

$A_1 + \emptyset = A_1$	Inclusion
$A_1 A_2 = A_2 A_1$	Commutative law
$A_1(A_2 A_3) = (A_1 A_2)A_3$	Associative law
$A_1 + (A_2 A_3) = (A_1 + A_2)(A_1 + A_3)$	Distributive law
$(A_1 + A_2)^c = A_1{}^c A_2{}^c$	DeMorgan's law

Table 2.3 Additional identities in the algebra of events.

show by induction for any finite number of events A_i, $i = 1, 2, \ldots, N$, that the events

$$\bigcup_{i=1}^{N} A_i = A_1 + A_2 + \cdots + A_N$$

and

$$\bigcap_{i=1}^{N} A_i = A_1 A_2 \cdots A_N$$

are also included in the algebra. In many cases it is important that the sum and product of a countably infinite number of events have a representation in the algebra. For example, suppose an experiment consists of measuring a random voltage, and the events A_i are defined as "$i - 1 \leq$ voltage $< i$; $i = 1, 2, \ldots$." Then the (infinite) sum of these events, which is the event "voltage ≥ 0," should be in the algebra. An algebra that includes the sum and product of an infinite number of events, that is,

$$\bigcup_{i=1}^{\infty} A_i = A_1 + A_2 + A_3 + \cdots$$

and

$$\bigcap_{i=1}^{\infty} A_i = A_1 A_2 A_3 \cdots$$

is called a sigma-algebra or a sigma-field. The algebra of events is defined to be such an algebra.

Since the algebra of events can be thought of as an algebra of sets, events are often represented as Venn diagrams. Figure 2.2 shows some typical Venn diagrams for a sample space and its events. The notation '$\subset$' is used to mean one event is "contained" in another and is defined by

$$A_1 \subset A_2 \iff A_1 A_2{}^c = \emptyset \tag{2.1}$$

2.1.2 Representation of the sample space

Since students of probability often have difficulty defining the an appropriate sample space for an experiment, it is worthwhile to spend a little more time on this concept.

Let us begin with two more ideas from the algebra of events. Let $A_1, A_2, A_3, \ldots$ be a finite or countably infinite set of events with the following properties:

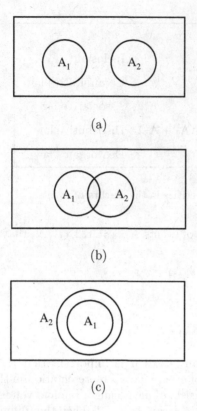

(a)

(b)

(c)

Figure 2.2 Venn diagram for events. (a) Events with no commonality $(A_1 A_2 = \emptyset)$. (b) Events with some commonality $(A_1 A_2 \neq \emptyset)$. (c) One event contained in another $(A_1 \subset A_2)$.

1. The events are *mutually exclusive*. This means that only one event can occur at a time, i.e., the occurrence of one event precludes the occurrence of other events. Equivalently,

$$A_i A_j = \emptyset \text{ for } i \neq j$$

2. The events are *collectively exhaustive*. In other words, one of the events A_i must always occur. That is,

$$A_1 + A_2 + A_3 + \cdots = \mathcal{S}$$

A set of events that has *both* properties is referred to as a *partition*.

Now, in defining the sample space it is assumed that the outcomes of the experiment satisfy these two properties. But there is more. For an experiment with discrete (countable) outcomes, the following provides a working definition of the sample space [1]:

The Sample Space is represented by the finest-grain, mutually exclusive, collectively exhaustive set of outcomes for an experiment.

You can see that the elements of the sample space have the properties of a partition; however, the outcomes defining the sample space must also be *"finest-grain."* While the choice of a finest-grain representation depends on the experiment and the questions to be answered, this property is very important. Without this property it may not be possible to represent all of the events of interest in the experiment. A Venn diagram is generally not sufficient to represent the sample space in solving problems because the representation usually needs to be more explicit.

A discrete sample space may be just a listing of the possible experimental outcomes as in Example 1.1 or it could take the form of some type of diagram. For example, in rolling of a pair of dice the sample space could be drawn as in Fig. 2.3, where the numbers 1 through 6 represent the number of dots showing on each individual die. The

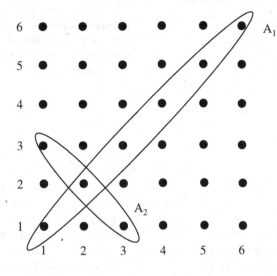

Figure 2.3 Sample space corresponding to a roll of the dice. A_1 is the event "rolling doubles"; A_2 is the event "rolling a '4'."

black dots represent the outcomes of the experiment, which are mutually exclusive and collectively exhaustive. Some more complex events, such as "rolling doubles" are also shown in the figure. It will be seen later that the probabilities of such more complicated events can be computed by simply adding the probabilities of the outcomes of which they are comprised.

As an example related to electrical and computer engineering, consider the transmission of N consecutive binary bits over a noisy communication channel (see Fig. 2.4). Because of the noise, the input and output bit sequences may not be identical. With $N = 2$ consecutive bits one can represent the four distinct symbols required in

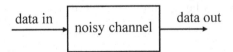

Figure 2.4 Binary transmission subject to noise.

the digital modulation schemes known as QPSK and QFSK.[2]

Figure 2.5 shows the sample space drawn for this case. Input symbols are shown along the horizontal axis and output symbols are shown along the vertical axis. The solid black points represent outcomes with no errors while the points denoted by open circles and squares represent outcomes with one and two errors respectively. As in the previous example, events of interest are represented by collections of points in the sample space. The important *event* "no errors" is represented by the encircled set of solid points in the figure.

The sample space for an experiment could also be continuous; i.e., it could consist

---

[2] Quadrature Phase Shift Keying and Quadrature Frequency Shift Keying.

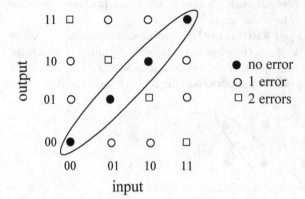

Figure 2.5 Sample space corresponding to transmission of two consecutive binary bits over a noisy channel. Each point in the sample space represents an outcome of the experiment. The set of encircled points represents the event "no errors."

of an uncountably infinite set of outcomes. For example, the phase of the sinusoidal carrier in a communication system may be random and take on values in the interval $[0, \pi)$. This semi-closed interval would represent the sample space for the experiment of measuring the phase of the received signal. The assignment of probability to outcomes in a continuous sample space, however, is not so straightforward as it is for a discrete sample space. Therefore most problems where the outcomes are real numbers are handled through the use of random variables where a complete set of tools and methods are available for solution. These tools and methods are discussed in Chapter 3.

2.2 Probability of Events

2.2.1 Defining probability

We have seen that probability represents the likelihood of occurrence of events. The probability model, when properly formulated, can be used to show that the *relative frequency* for the occurrence of an event in a large number of repetitions of the experiment, defined as

$$\text{relative frequency} = \frac{\text{number of occurrences of the event}}{\text{number of repetitions of the experiment}}$$

converges to the probability of the event. Although probability could be *defined* in this way, it is more common to use the axiomatic development given below.

Probability is conveniently represented in a Venn diagram if you think of the area covered by events as measures of probability. For example, if the area of the sample space S is normalized to one, then the area of overlap of events A_1 and A_2 in Fig. 2.2(b) can be thought of as representing the probability of the event $A_1 A_2$. If the probability of this joint event were larger, the events might be drawn to show greater overlap.

Probability can be defined formally by the following axioms:

(I) The probability of any event is nonnegative.

$$\Pr[A] \geq 0 \qquad (2.2)$$

(II) The probability of the universal event (i.e., the entire sample space) is 1.

$$\Pr[\mathcal{S}] = 1 \qquad (2.3)$$

(III) If A_1 and A_2 are mutually exclusive, then

$$\Pr[A_1 + A_2] = \Pr[A_1] + \Pr[A_2] \quad (\text{if } A_1 A_2 = \emptyset) \qquad (2.4)$$

(IV) If $\{A_i\}$ represent a countably infinite set of mutually exclusive events, then

$$\Pr\left[\bigcup_{i=1}^{\infty} A_i\right] = \sum_{i=1}^{\infty} \Pr[A_i] \quad (\text{if } A_i A_j = \emptyset \quad i \neq j) \qquad (2.5)$$

Although the additivity of probability for any finite set of disjoint events follows from (III), the property has to be stated explicitly for an infinite set in (IV). These axioms and the algebra of events can be used to show a number of other properties, some of which are discussed below.

From axioms (II) and (III), the probability of the complement of an event is

$$\Pr[A^c] = 1 - \Pr[A] \qquad (2.6)$$

Since by (I) the probability of any event is greater than or equal to zero, it follows from (2.6) that $\Pr[A] \leq 1$; thus

$$0 \leq \Pr[A] \leq 1 \qquad (2.7)$$

for any event A.

If A_1 is contained in A_2 ($A_1 \subset A_2$), then A_2 can be written as $A_2 = A_1 + A_1{}^c A_2$ (see Fig. 2.2(c)). Since the events A_1 and $A_1{}^c A_2$ are mutually exclusive, it follows from (III) and (I) that

$$\Pr[A_2] \geq \Pr[A_1]$$

From (2.6) and axiom (II), it follows that the probability of the null event is zero:

$$\Pr[\emptyset] = 0 \qquad (2.8)$$

Thus it also follows that if A_1 and A_2 are mutually exclusive, then $A_1 A_2 = \emptyset$ and consequently

$$\Pr[A_1 A_2] = 0$$

If events A_1 and A_2 are not mutually exclusive, i.e., they may occur together, then the following general relation applies

$$\Pr[A_1 + A_2] = \Pr[A_1] + \Pr[A_2] - \Pr[A_1 A_2] \qquad (2.9)$$

This is not an addition property; rather it can be derived using axioms (I) through (IV) and the algebra of events. It can be intuitively justified on the grounds that in summing the probabilities of the event A_1 and the event A_2, one has counted the common event "$A_1 A_2$" twice (see Fig. 2.2(b)). Thus the probability of the event "$A_1 A_2$" must be subtracted to obtain the probability of the event "$A_1 + A_2$."

These various derived properties are summarized in Table 2.4 below. It is a useful exercise to depict these properties (and the axioms as well) as Venn diagrams.

As a final consideration, let $A_1, A_2, A_3, \ldots$ be a finite or countably infinite set of

$$\Pr[A^c] = 1 - \Pr[A]$$

$$0 \leq \Pr[A] \leq 1$$

If $A_1 \subseteq A_2$ then $\Pr[A_1] \leq \Pr[A_2]$

$$\Pr[\emptyset] = 0$$

If $A_1 A_2 = \emptyset$ then $\Pr[A_1 A_2] = 0$

$$\Pr[A_1 + A_2] = \Pr[A_1] + \Pr[A_2] - \Pr[A_1 A_2]$$

Table 2.4 Some corollaries derived from the axioms of probability.

mutually exclusive and collectively exhaustive events.[3] Recall that such a set of events is referred to as a *partition*; the events cover the sample space like pieces in a puzzle (see Fig. 2.6). The probabilities of the events in a partition satisfy the relation

$$\sum_i \Pr[A_i] = 1 \tag{2.10}$$

and, if B is any other event, then

$$\sum_i \Pr[A_i B] = \Pr[B] \tag{2.11}$$

The latter result is referred to as the *principle of total probability*; it is frequently used in solving problems. The relation (2.11) is illustrated by a Venn diagram in Fig. 2.6. The event B is comprised of all of the pieces that represent intersections or overlap of event B with the events A_i.

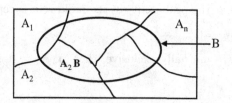

Figure 2.6 Venn diagram illustrating the principle of total probability.

Let us consider the following example to illustrate the formulas in this section.

Example 2.1: Simon's Surplus Warehouse has large barrels of mixed electronic components (parts) that you can buy by the handful or by the pound. You are not allowed to select parts individually. Based on your previous experience, you have determined that in one barrel, 29% of the parts are bad (faulted), 3% are bad resistors, 12% are good resistors, 5% are bad capacitors, and 32% are diodes. You decide to assign probabilities based on these percentages. Let us define the following events:

[3] see Section 2.1.2.

Event	Symbol
Bad (faulted) component	F
Good component	G
Resistor	R
Capacitor	C
Diode	D

A Venn diagram representing this situation is shown below along with probabilities of various events as given:

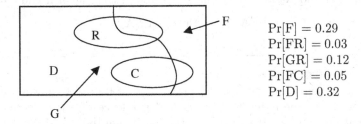

$$Pr[F] = 0.29$$
$$Pr[FR] = 0.03$$
$$Pr[GR] = 0.12$$
$$Pr[FC] = 0.05$$
$$Pr[D] = 0.32$$

We can now answer a number of questions.

1. *What is the probability that a component is a resistor (either good or bad)?*
   Since the events F and G form a partition of the sample space, we can use the principle of total probability (2.11) to write
   $$Pr[R] = Pr[GR] + Pr[FR] = 0.12 + 0.03 = 0.15$$

2. Assume you have no use for either defective parts or resistors. *What is the probability that a part is either defective and/or a resistor?*
   Using (2.9) and the previous result we can write
   $$Pr[F + R] = Pr[F] + Pr[R] - Pr[FR] = 0.29 + 0.15 - 0.03 = 0.41$$

3. *What is the probability that a part is useful to you?*
   Let U represent the event that the part is useful. Then (see (2.6))
   $$Pr[U] = 1 - Pr[U^c] = 1 - 0.41 = 0.59$$

4. *What is the probability of a bad diode?*
   Observe that the events R, C, and G form a partition, since a component has to be one and only one type of part. Then using (2.11) we write
   $$Pr[F] = Pr[FR] + Pr[FC] + Pr[FD]$$
   Substituting the known numerical values and solving yields
   $$0.29 = 0.03 + 0.05 + Pr[FD] \quad \text{or} \quad Pr[FD] = 0.21$$

□

It is worthwhile to consider what an appropriate representation of the sample space would be for this example. While the Venn diagram shown above represents the sample space in an abstract way, a more explicit representation is most useful. In this case, since a part can be a bad resistor, good resistor, bad capacitor and so on, a suitable representation is the list of outcomes:

sample space: { FR   GR   FC   GC   FD   GD }

(You may want to check that this satisfies the requirements discussed in Section 2.1.2.) In later sections of this chapter, you will see that the answers to the four questions posed in this example can be easily obtained if the probabilities of these six outcomes were specified or could be computed. In the example however, the probabilities of all of these experimental outcomes are not known, i.e., only partial information is given.

2.2.2 *Statistical independence*

There is one more concept that is frequently used when solving basic problems in probability, but cannot be derived from either the algebra of events or any of the axioms. Because of its practical importance in solving problems, we introduce this concept early in this discussion of probability:

> Two events A_1 and A_2 are said to be *statistically independent* if and only if
>
> $$\Pr[A_1 A_2] = \Pr[A_1] \cdot \Pr[A_2] \tag{2.12}$$

That is, for two independent events, the probability of both occuring is the product of the probabilities of the individual events. Independence of events is not generally something that you are asked to *prove* (although it may be). More frequently it is an assumption made when the conditions of the problem warrant it. The idea can be extended to multiple events. For example, if A_1, A_2 and A_3 are said to be *mutually independent* if and only if

$$\Pr[A_1 A_2 A_3] = \Pr[A_1] \Pr[A_2] \Pr[A_3]$$

Note also that for independent events, (2.9) becomes

$$\Pr[A_1 + A_2] = \Pr[A_1] + \Pr[A_2] - \Pr[A_1] \Pr[A_2]$$

so the computation of probability for the union of two events is also simplified.

As mentioned, the concept of statistical independence cannot be derived from anything presented so far, and does not have a convenient interpretation in terms of a Venn diagram. It can be justified, however, in terms of the relative frequency interpretation of probability. Suppose two events are "independent" in that they arise from two different experiments that have nothing to do with each other. Suppose further that in N_1 repetitions of the first experiment there are k_1 occurrences of the event A_1 and in N_2 repetitions of the second experiment there are k_2 occurrences of the event A_2. If N_1 and N_2 are sufficiently large, then the relative frequencies k_i/N_i remain approximately constant as N_1 and N_2 are increased. Let us now perform both experiments together a total of $N_1 N_2$ times. Consider the event A_1. Since it occurs k_1 times in N_1 repetitions of the experiment it will occur $N_2 k_1$ times in $N_1 N_2$ repetitions of the experiment. Now consider those $N_2 k_1$ cases where A_1 occured. Since event A_2 occurs k_2 times in N_2 repetitions, it will occur $k_1 k_2$ times in these $N_2 k_1$ cases where A_1 has occured. In other words, the two events occur together $k_1 k_2$ times in all of these $N_1 N_2$ repetitions of the experiments. The relative frequency for the occurrence of the two events together is therefore

$$\frac{k_1 k_2}{N_1 N_2} = \frac{k_1}{N_1} \cdot \frac{k_2}{N_2}$$

which is the product of the relative frequencies of the individual events. So given the relative frequency interpretation of probability, the definition (2.12) makes good sense.

2.3 *Some Applications*

Let us continue with some examples in which many of the ideas discussed so far in this chapter are illustrated.

2.3.1 Repeated independent trials

Many problems involve a repetition of independent events. As a typical example of this, consider the experiment of tossing a coin three times in succession. The result of the second toss is independent of the result of the first toss; likewise the result of the third toss is independent of the result of the first two tosses. Let us denote the probability of a "head" (H) on any toss by p and the probability of a "tail" (T) by $q = 1 - p$. (For a fair coin, $p = q = 1/2$, but let us be more general.) Since the results of the tosses are independent, the probability of any experimental outcome such as HHT is simply the product of the probabilities: $p \cdot p \cdot q = p^2 q$. The sequence HTH has the same probability: $p \cdot q \cdot p = p^2 q$. Experiments of this type are said to involve *repeated independent trials*.

An application which is familiar to electrical and computer engineers is the transmission of a binary sequence over a communication channel. In many practical cases the occurrence of bits (1 or 0) can be modeled as independent events. Thus the probability of a bit sequence of any length $101101\ldots$ is simply equal to the product of probabilities: $p \cdot q \cdot p \cdot p \cdot q \cdot p \cdots$. This considerably simplifies the analysis of such systems.

An example is given below where the outcomes of the experiment are based on repeated independent trials. Once the sample space and probabilities of the outcomes have been specified, a number of other probabilistic questions can be answered.

Example 2.2: CDs selected from the bins at Simon's Surplus are as likely to be good as to be bad. If three CDs are selected independently and at random, what is the probability of getting exactly *three* good CDs? Exactly *two* good CDs? How about *one* good CD?

Evidently buying a CD at Simon's is like tossing a coin. The sample space is represented by the listing of outcomes shown below: where G represents a good CD and B

BBB	BBG	BGB	BGG	GBB	GBG	GGB	GGG
A_1	A_2	A_3	A_4	A_5	A_6	A_7	A_8

represents a bad one. Each outcome is labeled as an event A_i; note that the events A_i are mutually exclusive and collectively exhaustive.

Three good CDs is represented by only the last event (A_8) in the sample space. Since the probability of selecting a good CD and the probability of selecting a bad CD are both equal to $\frac{1}{2}$, and the selections are independent, we can write

$$\Pr[3 \text{ good CDs}] = \Pr[A_8] = \Pr[G]\Pr[G]\Pr[G] = \left(\tfrac{1}{2}\right)^3 = \tfrac{1}{8}$$

(see Section 2.2.2).

The result of two good CDs is represented by the events A_4, A_6, and A_7. By a procedure similar to the above, each of these events has probability $\frac{1}{8}$. Since these three events are mutually exclusive, their probabilities add (see Section 2.2.1). That is,

$$\Pr[2 \text{ good CDs}] = \Pr[A_4 + A_6 + A_7] = \Pr[A_4] + \Pr[A_6] + \Pr[A_7] = \tfrac{1}{8} + \tfrac{1}{8} + \tfrac{1}{8} = \tfrac{3}{8}$$

Finally, a single good CD is represented by the events A_2, A_3, and A_5. By an identical procedure it is found that this result also occurs with probability $\frac{3}{8}$.

$\square$

To be sure you understand the steps in this example, you should repeat this example for the case where the probability of selecting a good CD is increased to 5/8. In this case the probability of all of the events A_i are not equal. For example, the probability of the event A_6 is given by $\Pr[G]\Pr[B]\Pr[G] = (5/8)(5/8)(5/8) = 75/512$. When you work through the example you will find that the probabilities of three and two good CDs is increased to $125/512$ and $225/512$ respectively while the probability of just one good CD is decreased to $135/512$.

2.3.2 Problems involving counting

Many important problems involve adding up the probabilities of a number of equally-likely events. These problems involve some basic combinatorial analysis, i.e., counting the number of possible events, configurations, or outcomes in an experiment.

Some discussion of combinatorial methods is provided in Appendix A. All of these deal with the problem of counting the number of pairs, triplets, or k-tuples of elements that can be formed under various conditions. Let us review the main results here.

> **Rule of product.** In the formation of k-tuples consisting of k elements where there are N_i choices for the ith element, the number of possible k-tuples is $\prod_{i=1}^{k} N_i$. An important special case is when there are the *same* number of choices N for each element. The number of k-tuples is then simply N^k.

> **Permutation.** A *permutation* is a k-tuple formed by selecting from a set of N *distinct* elements, where each element can only be selected once. (Think of forming words from a finite alphabet where each letter can be used only once.) There are N choices for the first element, $N - 1$ choices for the second element, $\ldots$, and $N - k + 1$ choices for the kth element. The number of such permutations is given by
> $$N \cdot (N - 1) \cdots (N - k + 1) = \frac{N!}{(N - k)!}$$
> For $k = N$ the result is simply N factorial.

> **Combination.** A *combination* is a k-tuple formed by selecting from a set of N distinct elements where the *order* of selecting the elements makes no difference. For example, the sequences ACBED and ABDEC would represent two different permutations, but only a single *combination* of the letters A through E. The number of combinations k from a possible set of N is given by the binomial coefficient
> $$\binom{N}{k} = \frac{N!}{k!(N - k)!}$$
> This is frequently read as "N *choose* k," which provides a convenient mnemonic for its interpretation.

Counting principles provide a way to assign or compute probability in many cases. This is illustrated in a number of examples below.

The following example illustrates the use of some basic counting ideas.

Example 2.3: In sequences of k binary digits, 1s and 0s are equally likely. What is the probability of encountering a sequence with a single 1 (in any position) and all other digits zero?

Imagine drawing the sample space for this experiment consisting of all possible sequences. Using the rule of product we see that there are 2^k events in the sample space and they are all equally likely. Thus we assign probability $1/2^k$ to each outcome in the sample space.

Now, there are just k of these sequences that have exactly one 1. Thus the probability of this event is $k/2^k$.

□

The next example illustrates the use of permutation.

Example 2.4:   IT technician Chip Gizmo has a cable with four twisted pairs running from each of four offices to the service closet; but he has forgotten which pair goes to which office. If he connects one pair to each of four telephone lines arbitrarily, what is the probability that he will get it right on the first try?

The number of ways that four twisted pairs can be assigned to four telephone lines is $4! = 24$. Assuming that each arrangement is equally likely, the probability of getting it right on the first try is $1/24 = 0.0417$.

□

The following example illustrates the use of permutations versus combinations.

Example 2.5:   Five surplus computers are available for adoption. One is an IBM (actually Lenovo); another is an HP (also made in China) and the rest are nondescript. You can request two of the surplus computers but cannot specify which ones. What is the probability that you get the IBM and the HP?

Consider first the experiment of randomly selecting two computers. Let's call the computers A, B, C, D, and E. The sample space is represented by a listing of pairs

$$A,B \quad B,A \quad A,C \quad C,A \quad \cdots$$

of the computers chosen. Each pair is a *permutation*, and there are $5!/(5-2)! = 5 \cdot 4 = 20$ such permutations that represent the outcomes in the sample space. Thus each outcome has a probability of $1/20$. We are interested in two of these outcomes, namely IBM,HP or HP,IBM. The probability is thus $2/20$ or $1/10$.

Another simpler approach is possible. Since we do not need to distinguish between ordering in the elements of a pair, we could choose our sample space to be

$$A,B \quad A,C \quad A,D \quad A,E \quad \cdots$$

where events such as B,A and C,A are not listed, since they are equivalent to A,B and A,C. The number of pairs in this new sample space is the number of *combinations* of 5 objects taken 2 at a time:

$$\binom{5}{2} = \frac{5!}{2!(5-2)!} = 10$$

Thus each outcome in this sample space has probability $1/10$. But only the single outcome IBM,HP is of interest. Therefore this probability is again $1/10$.

□

The final example for this section illustrates a more advanced use of the combinatoric ideas.

Example 2.6:   DEAL Computers Incorporated manufactures some of their computers in the US and others in Lower Slobbovia. The local DEAL factory store has a stock of 10 computers that are US made and 15 that are foreign made. You order five computers from the DEAL store which are randomly selected from this stock. What is the probability that two or more of them are US-made?

The number of ways to choose 5 computers from a stock of 25 is

$$\binom{25}{5} = \frac{25!}{5!(25-5)!} = 53130$$

This is the total number of possible outcomes in the sample space.

Now consider the number of outcomes where there are *exactly* 2 US-made computers in a selection of 5. Two US computers can be chosen from a stock of 10 in $\binom{10}{2}$ possible ways. For each such choice, three non-US computers can be chosen in $\binom{15}{3}$ possible ways. Thus the number of outcomes where there are exactly 2 US-made computers is given by

$$\binom{10}{2} \cdot \binom{15}{3}$$

Since the problem asks for "two or more" we can continue to count the number of ways there could be exactly 3, exactly 4, and exactly 5 out of a selection of five computers. Therefore the number of ways to choose 2 *or more* US-made computers is

$$\binom{10}{2}\binom{15}{3} + \binom{10}{3}\binom{15}{2} + \binom{10}{4}\binom{15}{1} + \binom{10}{5} = 36477$$

The probability of two or more US-made computers is thus the ratio $36477/53130 = 0.687$.

$\square$

2.3.3 Network reliability

Consider the set of communication links shown in Fig. 2.7 (a) and (b). In both cases

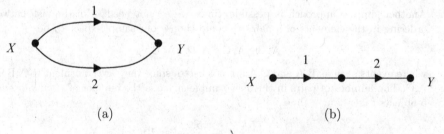

Figure 2.7 Connection of communication links. (a) Parallel. (b) Series.

it is desired to communicate between points X and Y. Let A_i represent the event that link i fails and F be the event that there is failure to communicate between X and Y. Further, assume that the link failures are *independent* events. Then for the parallel connection (Fig. 2.7(a))

$$\Pr[F] = \Pr[A_1 A_2] = \Pr[A_1]\Pr[A_2]$$

where the last equality follows from the fact that events A_1 and A_2 are independent. For the series connection (Fig. 2.7(b))

$$\Pr[F] = \Pr[A_1 + A_2] = \Pr[A_1] + \Pr[A_2] - \Pr[A_1 A_2] = \Pr[A_1] + \Pr[A_2] - \Pr[A_1]\Pr[A_2]$$

where we have applied (2.9) and again used the fact that the events are independent.

The algebra of events and the rules for probability can be used to solve some additional simple problems such as in the following example.

Example 2.7: In the simple communication network shown below, link failures occur in-

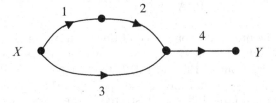

dependently with probability p. What is the largest value of p that can be tolerated if the overall probability of failure of communication between X and Y is to be kept less than 10^{-3}?

Let F represent the failure of communication; this event can be expressed as $F = (A_1 + A_2)A_3 + A_4$. The probability of this event can then be computed as follows:

$$\begin{aligned}
\Pr[F] &= \Pr[(A_1 + A_2)A_3 + A_4] \\
&= \Pr[A_1 A_3 + A_2 A_3] + \Pr[A_4] - \Pr[A_1 A_3 A_4 + A_2 A_3 A_4] \\
&= \Pr[A_4] + \Pr[A_1 A_3] + \Pr[A_2 A_3] - \Pr[A_1 A_2 A_3] \\
&\quad - \Pr[A_1 A_3 A_4] - \Pr[A_2 A_3 A_4] + \Pr[A_1 A_2 A_3 A_4] \\
&= p + 2p^2 - 3p^3 + p^4
\end{aligned}$$

The desired value of p can be found by setting this expression equal to 0.001 and searching for the roots of the polynomial

$$p^4 - 3p^3 + 2p^2 + p - 0.001$$

By using a computer program such as MATLAB, this polynomial can be found to have two complex conjugate roots, one real negative root, and one real positive root $p = 0.001$, which is the desired value.

□

An alternative method can be used to compute the probability of failure in this example. We list the possible outcomes (the sample space) and their probabilities as shown in the table below and put a check ($\checkmark$) next to each outcome that results in failure to communicate.

outcome	probability	F
$A_1 A_2 A_3 A_4$	p^4	$\checkmark$
$A_1 A_2 A_3 A_4{}^c$	$p^3(1-p)$	$\checkmark$
$A_1 A_2 A_3{}^c A_4$	$p^3(1-p)$	$\checkmark$
$A_1 A_2 A_3{}^c A_4{}^c$	$p^2(1-p)^2$	
$\vdots$	$\vdots$	
$A_1{}^c A_2{}^c A_3{}^c A_4{}^c$	$(1-p)^4$	

Then we simply add the probabilities of the outcomes that comprise the event F. The procedure is straightforward but slightly tedious because of the algebraic simplification required to get to the answer (see Problem 2.23).

2.4 Conditional Probability and Bayes' Rule

2.4.1 Conditional probability

If A_1 and A_2 are two events, then the probability of the event A_1 when it is known that the event A_2 has occurred is defined by the relation

$$\Pr[A_1|A_2] = \frac{\Pr[A_1 A_2]}{\Pr[A_2]} \tag{2.13}$$

$\Pr[A_1|A_2]$ is called the probability of "A_1 conditioned on A_2" or simply the probability of "A_1 *given* A_2." Note that in the special case that A_1 and A_2 are statistically independent, it follows from (2.13) and (2.12) that $\Pr[A_1|A_2] = \Pr[A_1]$. Thus when two events are independent, conditioning one upon the other has no effect.

The use of conditional probability is illustrated in the following simple example.

Example 2.8:   Remember Simon's Surplus and the CDs? A CD bought at Simon's is equally likely to be good or bad. Simon decides to sell them in packages of two and guarantees that in each package, at least one will be good. What is the probability that when you buy a single package, you get two good CDs?

Define the following events:

A_1 :   Both CDs are good.
A_2 :   At least one CD is good.

The sample space and these events are illustrated below:

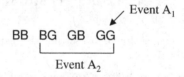

The probability we are looking for is

$$\Pr[A_1|A_2] = \frac{\Pr[A_1 A_2]}{\Pr[A_2]}$$

Recall that since all events in the sample space are equally likely, the probability of A_2 is 3/4. Also, since A_1 is included in A_2, it follows that $\Pr[A_1 A_2] = \Pr[A_1]$, which is equal to 1/4. Therefore

$$\Pr[A_1|A_2] = \frac{1/4}{3/4} = \frac{1}{3}$$

□

It is meaningful to interpret conditional probability as the Venn diagram of Fig. 2.8. *Given* the event A_2, the only portion of A_1 that is of concern is the intersection of A_1 with A_2. It is as if A_2 becomes the new sample space. In defining conditional probability $\Pr[A_1|A_2]$, the probability of event $A_1 A_2$ is therefore "renormalized" by dividing by the probability of A_2.

Getting back to the definition (2.13) can be rewritten as

$$\Pr[A_1 A_2] = \Pr[A_1|A_2]\Pr[A_2] = \Pr[A_2|A_1]\Pr[A_1] \tag{2.14}$$

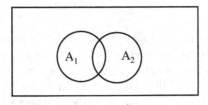

Figure 2.8 Venn diagram illustrating conditional probability.

where the second equality follows from the fact that $\Pr[A_1A_2] = \Pr[A_2A_1]$. Thus *the joint probability of two events can always be written as the product of a conditional probability and an unconditional probability.* Now let $\{A_i\}$ be a (finite or countably infinite) set of mutually exclusive collectively exhaustive events (a partition), and let B be some other event of interest. Recall the *principle of total probability* introduced in Section 2.2 and expressed by (2.11). This equation can be rewritten using (2.14) as

$$\Pr[B] = \sum_i \Pr[B|A_i]\,\Pr[A_i] \qquad\qquad (2.15)$$

Although both forms are equivalent, (2.15) is likely the more useful one to remember. This is because the information given in problems is more frequently in terms of conditional probabilities rather than joint probabilities. (And you must be able to recognize the difference!)

Let us consider one final fact about conditional probability before moving on. Again let $\{A_i\}$ be a partition of the sample space. Then the probabilities of the A_i conditioned on *any* event B sum to one. That is,

$$\sum_i \Pr[A_i|B] = 1 \qquad\qquad (2.16)$$

The proof of this fact is straightforward using the definition (2.13):

$$\sum_i \Pr[A_i|B] = \sum_i \frac{\Pr[A_iB]}{\Pr[B]} = \frac{\sum_i \Pr[A_iB]}{\Pr[B]} = \frac{\Pr[B]}{\Pr[B]} = 1$$

where in the next to last step the principle of total probability was used in the form (2.11).

As an illustration of this result, consider the following brief example.

Example 2.9: Consider the situation in Example 2.8. What is the probability that when you buy a package of two CDs, only one is good?

Since Simon guarantees that there will be *at least* one good CD in each package, we have the event A_2 defined in Example 2.8. Let A_3 represent the set of outcomes {BG GB} (only one good CD). This event has probability 1/2. The probability of only one good CD given the event A_2 is thus

$$\Pr[A_3|A_2] = \frac{\Pr[A_3A_2]}{\Pr[A_2]} = \frac{\Pr[A_3]}{\Pr[A_2]} = \frac{1/2}{3/4} = \frac{2}{3}$$

The events A_3 and A_1 are mutually exclusive and collectively exhaustive given the event A_2. Hence their probabilites, 2/3 and 1/3, sum to one.

□

2.4.2 Event trees

As stated above, the information for problems in probability is often stated in terms of *conditional* probabilities. An important technique for solving some of these problems is to draw the sample space by constructing a tree of dependent events and to use the information in the problem to determine the probabilities of compound events.

The idea is illustrated in Fig. 2.9. In this figure, A is assumed to be an event whose

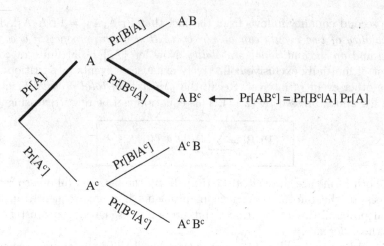

Figure 2.9 Sample space constructed using an event tree.

probability is known and does not depend on any other event. The probability of event B however, depends on whether or not A occurred. These conditional probabilities are written on the branches of the tree. The endpoints of the tree comprise the sample space for the problem; they form a partition of finest grain events which represent the outcomes of the experiment. The probabilities of these events are computed using (2.14), which is equivalent to multiplying probabilities along the branches of the tree that form a path to the event (see Fig. 2.9). Once the probabilities of these elementary events are determined, the probabilities for other compound events can be found by adding the appropriate probabilities. The technique is best illustrated by an example.

Example 2.10: You listen to the morning weather report. If the meteorologist forecasts rain, then the probability of rain is 0.75. If the meteorologist forecasts "no rain," then the probability of rain is 0.15. You have listened to this report for well over a year now and have determined that the meteorologist forecasts rain 1 out of every 5 days regardless of the season. What is the probability that the weather report is wrong? Suppose you take an umbrella if and only if the weather report forecasts rain. What is the probability that it rains and you are caught without an umbrella?

To solve this problem, define the events:

$$F = \text{``rain is forecast''} \qquad R = \text{``it actually rains''}$$

The problem statement then provides the following information:

$$\Pr[R|F] = 0.75 \quad \Pr[R|F^c] = 0.15$$

$$\Pr[F] = 1/5 \qquad \Pr[F^c] = 4/5$$

The events and conditional probabilities are depicted in the event tree below.

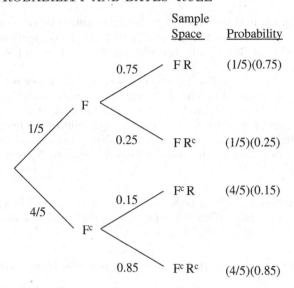

		Sample Space	Probability

The event that the weather report is wrong is represented by the two elementary events FR^c and F^cR. Since these events are mutually exclusive, their probabilities can be added to find

$$\Pr[\text{wrong report}] = \tfrac{1}{5}(0.25) + \tfrac{4}{5}(0.15) = 0.17$$

The probability that it rains and you are caught without an umbrella is the event F^cR. The probability of this event is $\tfrac{4}{5}(0.15) = 0.12$.

□

2.4.3 Bayes' rule

One of the most important uses of conditional probability was developed by Bayes in the late 1800s. It follows if (2.14) is rewritten as

$$\Pr[A_1|A_2] = \frac{\Pr[A_2|A_1] \cdot \Pr[A_1]}{\Pr[A_2]} \tag{2.17}$$

This allows one conditional probability to be computed from the other. A particularly important case arises when $\{A_j\}$ is a partition and B is some other event of interest. From (2.17) the probability of one of these events conditioned on B is given by

$$\Pr[A_i|B] = \frac{\Pr[B|A_i] \cdot \Pr[A_i]}{\Pr[B]} \tag{2.18}$$

But since the $\{A_j\}$ are a set of mutually exclusive collectively exhaustive events, the principle of total probability can be used to express the probability of the event B. Thus, substituting (2.15) into the last equation yields

$$\boxed{\Pr[A_i|B] = \frac{\Pr[B|A_i] \cdot \Pr[A_i]}{\sum_j \Pr[B|A_j] \Pr[A_j]}} \tag{2.19}$$

This result is known as *Bayes' theorem* or *Bayes' rule*. It is used in a number of problems that commonly arise in communications or other areas where decisions or inferences are to be made from some observed signal or data. Because (2.19) is a more

complicated formula, it is sufficient in problems of this type to remember the simpler result (2.18) and to know how to compute Pr[B] using the principle of total probability.
  As an illustration of Bayes' rule, consider the following example.

Example 2.11:   The US Navy is involved in a service-wide program to update memory in computers onboard ships. The Navy will buy memory modules only from American manufacturers known as A_1, A_2, and A_3. The Navy has studied the government procurement process and determined that the probabilities of buying from A_1, A_2, and A_3 are 1/6, 1/3, and 1/2 repectively. The Navy doesn't realize, however, that the probability of failure for the modules from A_1, A_2, and A_3 is 0.006, 0.015, and 0.02 (respectively).

Back in the fleet, an enlisted technician upgrades the memory in a particular computer and finds that it fails. What is the probability that the failed module came from A_1? What is the probability that it came from A_3?

Let F represent the event that a memory module fails. Using (2.18) and (2.19) we can write

$$Pr[A_1|F] = \frac{Pr[F|A_1] \cdot Pr[A_1]}{Pr[F]} = \frac{Pr[F|A_1] \cdot Pr[A_1]}{\sum_{j=1}^{3} Pr[F|A_j] Pr[A_j]}$$

Then substituting the known probabilities yields

$$Pr[A_1|F] = \frac{(0.006)1/6}{(0.006)1/6 + (0.015)1/3 + (0.02)1/2} = \frac{0.001}{0.016} = 0.0625$$

and likewise

$$Pr[A_3|F] = \frac{(0.02)1/2}{(0.006)1/6 + (0.015)1/3 + (0.02)1/2} = \frac{0.01}{0.016} = 0.625$$

Thus failed modules are ten times more likely to have come from A_3 than from A_1. In fact, A_3 accounts for almost two-thirds of all the failed modules.

  □

2.5 More Applications

This section illustrates the use of probability as it occurs in two problems involving digital communication. A basic digital communication system is shown in Fig. 2.10. The system has three basic parts: a transmitter which codes the message into some

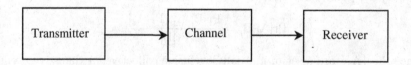

Figure 2.10 Digital communication system.

representation of binary data; a channel over which the binary data is transmitted; and a receiver, which decides whether a 0 or a 1 was sent and decodes the data. The simple binary communication channel discussed in Section 2.5.1 below is a probabilistic model for the binary communication system, which models the effect of sending one bit at a time. Section 2.5.2 discusses the digital communication system using the formal concept of "information," which is also a probabilistic idea.

2.5.1 The binary communication channel

A number of problems naturally involve the use of conditional probability and/or Bayes' rule. The binary communication channel, which is an abstraction for a communication system involving binary data, uses these concepts extensively. The idea is illustrated in the following example.

Example 2.12:   The transmission of bits over a binary communication channel is represented in the drawing below, where notation like 0_S, 0_R ... is used to denote events "0

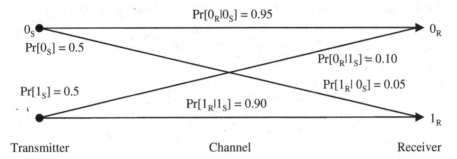

Transmitter   Channel   Receiver

sent," "0 received," etc. When a 0 is transmitted, it is correctly received with probability 0.95 or incorrectly received with probability 0.05. That is $\Pr[0_R|0_S] = 0.95$ and $\Pr[1_R|0_S] = 0.05$. When a 1 is transmitted, it it is correctly received with probability 0.90 and incorrectly received with probability 0.10. The probabilities of sending a 0 or a 1 are denoted by $\Pr[0_S]$ and $\Pr[1_S]$ and are known as the *prior* probabilities. It is desired to compute the *probability of error* for the system.

This is an application of the principle of total probability. If two events A_1 and A_2 form a partition, then (2.15) can be written as

$$\Pr[B] = \Pr[B|A_1]\Pr[A_1] + \Pr[B|A_2]\Pr[A_2]$$

Since the two events 0_S and 1_S are mutually exclusive and collectively exhaustive, they can be identified with the events A_1 and A_2 and the event B can be taken to be the event that an error occurs. It then follows that

$$
\begin{aligned}
\Pr[\text{error}] &= \Pr[\text{error}|0_S]\Pr[0_S] + \Pr[\text{error}|1_S]\Pr[1_S] \\
&= \Pr[1_R|0_S]\Pr[0_S] + \Pr[0_R|1_S]\Pr[1_S] \\
&= (0.05)(0.5) + (0.10)(0.5) = 0.075
\end{aligned}
$$

□

The probability of error is an overall measure of performance that is frequently used for a communication system. Notice that it involves not just the conditional error probabilities $\Pr[1_R|0_S]$ and $\Pr[0_R|1_S]$ but also the prior probabilities $\Pr[0_S]$ and $\Pr[1_S]$ for transmission of a 0 or a 1. One criterion for optimizing a communication system is to minimize the probability of error.

If a communication system is correctly designed, then the probability that a 1 was sent given that a 1 is received should be greater than the probability that a 0 was sent given that a 1 is received. In fact, as shown later in the text, applying this condition to both 0s and 1s leads to *minimizing* the probability of error. The computation of these "inverse" probabilities is illustrated in the next example.

Example 2.13: Assume for the communication system illustrated in the previous example that a 1 has been received. What is the probability that a 1 was sent? What is the probability that a 0 was sent?

This is an application of conditional probability and Bayes rule. For a 1, the conditional probability is

$$\Pr[1_S|1_R] = \frac{\Pr[1_R|1_S]\Pr[1_S]}{\Pr[1_R]} = \frac{\Pr[1_R|1_S]\Pr[1_S]}{\Pr[1_R|1_S]\Pr[1_S] + \Pr[1_R|0_S]\Pr[0_S]}$$

Substituting the numerical values from the figure in Example 2.12 then yields

$$\Pr[1_S|1_R] = \frac{(0.9)(0.5)}{(0.9)(0.5) + (0.05)(0.5)} = 0.9474$$

For a 0, the result is

$$\Pr[0_S|1_R] = \frac{\Pr[1_R|0_S]\Pr[0_S]}{\Pr[1_R|1_S]\Pr[1_S] + \Pr[1_R|0_S]\Pr[0_S]}$$

$$= \frac{(0.05)(0.5)}{(0.9)(0.5) + (0.05)(0.5)} = 0.0526$$

Note that $\Pr[1_S|1_R] > \Pr[0_S|1_R]$ as would be expected, and also that $\Pr[1_S|1_R] + \Pr[0_S|1_R] = 1$.

□

2.5.2 Measuring information and coding

The study of *Information Theory* is fundamental to understanding the trade-offs in design of efficient communication systems. The basic theory was developed by Shannon [2, 3, 4] and others in the late 1940s and '50s and provides fundamental results about what a given communication system can or cannot do. This section provides just a taste of the results which are based on a knowledge of basic probability.

Consider the digital communication system depicted in Fig. 2.10 and let the events A_1 and A_2 represent the transmission of two codes representing the symbols 0 and 1. To be specific, assume that the transmitter, or source, outputs the symbol to the communication channel with the following probabilities: $\Pr[A_1] = 1/8$ and $\Pr[A_2] = 7/8$. The *information* associated with the event A_i is defined as

$$I(A_i) = -\log \Pr[A_i]$$

The logarithm here is taken with respect to the base 2 and the resulting information is expressed in *bits*.[4] The information for each of the two symbols is thus

$$I(A_1) = -\log \Pr[\tfrac{1}{8}] = 3 \text{ (bits)}$$
$$I(A_2) = -\log \Pr[\tfrac{7}{8}] = 0.193 \text{ (bits)}$$

Observe that events with *lower* probability have *higher* information. This corresponds to intuition. Someone telling you about an event that almost always happens provides little information. On the other hand, someone telling you about the occurrence of a very rare event provides you with much more information. The news media works on this principle in deciding what news to report and thus tries to maximize information.

---

[4] Other less common choices for the base of the logarithm are 10, in which case the units of information are Hartleys, and e in which case the units are called nats.

The *average* information[5] H is given by the weighted sum

$$H = \sum_{i=1}^{2} \Pr[A_i] I(A_i) = \tfrac{1}{8} \cdot 3 + \tfrac{7}{8} \cdot 0.193 = 0.544$$

Notice that this average information is less than one bit, although it is not possible to transmit two symbols with less than one binary digit (bit).

Now consider the following scheme. Starting anywhere in the sequence, group together two consecutive bits and assign this pair to one of four possible codewords. Let the corresponding events (or codewords) be denoted by B_j for $j = 1, 2, 3, 4$ as shown in the table below and assume that two consecutive symbols are independent.

codeword	symbols	probability	information (bits)
B_1	00	$\tfrac{1}{8} \cdot \tfrac{1}{8} = \tfrac{1}{64}$	6.000
B_2	01	$\tfrac{1}{8} \cdot \tfrac{7}{8} = \tfrac{7}{64}$	3.193
B_3	10	$\tfrac{7}{8} \cdot \tfrac{1}{8} = \tfrac{7}{64}$	3.193
B_4	11	$\tfrac{7}{8} \cdot \tfrac{7}{8} = \tfrac{49}{64}$	0.386

Notice that since the probabilities of two consecutive symbols multiply, the corresponding information adds. For example, the symbol 0 by itself has information of 3 bits, while the pair of symbols 00 shown in the table has information of 6 bits. The average information for this scheme is given by the weighted sum

$$H = \tfrac{1}{64} \cdot 6 + \tfrac{7}{64} \cdot 3.193 + \tfrac{7}{64} \cdot 3.193 + \tfrac{49}{64} \cdot 0.386 = 1.087$$

The average information per bit is still $1.087/2 = 0.544$.

The previous analysis does not result in any practical savings, since the average information H is still more than one bit and therefore in a binary communication system it will still require a minimum of two bits to send the four codewords. A continuation of this procedure using larger groups of binary symbols mapped to codewords, however, does lead to some efficiency. The table below lists the average information with increasing numbers of symbols per codeword.

No. symbols	Avg. Information	Avg. Inf. / bit
2	1.087	0.544
3	1.631	0.544
4	2.174	0.544
5	2.718	0.544
8	4.349	0.544
10	5.436	0.544
12	6.523	0.544

From this table, it is seen that when three symbols are grouped together, the average information is 1.631 bits. It would therefore seem that only two binary digits should be theoretically required to transmit the codewords, since the information $I = 1.631$ is less than 2 bits. Likewise, when 12 symbols are grouped together, it should require

---

[5] Average information is also known as the source *entropy* and is discussed further in Chapter 4.

no more than 7 binary digits on average to code the message ($I = 6.523 < 7$). How to achieve such efficiency in practice has led to various coding algorithms such as Huffman coding and Shannon-Fano coding. The basic idea is to use variable length codes and assign fewer binary digits to codewords that occur more frequently. This reduces the average number of bits that are needed to transmit the message. The example below illustrates the technique using the Shannon-Fano algorithm.

Example 2.14: It is desired to code the message "ELECTRICAL ENGINEERING" in an efficient manner using Shannon-Fano coding. The probabilities of the letters (excluding the space) are represented by their relative frequency of occurrence in the message. The letters { E, L, C, T, R, I, A, N, G } are thus assigned the probabilities

$$\left\{ \frac{5}{21}, \frac{2}{21}, \frac{2}{21}, \frac{1}{21}, \frac{2}{21}, \frac{3}{21}, \frac{1}{21}, \frac{3}{21}, \frac{2}{21} \right\}.$$

The codewords are assigned as illustrated in the steps below. The letters are arranged

Code assignment

E	5/21	0
I	3/21	0
N	3/21	0
L	2/21	1
C	2/21	1
R	2/21	1
G	2/21	1
T	1/21	1
A	1/21	1

in order of decreasing probability; any ties are broken arbitrarily. The letters are then partitioned into two groups of approximately equal probability (as closely as possible). This is indicated by the partition labeled 1. Those letters in the first group are assigned a codeword beginning with 0 while those in the second group are assigned a codeword beginning with 1.

Within each group, this procedure is repeated recursively to determine the second, third, and fourth digit of the codeword.

Code assignment

E	5/21	0	0
I	3/21	0	1
N	3/21	0	1
L	2/21	1	0
C	2/21	1	0
R	2/21	1	0
G	2/21	1	1
T	1/21	1	1
A	1/21	1	1

Code assignment

E	5/21	0	0	
I	3/21	0	1	0
N	3/21	0	1	1
L	2/21	1	0	0
C	2/21	1	0	1
R	2/21	1	0	1
G	2/21	1	1	0
T	1/21	1	1	1
A	1/21	1	1	1

The final result is as shown on the next page.

An inherent and necessary property (for decoding) of any variable-length coding scheme is that no codeword is a prefix of any longer codeword. Thus, for example,

		Code assignment				Codeword	Length
E	5/21	0	0			00	2
I	3/21	0	1	0		010	3
N	3/21	0	1	1		011	3
L	2/21	1	0	0		100	3
C	2/21	1	0	1	0	1010	4
R	2/21	1	0	1	1	1011	4
G	2/21	1	1	0		110	3
T	1/21	1	1	1	0	1110	4
A	1/21	1	1	1	1	1111	4

upon finding the sequence 011, we can uniquely determine that this sequence corresponds to the letter N, since there is no codeword of length 4 that has 011 as its first three binary digits.

Now consider the efficiency achieved by this coding scheme. The average number of bits used in coding of the message is the sum of the lengths of the codewords weighted by the probability of the codeword. For this example, the average length is given by (see final figure)

$$2 \cdot 5/21 + 3 \cdot 3/21 + \cdots + 4 \cdot 1/21 = 3.05 \text{ (bits)}$$

On the other hand, if a fixed-length coding scheme were used then the length of each codeword would be 4 bits. (Since there are nine letters, three bits are insufficient and four bits are needed to code each letter.) Thus the variable-length coding scheme, which is based on estimating the average information in the message, reduces the communication traffic by about 24%.

□

2.6 Summary

The study of probability deals with the occurrence of random "events." Such events occur as outcomes or collections of outcomes from an experiment. The complete set of outcomes from an experiment comprise the Sample Space. The outcomes in this set must be mutually exclusive and collectively exhaustive and be of finest grain in representing the conditions of the experiment. Events are defined in the sample space.

The algebra of events is a form of set algebra that provides rules for describing arbitarily complex combinations of events in an unambiguous way. Venn diagrams are useful as a complementary geometric method for depicting relationships among events.

Probability is a number between 0 and 1 assigned to an event. Several rules and formulae permit computing the probabilities of intersections, unions, and other more complicated expressions in the algebra of events when the probabilities of the other events in the expression are known. An important special rule applies to independent events: to compute the probability of the combined event, simply multiply the individual probabilities.

Conditional probability is necessary when events are not independent. The formulas developed in this case provide means for computing joint probabilities of sets of events that depend on each other in some way. These conditional probabilities are also useful in developing a "tree diagram" from the facts given in a problem, and an associated representation of the sample space. Bayes' rule is an especially important use of conditional probability, since it allows you to "work backward" and compute the probability of unknown events from related observed events. Bayes' rule forms the basis for methods of "statistical inference" and finds much use (as seen later) in engineering problems.

A number of examples and applications of the theory are discussed in this chapter. These are important to illustrate how the theory applies in specific situations. The applications also discuss some well-established models, such as the binary communication channel, which are important to electrical and computer engineering.

References

[1] Alvin W. Drake. *Fundamentals of Applied Probability Theory*. McGraw-Hill, New York, 1967.

[2] Claude E. Shannon. A mathematical theory of communication. *Bell System Technical Journal*, 27(3):379–422, July 1948. (See also [4].)

[3] Claude E. Shannon. A mathematical theory of communication (concluded). *Bell System Technical Journal*, 27(4):623–656, October 1948. (See also [4].)

[4] Claude E. Shannon and Warren Weaver. *The Mathematical Theory of Communication*. University of Illinois Press, Urbana, IL, 1963.

Problems

Algebra of events

2.1 Draw a set of Venn diagrams to illustrate each of the following identities in the algebra of events.

    (a) $A \cdot (B + C) = AB + AC$

    (b) $A + (BC) = (A + B)(A + C)$

    (c) $(AB)^c = A^c + B^c$

    (d) $(A + B)^c = A^c B^c$

    (e) $A + A^c B = A + B$

    (f) $AB + B = B$

    (g) $A + AB + B = A + B$

2.2 A sample space S is given to be $\{a_1, a_2, a_3, a_4, a_5, a_6\}$. The following events are defined on this sample space: $A_1 = \{a_1, a_2, a_4\}$, $A_2 = \{a_2, a_3, a_6\}$, and $A_3 = \{a_1, a_3, a_5\}$.

    (a) Find the following events:

        (i) $A_1 + A_2$
        (ii) $A_1 A_2$,
        (iii) $(A_1 + A_3^c)A_2$

    (b) Show the following identities:

        (i) $A_1(A_2 + A_3) = A_1 A_2 + A_1 A_3$
        (ii) $A_1 + A_2 A_3 = (A_1 + A_2)(A_1 + A_3)$
        (iii) $(A_1 + A_2)^c = A_1^c A_2^c$

Probability of events

2.3 By considering probabilities to be represented by areas in a Venn diagram, show that the four axioms of probability and the results listed in Table 2.4 "make sense." If the result cannot be shown by Venn diagram, say so.

2.4 Starting with the expression for $\Pr[A_1 + A_2]$, show that for *three* events

$$\Pr[A_1 + A_2 + A_3] = \Pr[A_1] + \Pr[A_2] + \Pr[A_3] - \Pr[A_1 A_2] \\ - \Pr[A_1 A_3] - \Pr[A_2 A_3] + \Pr[A_1 A_2 A_3]$$

2.5 Consider the sample space in Prob. 2.2 in which all the outcomes are assumed equally likely. Find the following probabilities:

(a) $\Pr[A_1 A_2]$

(b) $\Pr[A_1 + A_2]$

(c) $\Pr[(A_1 + A_3^c)A_2]$

2.6 A signal has been sampled, quantized (8 levels), and encoded into 3 bits. These bits are sequentially transmitted over a wire.

(a) Draw the sample space corresponding to the received bits. Assume that no errors occur during transmission.

(b) Determine the probability that the received encoded sample has a value greater than 5.

(c) What is the probability that the value of the received sample is between 3 and 6 (inclusive)?

2.7 For purposes of efficient signal encoding, a given speech signal is subject to a companding operation. The μ-law companding used in commercial telephony in North America is given by

$$|y| = \frac{\ln(1 + \mu|x|)}{\ln(1 + \mu)}$$

where $\mu \approx 100$, x is the input signal value and y is the output signal value. Assume that the input is in the range of $0 \le x \le 5$ and that all values are equally likely.

(a) Draw the sample spaces that could be used to represent events involving the input and the output signals.

(b) What is the probability that the output value is less than 1?

(c) Find these probabilities: $\Pr[\frac{1}{2} \le x \le 1]$ and $\Pr[\frac{1}{2} \le y \le 1]$.

2.8 The input-output relationship in a system is described by

$$y = 16e^{-1.6x}$$

where x is the input and y is the output. The input is in the range of $-1 \le x \le 2$, and all values in this range occur in an equally likely fashion.

(a) Draw the sample spaces of the input and the output.

(b) Find these probabilities: $\Pr[Y \le 16]$ and $\Pr[2 \le Y < 20]$.

2.9 The following events and their probabilities are listed below.

event:	A	B	C	D
probability:	1/4	1/3	1/4	1/4

(a) Compute $\Pr[A + B]$ assuming $AB = \emptyset$.

(b) Compute $\Pr[A + B]$ assuming A and B are *independent*.

(c) Is it possible that all of A, B, C, and D are mutually exclusive? Tell why or why not.

2.10 Consider a 12-sided (dodecahedral) unfair die. In rolling this die, the even numbered sides are twice as likely as the odd numbered sides. Define the events: $A = \{\text{odd numbered side}\}$ and $B = \{4, 5, 6, 7, 8\}$.

(a) Find the probability $\Pr[A]$.

(b) Find the probability $\Pr[B]$.

(c) Find the probability $\Pr[AB]$.

2.11 An experiment consists of rolling two fair dice: an eight-sided die and a six-sided die. Define I to be the difference of the dots on the faces of eight-sided and six-sided, respectively. Find the probability of the union of events $\{I = -1\}$ and $\{I = 5\}$, i.e., $\Pr\left[\{I = -1\} \cup \{I = 5\}\right]$.

2.12 Computers purchased from *Simon's Surplus* will experience hard drive failures with probability 0.3 and memory failures with probability 0.2 and will experience both types of failures simultaneously with probability 0.1

(a) What is the probability that there will be one type of failure but not the other?

(b) What is the probability that there will be no failures of either kind?

Use the algebra of events to write expressions for the desired events and the rules of probability to compute probabilities for these event expressions.

2.13 In the 2003 race for Governor of California the alphabet was "reordered" so that none of the 135 candidates would feel discriminated against by the position of their name on the ballot.

(a) The week of the election it was announced that the first four letters of the reordered alphabet were R, W, Q, O. Assuming that all letters were initially equally-likely, what was the probability that this *particular* sequence would be chosen? (There are 26 letters in the English alphabet.)

(b) The letter S (for Schwarzenegger) turned up at position 10 in the new alphabet. What is the probability that S would end up in the 10th position?

Applications

2.14 Consider the problem of purchasing CDs described in Example 2.2.

(a) Assuming that the probabilities of good CDs and bad CDs are equal, what is the probability that you have one or more good CDs?

(b) If the probability of a good disk is 5/8, what is the probability that you have one or more good CDs?

2.15 Simon has decided to improve the quality of the products he sells. Now only one out of five CDs selected from the bins is defective (i.e., the probability that a diskette is bad is only 0.2). If three CDs are chosen at random, what is the probability

    (a) that all three CDs are good?

    (b) that all three CDs are bad?

    (c) that [exactly] one of the three CDs is bad?

    (d) that there are more good CDs than bad ones?

2.16 In a certain digital control system, the control command is represented by a set of four hexadecimal characters.

    (a) What is the total number of control commands that are possible?

    (b) If each control command must have a unique prefix, i.e., starting from left to right no hex character must be repeated, how many control commands are possible?

2.17 A digital transmitter sends groups of 8 bits over a communication channel sequentially. The probability of a single bit error in the channel is p.

    (a) How many different ways can two errors occur in 8 bits? (We will call each of these a 2-bit error.)

    (b) Assume that a particular 2-bit error has occured: $\Phi x\Phi\Phi\Phi x\Phi\Phi$, where x indicates an error bit and Φ the correct bit (0 or 1). What is the probability of this particular event?

    (c) Determine the probability of 2-bit errors in this system.

    (d) Plot the probability of 2-bit errors for $p = 0.1,\ 0.01,\ 0.001,\ 0.0001$.

2.18 In the Internet, the TCP/IP protocol suite is used for transmitting packets of information. The receiving TCP entity checks the packets for errors. If an error is detected, it requests retransmission of the packet that is found to be in error. Suppose that the packet error probability is 10^{-2}.

    (a) What is the probability that the third packet requires retransmission?

    (b) What is the probability that the tenth packet requires retransmission?

    (c) What is the probability that there is a packet retransmission within the first 5 packets?

    (d) Following on the lines of (c), how many packets are required in order for the probability of retransmission to be equal to or greater than 0.1?

2.19 In the serial transmission of a byte (such as over a modem) errors in each bit occur independently with probability Q. For every 8-bit byte sent, a parity check bit is appended so that each byte transmitted actually consists of 9 bits. The parity bit is chosen so that the group of 9 bits has "even parity," i.e., the number of 1's, is even. Errors can be detected by checking the parity of the received 9-bit sequence.

    (a) What is the probability that a *single* error occurs?

    (b) What is the probability that errors occur but are not detected? Write your answer as an expression involving the error probability Q.

2.20 The diagram below represents a communication network where the source S communicates with the receiver R. Let A represent the event "link a fails," and B represent the event "link b fails," etc. Write an expression in the algebra of events for the event F = "S fails to communicate with R."

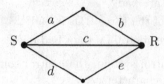

2.21 Repeat Prob. 2.20 for the following network.

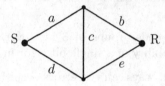

2.22 In Prob. 2.20 assume that link failures are independent and that each link can fail with probability 0.1. Compute the probability of the event F = "S fails to communicate with R."

2.23 By using the alternate procedure described on page 25, compute the probability of failure for the network of Example 2.7.

2.24 A communication network is shown below. Define the following events:

A   link a fails
B   link b fails
C   link c fails
F   S cannot communicate with R

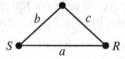

(a) Assume link failures are independent events. Write an expression in the Algebra of Events for the event F. Your expression should be in terms of the events A, B, and C.

(b) Repeat part (a) for the following network, shown below

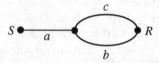

(c) If every link has the same probability of failure $Pr[A] = Pr[B] = Pr[C] = p = 0.1$, then which network has the lowest $Pr[F]$? Justify your answer.

2.25 The following communication network connects the source S with the receiver R. The probability of failure of any link in the network is $p = 0.01$, and the link failures are independent.

(a) Write an algebraic expression for the failure F to connect between S and R.

(b) If $p = 0.01$, calculate the probability that the path between S and R is established.

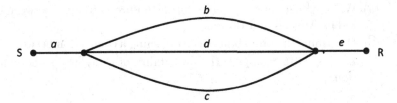

2.26 The diagram below is meant to represent a communication network with links *a*, *b*, *c*, and *d*. Let A denote the event "Link *a* is OK." Events B, C, and D are defined similarly. Each link has a probability of failure of 0.5 (in other words, the probability that each link is OK is 0.5) and the failure of any link is *independent* of the failure of any other link.

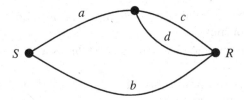

(a) Write an expression in the Algebra of Events that represents the event "S communicates with R." The expression should involve only the events A, B, C, and D. Do not use any probabilities here.

(b) What is the probability that *c* or *d* or both links are OK (i.e., what is the probability of the event CD)?

(c) What is the probability of the event "S communicates with R"?

(d) If S fails to communicate with R, which link has the highest probability of having failed? Which link has the next highest probability of having failed? What are these probabilities? In what order should we test the links (according to their probability of failure) to determine if each is functioning?

2.27 Consider the probability of failure of a communication network as described in Example 2.7.

(a) What is the largest value of *p* that can be tolerated if the probability of failure must be less than 0.001?

(b) What is the largest value of *p* if the probability of failure must be less than 0.05?

2.28 In the early days of its development the ARPA network looked something like the diagram shown below. Assume the probability of failure of any link in the network is *p*, and the link failures are independent.

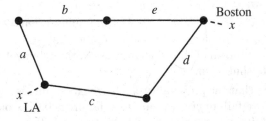

Let A, B, C, D, E be the events that the corresponding link fails.

(a) Write an algebraic expression for the failure to connect between subscribers in Los Angeles and Boston.

(b) Determine the probability of establishing a connection between these users.

(c) Let $p = 0.01$. Compute the probability of establishing the above connection.

Conditional probability and Bayes' rule

2.29 An analog to digital (A/D) converter generates 3-bit digital words corresponding to eight levels of the input signal (0, 1, ..., 7) with the following probabilities: $\Pr[0] = \Pr[1] = \Pr[6] = \Pr[7] = 1/16$, $\Pr[2] = \Pr[5] = 1/8$, and $\Pr[3] = \Pr[4] = 1/4$.

(a) Find the probability that the input has a value > 5 given that the first bit of the output is a 1.

(b) Find the probability that the first bit is a 1 given that the input value is greater than 5.

(c) What is the probability of a binary 1? <u>Hint:</u> Use the principle of total probability approach.

(d) Determine the probability of a binary 0.

2.30 In a certain computer, the probability of a memory failure is 0.01, while the probability of a hard disk failure is 0.02. If the probability that the memory and the hard disk fail simultaneously is 0.0014, then

(a) Are memory failures and hard disk failures independent events?

(b) What is the probability of a memory failure, given a hard disk failure?

2.31 Repeat Prob. 2.30 if the probability of a memory failure is 0.02, the probability of a disk failure is 0.015, and the probability that both fail simultaneously is 0.0003.

2.32 In the triple-core computers that Simon's Surplus has just sold back to the government, the probability of a memory failure is 0.02, while the probability of a hard disk failure is 0.015. If the probability that the memory and the hard disk fail simultaneously is 0.0003, then

(a) Are memory failures and hard disk failures independent events?

(b) What is the probability of a memory failure, given a hard disk failure?

2.33 In the process of transmitting binary data over a certain noisy communication channel, it is found that the errors occur in bursts. Given that a single bit error occurs, the probabilty that the next bit is in error is *twice* the probability of a single bit error. If it is known that two consecutive errors occur with probability 2×10^{-4}, what is the probability of a single bit error?

2.34 In the Navy's new advanced YK/2 (pronounced "yuk-2") computer, hardware problems are not necessarily catastrophic, but software problems will cause complete shutdown of ship's operations. Nothing but software failures can cause a complete shutdown.

It is known that the probability of a hardware failure is 0.001 and the probability of a software failure *given* a hardware failure is 0.02. Complete shutdowns (i.e., software failures) occur with probability 0.005. If a shutdown occurs, what is the probability that there was a hardware failure?

2.35 A random hexadecimal character in the form of four binary digits is read from a storage device.

(a) Draw the tree diagram and the sample space for this experiment.

(b) Given that the first bit is a zero, what is the probability of more zeros than ones?

(c) Given that the first two bits are 10, what is the probability of more zeros than ones?

(d) Given that there are more zeros than ones, what is the probability that the first bit is a zero?

2.36 Beetle Bailey has a date with Miss Buxley, but Beetle has an old jeep which will break down with probability 0.4. If his jeep breaks down he will be late with probability 0.9. If it does not break down he will be late with probability 0.2. What is the probability that Beetle will be late for his date?

2.37 Rudy is an astronaut and the engineer for Project Pluto. Rudy has determined that the mission's success or failure depends on only three major systems. Further, Rudy decides that the mission is a failure if and only if two or more of the major systems fail. The following is known about these systems. System I, the auxiliary generator, fails with probability 0.1 and does not depend on the other systems. System II, the fuel system, fails with probability 0.5 if at least one other system fails. If no other system fails, the probability that the fuel system fails is 0.1. System III, the beer cooler, fails with probability 0.5 if the generator system fails. Otherwise the beer cooler cannot fail.

(a) Draw the event space (sample space) for this problem by constructing a tree of possible events involving failure or nonfailure of the major systems. Use the notation G^c, B^c, F^c to represent failure of the generator, beer cooler, and fuel systems and G, B, F to represent nonfailure of those systems. The ends of your tree should be labeled with triples such as G^cBF and their probabilities.

Hint: Start with the Generator. What system should go on the next level of the tree?

(b) Answer the following questions about the mission.

(i) What is the probability that the mission fails?
(ii) What is the probability that all three systems fail?

(c) Given that more than one system failed, what is the probability:

(i) that the generator did not fail?
(ii) that the beer cooler failed?
(iii) that both the generator and the fuel system failed?

(d) Given that the beer cooler failed, what is the probability that the mission succeeded?

[This problem is a modified version of a problem from Alvin W. Drake, *Fundamentals of Applied Probability Theory*, McGraw-Hill, New York, 1967. Reproduced by permission.]

2.38 Microsoft's Internet Explorer (IE) and Mozilla's FireFox (FF) are two popular Web browsers. IE supports certain features that are present at a given site, such as security, that FF does not and vice versa. Assume that all features on all sites are exploitable by either IE or FF. The probability that IE supports a given feature is $1 - \epsilon$, and the probability that FF fails to support a feature is δ.

(a) A user selects a browser at random and attempts to access a site at random. What is the probability that the user will be forced to change the browsers?

(b) Given that the browser fails to access a site, what is the probability that it is IE?

2.39 A box contains 16 integrated circuits (ICs) of which 3 are known to be defective.

(a) You are asked to randomly select two ICs from the box. What is probability that at least one IC is good?

(b) If you randomly select one IC from the box, what is the probability that it is good?

2NB3 Components from three different manufacturers have been procured; 100 from Manufacturer A, of which 10% are defective; 300 from Manufacturer B, of which 5% are defective; and 500 from Manufacturer C, of which 20% are defective. We *randomly* select the shipment from one of the manufacturers and then *randomly* pick a component from it.

(a) What is the probability that the selected component is defective?

(b) If the selected component is found to be defective, what is the probability that it came from Manufacturer A?

More applications

2.40 In a certain binary communication channel, it is equally likely to send a 1 or a 0 (both probabilities are equal to 1/2). The probability of an error given that a 1 is sent is 2/9, while the probability of an error given a 0 is sent is 1/9.

(a) What is the probability that a 1 is received?

(b) What is the (unconditional) probability that an error occurs?

(c) What is the probability that a 1 was sent, given a 1 was received?

2.41 Consider the binary communication channel depicted below:

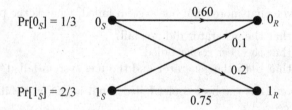

$\Pr[0_S] = 1/3 \quad 0_S$ 0.60 0_R
0.1
0.2
$\Pr[1_S] = 2/3 \quad 1_S$ 0.75 1_R

(a) What is $\Pr[1_R|0_S]$?

(b) What is $\Pr[1_R]$?

(c) What is $\Pr[\text{error}]$?

(d) What is $\Pr[0_S|1_R]$?

2.42 A binary symmetric communication channel is shown below.

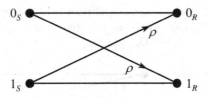

An "Error" occurs if 0 is sent and 1 is received or 1 is sent and 0 is received. Assume that $\Pr[1_R|0_S] = \Pr[0_R|1_S] = \rho$.

(a) Assume $\Pr[0_S] = \Pr[1_S] = \frac{1}{2}$. Find ρ such that $\Pr[\text{Error}] = 0.001$.

(b) Repeat part (a) for $\Pr[0_S] = 0.2$ and $\Pr[1_S] = 0.8$.

(c) What is the probability that a 0 was sent, given that a 0 is received? Assume the same conditions as part (a).

2.43 A binary communication channel is depicted below. Assume that the random experiment consists of transmitting a single binary digit and that the probability of transmitting a 0 or a 1 is the same.

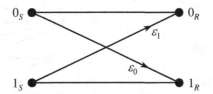

(a) Draw the sample space for the experiment and label each elementary event with its probability.

(b) What is the probability of an error?

(c) Given that an error occurred, what is the probability that a 1 was sent?

(d) What is the probability a 1 was sent given that a 1 was received?

2.44 In a digital communication system known as QPSK[6], one of four symbols $\{A,\ B,\ C,\ D\}$ is transmitted at a time. The channel characteristics are indicated in the figure below. The *prior* probabilities of the input symbols are given as: $\Pr[A_S] = \Pr[B_S] = \Pr[C_S] = \Pr[D_S] = 1/4$.

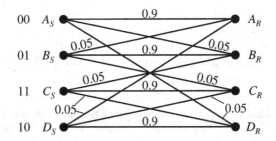

(a) What is the probability of receiving an A?

(b) Determine the probability of error in this channel.

(c) Given that D is received, find the probability that (i) A was sent, (ii) B was sent, (iii) C was sent, and (iv) D was sent.

---

[6] Quadrature Phase Shift Keying.

2.45 Repeat Prob. 2.44 for the channel shown in the figure below. The *prior* probabilities of the input symbols are given to be: $\Pr[A_S] = 3/16$, $\Pr[B_S] = 1/4$, $\Pr[C_S] = 5/16$, $\Pr[D_S] = 1/4$.

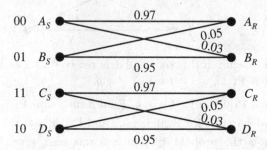

2.46 Never-Ready Wireless has just invented a new signaling system known as 3PSK. The system involves transmitting one of three symbols $\{0, 1, 2\}$. The conditional probabilities for their network are given in the following table:

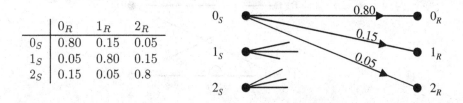

	0_R	1_R	2_R
0_S	0.80	0.15	0.05
1_S	0.05	0.80	0.15
2_S	0.15	0.05	0.8

For example, the probability that a 2 is received, given a 0 was transmitted is $\Pr[2_R|0_S] = 0.05$. A partial sketch of the communication channel is also provided above. The *prior* probabilities of transmitting each of the symbols are given by $\Pr[0_S] = 2/5$, $\Pr[1_S] = 2/5$ and $\Pr[2_S] = 1/5$.

(a) What is the probability of an error given that 0 was sent?

(b) What is the (unconditional) probability of error?

(c) Given that 1 is received, what is the probability that 1 was sent?

(d) Given that 1 is received, what is the probability that this is an error?

2.47 Consider an asymmetric binary communication channel characterized by conditional probabilities $\Pr[1_R|0_S] = 0.03$ and $\Pr[0_R|1_S] = 0.02$. The *a priori* probability of sending a binary 1 is given by $\Pr[1_S] = 0.56$. We now transmit 25 bits of binary information over this channel. What is the probability that there are less than three bit errors?

2.48 A binary sequence is transmitted from a source S to a destination D via a repeater station R. The communication channels on either side of the repeater station have different characteristics. The *a priori* probabilities at S are: $\Pr[1_S] = 0.52$ and $\Pr[0_S] = 0.48$. The conditional probabilities describing the channel between S and R are: $\Pr[1_R|1_S] = 1$, $\Pr[1_R|0_S] = 0.10$, and $\Pr[0_R|0_S] = 0.90$; and between R and D are: $\Pr[1_D|1_R] = 0.95$, $\Pr[0_D|1_R] = 0.05$, and $\Pr[0_D|0_R] = 1$.

(a) Draw the network diagram and label the branches.

(b) Determine $\Pr[1_R]$, i.e., the probability of a binary 1 at the repeater.

(c) Find the probability $\Pr[0_R|0_D]$.

2.49 Two symbols A and B are transmitted over a binary communications channel shown below. The two received symbols are denoted by α and β. The numbers on the branches in the graph indicate *conditional probabilities* that α or β is received given that A or B is transmitted. For example, $\Pr[\alpha|A] = 0.8$.

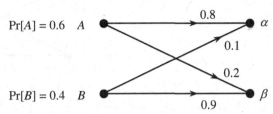

The symbols A and B are transmitted with probabilities indicated above.

(a) What is the probability that the received symbol is α (i.e., what is $\Pr[\alpha]$)?

(b) What decision rule should be used to minimize the probability of error? In other words, tell whether A or B should be chosen when α is received and tell whether A or B should be chosen when β is received.

(c) Someone has "switched the wires" so now the channel looks like:

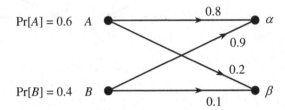

Answer the question in part (b) for this new situation. Also compute the *probability of error* corresponding to this decision rule.

2.50 Consider the following message: FORMULA FOR SUCCESS

(a) Estimate the probabilities of the letters in the message using the relative frequency approach.

(b) Find the average information in the message.

2.51 The secure data communication line running into Admiral Ennud's office can be modeled on a bit level by a binary communication channel. What the admiral doesn't know and even his closest associates won't tell him however, is that the line does not go directly from the server to the admiral's office. The line from the server goes to a patch panel in the basement where the signal is detected and regenerated and sent to the admiral's office via another line. The overall channel can be represented by the concatenation of two binary communication channels as shown below:

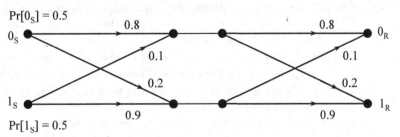

(a) Find the parameters of the single binary communication channel which is equivalent to the above. Label the branch probabilities below.

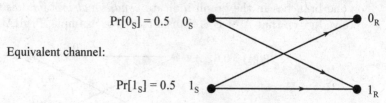

$$Pr[0_S] = 0.5 \quad 0_S$$

Equivalent channel:

$$Pr[1_S] = 0.5 \quad 1_S$$

(b) Compute the probability of error, Pr[error], for the equivalent channel.

(c) An IT technician has decided that the probability of error can be reduced if she inverts the connection to the second channel. In other words, the situation would look like this:

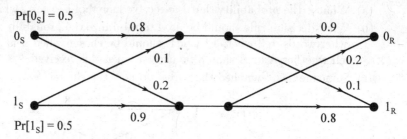

Is she correct? Justify your answer.

2.52 Consider the following seemingly meaningless sentence:

MAT CAT MAN MITTEN MAM MEN EATEN.

(a) Estimate the probabilities of each of the symbols in the sentence by computing the relative frequency of the symbols. You may ignore spaces.

(b) Compute the average information for the sentence.

(c) Obtain the Huffman code for each of the symbols in the sentence.

(d) Determine the average length of the codewords.

HUFFMAN ALGORITHM

Step 1: Arrange the symbols in order of decreasing probability. If the probabilities are equal, break the tie randomly.

Step 2: Combine the two smallest values in the list to form a new entry whose probability is the sum of the two original entries.

Step 3: Continue combining two smallest values in the modified list (original untouched values and the new entry) until the root of the tree is reached.

Step 4: Assign bits from left to right on the tree: at each branch, assign a 0 to the higher valued branch and a 1 to the other.

Step 5: Form the codewords by reading from the root to the node of the symbol.

Computer Projects

Project 2.1

In this project, you are to simulate a simple binary communication channel characterized by appropriate conditional and prior probabilities and estimate the probability of error as well as the probability of receiving either a 1 or a 0.

We start out with a symmetric binary communication channel characterized by the conditional probabilities $\Pr[0_R|0_S] = \Pr[1_R|1_S] = 0.975$ and $\Pr[0_R|1_S] = \Pr[1_R|0_S] = 0.025$. The prior probabilities of a 0 or a 1 are given by $\Pr[0_S] = 0.512$ and $\Pr[1_S] = 0.488$. The input to the binary communication channel is to be a sequence of 0s and 1s. Each 0 or 1 in the sequence is statistically independent of the others and is generated according to the probabilities $\Pr[0_S]$ and $\Pr[1_S]$ given above.

1. Generate the data input sequence of 0s and 1s according to the required probabilities. The size of the sequence is your choice; however, to obtain meaningful results, it should be at least 5000 points long.

2. Simulate the channel by writing a computer program to do the following:

   (a) When a 0 is presented as input to the channel, the channel should generate an output of 0 with probability $\Pr[0_R|0_S]$ and an output of 1 with probability $\Pr[1_R|0_S]$ (where these numerical values are given above).

   (b) When a 1 is presented to the channel, the channel should generate a 0 with probability $\Pr[0_R|1_S]$ and a 1 with probability $\Pr[1_R|1_S]$.

3. Compute the theoretical values for the following probabilities: $\Pr[0_S|0_R]$, $\Pr[1_S|1_R]$, and $\Pr[\text{error}]$ (see Section 2.5.1).

4. Apply the input data sequence generated in Step 1 to the channel in Step 2, estimate the probabilities in Step 3. To estimate the probability, use relative frequency; for example, to estimate $\Pr[0_S|0_R]$ you would compute

$$\frac{\#\text{ times 0 sent and 0 received}}{\#\text{ times 0 received}}$$

5. Compare the estimated values to the theoretical values. If necessary, repeat the experiment using a longer input sequence. You may also wish to compare the results for *various* length input sequences.

Repeat Steps 1 through 5 for a nonsymmetric binary communication channel with conditional probabilities $\Pr[0_R|0_S] = 0.975$, $\Pr[1_R|1_S] = 0.9579$, $\Pr[1_R|0_S] = 0.025$, $\Pr[0_R|1_S] = 0.0421$. Let the prior probabilities be $\Pr[0_S] = 0.5213$ and $\Pr[1_S] = 0.4787$.

MATLAB programming notes

You can generate a binary random number with the following statement:

```
x=rand(1)<P;
```

where P is the desired probability of a 1. Note that the expression to the right of the equal sign is a boolean expression; thus MATLAB assigns a 0 or 1 to the variable x. You can replace the argument '1' of the 'rand' function with the size of a vector or matrix and generate a whole vector or matrix of binary random numbers.

Project 2.2

The Huffman coding algorithm is described in Prob. 2.52 (above). In this project, you will measure the average information, code the given message using the Huffman algorithm, transmit the coded message over a simple binary channel, reconstruct the message, and compare the performance to that without coding. The following schematic diagram illustrates the sequence for implementing these operations.

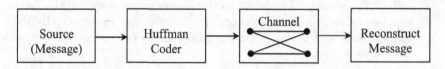

1. Consider the following message for source coding and transmission:

   APPARENTLY NEUTRAL'S PROTEST IS THOROUGHLY DISCOUNTED AND IGNORED. ISMAN HARD HIT. BLOCKAGE ISSUE AFFECTS PRE-TEXT FOR EMBARGO ON BY-PRODUCTS, EJECTING SUETS AND VEGETABLE OILS.

   This message is included in the data package for this book (msg.txt).

2. Following the procedure of Section 2.5.2, determine the average information for this message. (You may ignore hyphens, spaces, and punctuation.)

3. Using the Huffman algorithm, code the above message.

4. Determine the average codeword length.

5. Transmit the codewords in binary form across the binary symmetric communication channel specified in Project 2.1.

6. Reconstruct the message from the received codewords.

7. Determine the average information of the received message. Would you expect it to be larger or smaller than that of the transmitted message?

3 Random Variables and Transformations

Situations involving probability do not always deal strictly with events. Frequently there are real-valued measurements or observations associated with a random experiment. Such measurements or observations are represented by *random variables*.

This chapter develops the necessary mathematical tools for the analysis of experiments involving random variables. It begins with discrete random variables, i.e., those random variables that take on only a discrete (but possibly countably infinite) set of possible values. Some common types of discrete random variables are described that are useful in practical applications. Moving from the discrete to the continuous, the chapter discusses random variables that can take on an uncountably infinite set of possible values and some common types of these random variables.

The chapter also develops methods to deal with problems where one random variable is described in terms of another. This is the subject of "transformations."

The chapter concludes with two important practical applications. The first involves the detection of a random signal in noise. This problem, which is fundamental to every radar, sonar, and communication system, can be developed using just the information in this chapter. The second application involves the classification of objects from a number of color measurements. Although the problem may seem unrelated to the detection problem, some of the underlying principles are identical.

3.1 Discrete Random Variables

Formally, a random variable is defined as a function $X(\cdot)$ that assigns a real number to each elementary event in the sample space. In other words, it is a mapping from the sample space to the real line (see Fig 3.1). A random variable therefore takes on

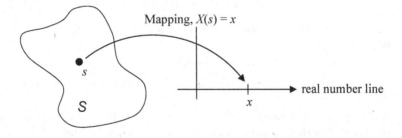

Figure 3.1 Illustration of a random variable.

a given numerical value with some specified probability. A simple example is useful to make this abstract idea clearer. Consider the experiment of rolling a pair of dice. The sample space showing the complete set of outcomes is illustrated in Chapter 2, Fig. 2.3. Given an outcome s, let us define the random variable K as follows:

$$K(s) = \text{the number rolled}$$

Now, suppose s is the point in the sample space representing a roll of $(3,1)$ on the dice; then $K(s) = 4$. Likewise, if s represents the roll of $(2,2)$ or $(1,3)$ then again $K(s) = 4$. Thus the mapping provided by the random variable is in general many-to-one.

In discussing random variables it is common to drop the argument s and simply write (for example) $K = 4$. In this example K is a *discrete* random variable, meaning that it takes on only a discrete set of values, namely the integer values 2 through 12. The probability assigned to these values of K is determined by the probability of the outcomes s in the sample space. If each outcome in the sample space has probability $1/36$, then the probability that $K = 4$ is $3/36$ or $1/12$. (This is the sum of the probabilities for the outcomes $(3,1)$, $(2,2)$, and $(1,3)$ all of which result in $K = 4$.) Similarly, the probability that $K = 2$ is $1/36$. (There is only one experimental outcome which results in this value of K.)

The probability associated with random variables is conveniently represented by a plot similar to a bar chart and illustrated in Fig. 3.2. The function $f_K[k]$ depicted in

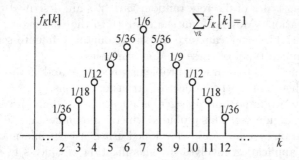

Figure 3.2 Probability mass function for rolling of dice.

the figure is called a *probability mass function* (PMF). The PMF shows the probability of the random variable for each possible value that it can take on. The sum of the probabilities represented in the PMF is therefore equal to 1.

Let us now establish some notation and a formal definition. It is common to represent a random variable by a capital letter such as K. The value that the random variable takes on is then denoted by the corresponding lower case letter k. The PMF is then a function $f_K[k]$ defined by

$$f_K[k] = \Pr[s \in \mathcal{S} : K(s) = k]$$

In other words, $f_K[k]$ is the probability that random variable K takes on the value k. Less formally, we can write

$$f_K[k] = \Pr[K = k] \tag{3.1}$$

where the precise meaning of the expression $\Pr[K = k]$ is given in the equation preceding (3.1).

The probability that the random variable takes on a set of values is obtained by summing the PMF over that whole set of values. For example, the probability of rolling a number greater than or equal to 10 is given by (see Fig. 3.2)

$$\Pr[K \geq 10] = \sum_{k=10}^{12} f_K[k] = \frac{1}{12} + \frac{1}{18} + \frac{1}{36} = \frac{1}{6}$$

Formally, the PMF must satisfy two conditions: First, since $f_K[k]$ represents a *probability*, it must be true that

$$0 \leq f_K[k] \leq 1 \tag{3.2}$$

for all values of k. Secondly, the sum of all the values is one:

$$\sum_{k} f_K[k] = 1 \qquad (3.3)$$

Any discrete function satisfying these two conditions can be a PMF.

To further illustrate the PMF, let us consider another example.

Example 3.1: The price of Simon's CDs is three for $1. With the sales tax this comes to $1.08 or $.36 per CD. Customers will always return bad CDs and Simon will exchange any bad CD for another one guaranteed to be good, but customers have determined that the cost of returning a CD in terms of trouble and aggravation is $.14. The net effect is that good CDs cost $.36 while bad ones cost the customer $.50. What is the discrete probability density function for the random variable C, the cost of buying three diskettes at Simon's? What is the probability that you will end up paying more than $1.25?

The events of the sample space and the cost associated with each event are shown below:

Events:	BBB	BBG	BGB	BGG	GBB	GBG	GGB	GGG
Cost:	$1.50	$1.36	$1.36	$1.22	$1.36	$1.22	$1.22	$1.08
Probability:	1/8	1/8	1/8	1/8	1/8	1/8	1/8	1/8

Since all of the events in the sample space are equally likely, we can construct the PMF by counting events in the sample space that result in the same cost. The resulting PMF is shown below.

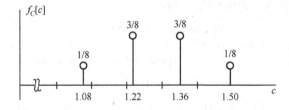

The probability of paying more than $1.25 is given by

$$f_C[1.36] + f_C[1.50] = \frac{3}{8} + \frac{1}{8} = \frac{1}{2}$$

□

This example illustrates that although the domain of the PMF is discrete, the random variable does not necessarily have to take on integer values. Integer values for a discrete random variable are the most common, however.

3.2 Some Common Discrete Probability Distributions

There are a number of common types of random variables that occur frequently in various problems. The PMFs that describe these random variables are sometimes referred to as "probability laws" or *distributions*. A few of these common distributions are discussed below. In all cases the random variable is assumed to take on values in the set of integers.

3.2.1 *Bernoulli random variable*

A Bernoulli random variable is a discrete random variable that takes on only one of
two discrete values (usually 0 and 1). The Bernoulli PMF is defined by

$$f_K[k] = \begin{cases} p & k = 1 \\ 1 - p & k = 0 \\ 0 & \text{otherwise} \end{cases} \tag{3.4}$$

and is illustrated in Fig. 3.3. The parameter p is the probability of a one.

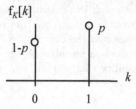

Figure 3.3 The Bernoulli
PMF.

Example 3.2:   A certain binary message $M[n]$ is represented by a sequence of zeros and
ones. If the sequence is sampled at a random time n_o the result $K = M[n_o]$ is a
Bernoulli variable. Unless there is evidence to the contrary, we would expect the
parameter of the distribution to be $p = 0.5$.

    □

3.2.2 *Binomial random variable*

Consider a binary sequence of length n, where each element of the sequence is a
Bernoulli random variable occurring independently. Let the random variable K be
defined as the number of 1's in the sequence. Consider any such sequence, say

$$1\,0\,0\,1\,1\,\dots\,0\,1$$

in which k 1's occur. The corresponding probability of this sequence is

$$p \cdot (1 - p) \cdot (1 - p) \cdot p \cdot p \cdots (1 - p) \cdot p = p^k (1 - p)^{(n-k)}$$

If you were to list all the possible sequences with k 1's, you would find that there
are $C_k^n = \binom{n}{k}$ such sequences (see Appendix A). Therefore the probability that one of
these sequences occurs is given by[1]

$$f_K[k] = \binom{n}{k} p^k (1 - p)^{(n-k)} \qquad 0 \le k \le n \tag{3.5}$$

This PMF is known as the binomial distribution. It is depicted in Fig. 3.4 for the
parameter values $n = 20$ and $p = 0.25$.

---

[1] Throughout this section, $f_K[k]$ may be assumed to be 0 for values of k not explicitly represented
in the formula.

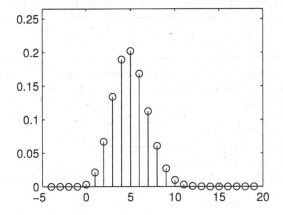

Figure 3.4 The binomial distribution.
$(n = 20, p = 0.25)$

Example 3.3:   A certain modem connection has a channel bit error rate of $p = 0.01$. Given that data are sent as packets of 100 bits, what is the probability that (a) just 1 bit is in error and (b) that 3 bits are in error?

The answers to the question are

$$\Pr[1 \text{ bit error}] = f_K[1] = \binom{100}{1} 0.01^1 \, 0.99^{99} = 0.3697$$

$$\Pr[3 \text{ bit errors}] = f_K[3] = \binom{100}{3} 0.01^3 \, 0.99^{97} = 0.0610$$

□

3.2.3 Geometric random variable

The geometric random variable is typically used to represent waiting times. Again, consider a sequence of binary digits being sent over some communication channel. Let us consider the number of bits that are observed before the first error occurs; this is a random variable K. (The *method* of error detection is irrelevant.) A tree diagram for this experiment is shown in Fig. 3.5. If the first error occurs at the kth bit, then $k - 1$

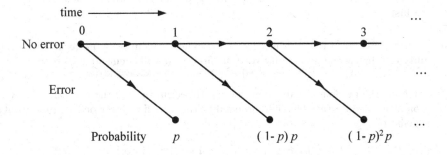

Figure 3.5 Tree diagram for error detection experiment.

bits are without error. The probability of this event is given by the geometric PMF:

$$f_K[k] = p(1-p)^{(k-1)} \qquad 1 \le k < \infty \qquad (3.6)$$

Actually, there are two common variations of this distribution. The other form corresponds to an error occuring at the $k+1^{st}$ bit and is given by

$$f_K[k] = p(1-p)^k \qquad 0 \le k < \infty \qquad (3.7)$$

The distributions (3.6) and (3.7) will be referred to as Type 1 and Type 0 respectively. A plot of the Type 0 geometric PMF for $p = 0.25$ is shown in Fig. 3.6. The function is a simple decaying exponential.

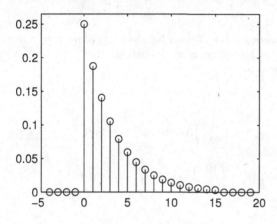

Figure 3.6 The geometric distribution (Type 0). $(p = 0.25)$

Since this is the first example of a random variable that can take on an *infinite* number of possible values, it is interesting to check that the sum of the PMF is equal to 1 as required. This is most easily done by considering the second form of the distribution and defining $q = 1 - p$. Then since $q < 1$ we can write

$$\sum_{k=0}^{\infty} f_K[k] = \sum_{k=0}^{\infty} (1-q)q^k = (1-q)\sum_{k=0}^{\infty} q^k = (1-q)\frac{1}{1-q} = 1$$

where the formula for the sum of an infinite geometric series has been used in the second to last step.[2]

Example 3.4: Errors in the transmission of a random stream of bytes occur with probability ϵ. What is the probability that the first error will occur after 16 bytes?

Let K be the random variable representing the occurrence of the first error. So K is a geometric random variable. The probability that the first error occurs after 16 bytes is given by

$$\Pr[K > 16] = \sum_{k=16}^{\infty} \epsilon(1-\epsilon)^k = (1-\epsilon)^{16}$$

---

[2] See, e.g., [1].

The last step is left as a challenge to you to evaluate the sum using the formula for the summation of an infinite geometric series. You can check that the answer is correct by the following argument. The event that the first error occurs after the 16th byte is the same as the product or union of events:

$$\bigcup_{k=1}^{16} (\text{"No error on the } k^{\text{th}} \text{ byte"})$$

Since the events in the union are mutually independent and each has probability $1 - \epsilon$, the probability of the overall event is $(1 - \epsilon)^{16}$.

$\square$

3.2.4 Poisson random variable

Consider the situation where events occur randomly in a certain time interval t. The "events" could be the arrival of print jobs at a network server, the number of "hits" on a particular web page, the number of telephone calls occuring in a busy office, or any similar set of events. The time interval could be measured in microseconds or hours or any other set of units appropriate to the problem. Let λ be a parameter that represents the *rate* of arrival of these events (e.g., $\lambda = 20$ hits/hour on a certain web page). Then, a useful expression for the probability of k events happening in time t that has been experimentally verified is

$$\Pr[k \text{ events in } t] = \frac{(\lambda t)^k}{k!} e^{-\lambda t}$$

(This formula is derived later in the text.) The number of events K is said to be a Poisson random variable with corresponding PMF

$$f_K[k] = \frac{\alpha^k}{k!} e^{-\alpha} \qquad 0 \le k < \infty \tag{3.8}$$

The single parameter $\alpha = \lambda t$ can be interpreted as the average number of events arriving in the time interval t. For example, if $\lambda = 20$ events/hour and t=0.2 hour (12 minutes), then $\alpha = (20)(0.2) = 4$. A plot of the PMF for $\alpha = 4$ is given in Fig. 3.7.

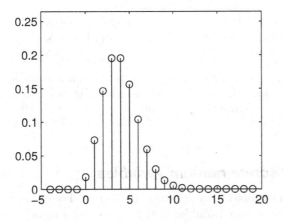

Figure 3.7 The Poisson PMF. ($\alpha = 4$)

Example 3.5:   Packets at a certain node on the internet arrive with a rate of 100 packets per minute. What is the probability that no packets arrive in 6 seconds? What about the probability that 2 or more packets arrive in the first 6 seconds?

The parameter α is given by

$$\alpha = \lambda t = 100/\text{min} \cdot \frac{6}{60} \text{ min} = 10 \text{ (arrivals)}$$

The probability that no packets arrive is thus given by

$$f_K[0] = \frac{10^0}{0!} e^{-10} = e^{-10} = 4.540 \times 10^{-5}$$

To answer the second question, let us first compute the probability of a *single* arrival in the first 6 seconds. This is given by

$$f_K[1] = \frac{10^1}{1!} e^{-10} = 10e^{-10} = 4.540 \times 10^{-4}$$

The probability of two or more arrivals is thus given by

$$1 - f_K[0] - f_K[1] = 0.9995$$

□

The Poisson PMF is also useful as an approximation to the binomial PMF. This approximation is given by

$$\binom{n}{k} p^k (1-p)^{(n-k)} \approx \frac{\alpha^k}{k!} e^{-\alpha}$$

where $\alpha = np$ in the limit as $n \to \infty$ and $p \to 0$. This relation is discussed further in Section 11.1 of Chapter 11.

3.2.5 Discrete uniform random variable

The final distribution to be discussed is the discrete uniform distribution defined by

$$f_K[k] = \begin{cases} \dfrac{1}{n-m+1} & m \leq k \leq n \\ 0 & \text{otherwise} \end{cases} \tag{3.9}$$

and illustrated in Fig. 3.8 for $m = 0$, $n = 9$. This PMF is useful as the formal description of a discrete random variable all of whose values over some range are equally likely.

Example 3.6:   Let K be a random variable corresponding to the outcome of the roll of a single die. K is a discrete uniform random variable with parameters $m = 1$ and $n = 6$. All of the nonzero values of the function are equal to $1/(n-m+1) = 1/6$.

□

3.2.6 Other types of discrete random variables

Several other discrete random variables arise elsewhere in the text. For example, Chapter 5 describes some discrete random variables that arise as sums of other random variables. A listing of discrete random variables that appear in the book can be found on the inside covers along with the sections in which they are described.

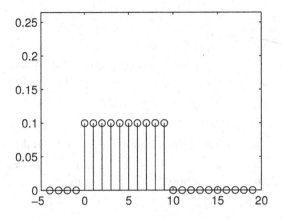

Figure 3.8 The discrete uniform distribution. ($m = 0$, $n = 9$)

3.3 Continuous Random Variables

Not all random variables take on a discrete set of values, even if the number of such values is infinite. For example, suppose an experiment is to monitor the power line during an electrical storm. The peak voltage measured from a power surge on the line can be represented by a random variable that takes on a continuum of values (over some possible range). In order to deal with this kind of problem, new tools are needed to work with continuous random variables.

3.3.1 Probabilistic description of continuous random variables

Consider the example of the surge voltage cited above. The event that the peak voltage is less than, say 2000 volts has some reasonable likelihood of occurrence. The event that the voltage is *precisely equal* to 2000 volts (or any other specific value) is extremely unlikely, however. A mathematical model involving continuous random variables should reflect these characteristics and assign probability to intervals of the real line. In order that these intervals have finite probability between 0 and 1, all (or at least most) of the infinite number of points within an interval should have probability approaching 0.

Let us begin with a random variable X and consider the probability

$$F_X(x) = \Pr[s \in \mathcal{S} : X(s) \leq x]$$

As with discrete random variables, we denote the random variable by a capital letter (X) and a value that it takes on by the corresponding lower case letter (x). To make the notation less cumbersome, reference to the sample space is frequently dropped and the previous equation is written simply as

$$F_X(x) = \Pr[X \leq x] \tag{3.10}$$

The quantity $F_X(x)$ is known as the *cumulative distribution function* (CDF) and has the general character depicted in Fig. 3.9. When plotted as a function of its argument x, this function starts at 0 and increases monotonically toward a maximum value of 1 (see Fig. 3.9). The following properties, which can be shown rigorously, make sense intuitively:

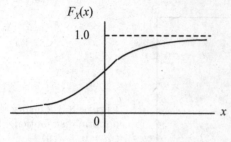

Figure 3.9 Cumulative distribution function.

1. Since no value of X can be less than minus infinity, $F_X(-\infty)$ should be 0.

2. If Δx is any positive value, then $\Pr[X \leq x + \Delta x]$ should not be less than $\Pr[X \leq x]$. Thus $F_X(x)$ should be monotonically increasing.

3. Since every value of X is less than infinity, $F_X(\infty)$ should be 1.

Next, consider the probability that X is in some more general interval[3] $a < X \leq b$ where a and b are any two values with $a < b$. We can write

$$\Pr[X \leq b] = \Pr[X \leq a] + \Pr[a < X \leq b]$$

This follows because the two intervals on the right are nonoverlapping and their union is the interval on the left ($X \leq b$). Then rearranging this equation and using the definition (3.10) yields

$$\Pr[a < X \leq b] = F_X(b) - F_X(a) \tag{3.11}$$

Now, *assume* for the moment that $F_X(x)$ is continuous and differentiable throughout its range, and let $f_X(x)$ represent the derivative

$$f_X(x) = \frac{dF_X(x)}{dx} \tag{3.12}$$

By using this definition, (3.11) can be written as

$$\boxed{\Pr[a < X \leq b] = \int_a^b f_X(x)dx} \tag{3.13}$$

The function $f_X(x)$ is known as the *probability density function* (PDF). The CDF and the PDF are the primary tools for dealing with continuous random variables.

3.3.2 More about the PDF

A typical generic PDF is shown in Fig. 3.10. The PDF for a continuous random variable

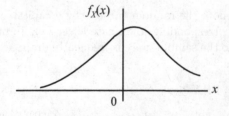

Figure 3.10 Typical probability density function for a continuous random variable.

has two fundamental properties. First, it is everywhere *nonnegative*:

$$f_X(x) \geq 0 \tag{3.14}$$

---

[3] The two different forms of inequality used here are deliberate and important in our development.

This follows directly from the monotonicity of $F_X(x)$ and (3.12). Secondly, the PDF must integrate to 1:

$$\int_{-\infty}^{\infty} f_X(x)dx = 1 \tag{3.15}$$

This follows from (3.13) by taking the interval $(a, b]$ to be the entire real line. These two conditions correspond to the two conditions (3.2) and (3.3) for the discrete probability density function. Note however, that since $f_X(x)$ is *not* a probability, but rather a probability *density*, is *not* required to be ≤ 1.

Let us explore this last point just a bit further. Consider (3.13) and let $b = a + \Delta_x$ where Δ_x is a small number. If $f_X(x)$ is continuous in the interval $(a, a + \Delta_x]$, then

$$\Pr[a < X \leq a + \Delta_x] = \int_a^{a+\Delta_x} f_X(x)dx \approx f_X(a) \cdot \Delta_x \tag{3.16}$$

The approximate equality follows from the mean value theorem of calculus and is consistent with our previous discussion. Equation 3.16 is a very important result because it shows that $f_X(x)$ by itself is *not* a probability, but it becomes a probability when multiplied by a small quantity Δ_x. *This interpretation of the probability density function for a continuous random variable will be used on several occasions in the text.*

3.3.3 A relation to discrete random variables

Let us to explore the relation between a continuous and discrete random variable in the context of a familiar application, namely that of analog-to-digital signal conversion. In the digital recording of audio signals such as speech or music, the analog signal has to be sampled and quantized. This operation is performed by an analog-to-digital coverter (ADC) as shown in Fig. 3.11. The ADC contains a quantizer which maps values of

$X \longrightarrow$ ADC $\longrightarrow K$

Figure 3.11 Analog-to-digital converter.

X into one of 2^N possible output values or "bins"; each of these bins is represented by a binary number with N bits. The number N is known as the bit "depth." It will further be assumed that the quantizer provides a signed binary integer k as output if the input x is in the interval $x_{k-1} < x \leq x_k$ where $x_k = x_{k-1} + \Delta$. The input/output characteristic of an ideal ADC is shown in Fig. 3.12.

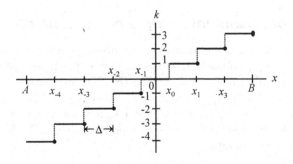

Figure 3.12 Input/output characteristic of a 3-bit ADC.

Suppose that the input signal at a particular point in time is a continuous random variable X taking values in the interval $A < X \leq B$. The PDF for the random variable is depicted in Fig. 3.13 (a). The output K of the quantizer is a discrete random variable that takes on the values $k = -2^{N-1}, \ldots, 0, 1, \ldots, 2^{N-1} - 1$. The PMF for the output K is depicted in Fig. 3.13 (b).

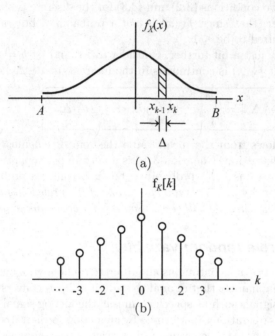

(a)

(b)

Figure 3.13 Discrete approximation for continuous random variable. (a) PDF for input random variable X. (b) PMF for output random variable K.

The probability that random variable K takes on a particular value k is given by the PMF $f_K[k]$. This probability is equal to the probability that X falls in the interval $x_{k-1} < X \leq x_k$. Thus we have

$$f_K[k] = \int_{x_{k-1}}^{x_k} f_X(x)dx = f_X(x'_k)\Delta$$

where x'_k, by the mean value theorem, is a point within the interval.

In a typical audio application, the bit depth N may be fairly large (e.g., $N = 16$ or $N = 20$). Therefore, except for the constant factor Δ, the PMF of the output is a very accurate discrete approximation to the PDF of the input.

3.3.4 Solving problems involving a continuous random variable

While solving problems for discrete random variables generally involves summing the PMF over various sets of points in its domain, solving problems involving continuous random variables usually involves integrating the PDF over some appropriate region or regions. In so doing, you must be careful to observe the limits over which various algebraic expressions may apply, and set up the limits of integration accordingly. In particular, the integral must *not* include regions where the PDF is 0. The following example illustrates these points.

Example 3.7: The time T (in seconds) for the completion of two sequential events is a continuous random variable with PDF given below:

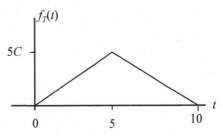

$$f_T(t) = \begin{cases} Ct & 0 < t \le 5 \\ C(10 - t) & 5 < t \le 10 \\ 0 & \text{otherwise} \end{cases}$$

What is the value of the constant C so that $f_T(t)$ is a legitimate probability density function? Find the probability that both events are completed in 3 seconds or less. What is the probability that both events are completed in 6 seconds or less?

To answer the first question, use the fact that the PDF must integrate to 1:

$$\int_{-\infty}^{\infty} f_T(t)dt = \text{ area of triangle } = 25C = 1$$

Therefore the constant C is equal to $1/25$. Notice how the problem of formally evaluating the integral by calculus was replaced by a simple geometric computation. This is a useful technique whenever it is possible.

The probability that the events are completed in no more than 3 seconds is found by integrating the density over the region shown shaded below.

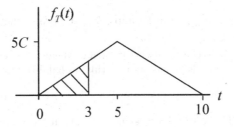

$$\Pr[T \le 3] = \int_{-\infty}^{3} f_T(t)dt = \int_{0}^{3} \frac{1}{25}t \, dt = \frac{t^2}{50}\Big|_0^3 = \frac{9}{50} = 0.18$$

Observe that since the PDF is 0 for negative values of t, when the expression $\frac{1}{25}t$ is substituted for $f_T(t)$ the lower limit on the integral becomes 0 (not $-\infty$!).

The probability that the events are completed in no more than 6 seconds is found by integrating the density over the following region:

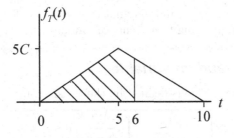

In this case, the region of integration has to be broken up into two regions where different algebraic expressions apply. The computation is as follows:

$$\Pr[T \le 6] = \int_{-\infty}^{6} f_T(t)dt = \int_{0}^{5} \frac{1}{25}t\,dt + \int_{5}^{6} \frac{1}{25}(10-t)\,dt$$

$$= \left.\frac{t^2}{50}\right|_{0}^{5} + \left.\frac{-(10-t)^2}{50}\right|_{5}^{6} = \frac{25}{50} + \frac{9}{50} = \frac{17}{25} = 0.68$$

□

When the intervals of concern are very small, the interpretation of probability density provided by (3.16) is useful. The next example explores this point.

Example 3.8:   Consider the time T (in seconds) for the completion of two sequential events as described by the PDF in Example 3.7. What is the most likely time of completion of the events? What is the (approximate) probability that the events are completed within 10 msec of the most likely time?

Since T is a continuous random variable, the probability for events to be completed at any *specific* time t is zero. We interpret the question however, to mean that the event is completed within a small interval of time $(t, t + \Delta_t]$. To find the most likely time of completion, notice that the probability that the event occurs in any such small interval is given by (see (3.16))

$$f_T(t)\Delta_t$$

From the sketch of $f_T(t)$ in Example 3.7, the time t that maximizes this probability is seen to be $t = 5$.

We interpret the second question to mean "What is the probability that T is between 4.99 and 5.01 seconds?" To compute this probability first use (3.16) to write

$$\Pr[5 < T \le 5 + 0.01] \approx f_T(5) \cdot 0.01 = \frac{1}{5}(0.01) = 0.002$$

Then, since the distribution is symmetric about the point $t = 5$, the probability that $4.99 < T \le 5$ is equal to the same value. Hence the probability that T is within 10 ms of the most likely value is approximately 0.004.

□

3.4 Some Common Continuous Probability Density Functions

A number of common forms for the PDF apply in certain applications. These probability density functions have closed form expressions, which can be found listed in several places. A few of these PDFs are described here. Expressions for the CDF are generally *not* listed in tables however, and may not even exist in closed form at all. This is one reason why most engineering treatments of random variables focus on using the PDF.

3.4.1 Uniform random variable

A uniform random variable is defined by the density function (PDF)

$$f_X(x) = \begin{cases} \dfrac{1}{b-a} & a \le x \le b \\[2mm] 0 & \text{otherwise} \end{cases} \qquad (3.17)$$

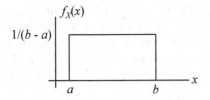

Figure 3.14 The uniform PDF.

and illustrated in Fig. 3.14. Like its discrete counterpart, this PDF is useful as the formal description of a random variable all of whose values over some range are equally likely.

Example 3.9:   An analog signal is hard limited to lie between -5 and $+5$ volts before digital-to-analog conversion. In the absence of any specific knowledge about the signal and its characteristics, the signal value can be modeled as a random variable having a uniform PDF with $a = -5$ and $b = +5$. (The signal is said to be "uniformly distributed" between -5 and $+5$ volts.)

□

3.4.2 Exponential random variable

The exponential random variable has the negative exponential PDF given by

$$f_X(x) = \begin{cases} \lambda e^{-\lambda x} & x \geq 0 \\ 0 & x < 0 \end{cases} \tag{3.18}$$

and depicted in Fig. 3.15. The exponential density is used to describe waiting times

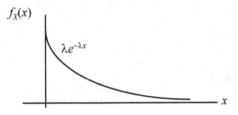

Figure 3.15 The exponential PDF.

for events, which usually follow this distribution in practice. The exponential density can be derived as a limiting form of the geometric distribution for discrete events (see e.g. [2]). It is also closely related to the Poisson random variable. In this context the parameter λ represents the arrival rate of any recurring events (electrons impinging on a cathode, jobs arriving at a printer server, buses arriving at a bus stop, etc.). While the Poisson PMF describes the *number* of arrivals in a fixed time interval t, the exponential density function $f_T(t)$ describes the waiting time T to the arrival of the *next* event.

Example 3.10:   In a printer queue, the waiting time W for the next job is an exponential random variable. Jobs arrive at an average rate of $\lambda = 30$ jobs per hour. What is the probability that the waiting time to the next job is between 2 and 4 minutes?

This probability is easily calculated:

$$\Pr\left[\tfrac{2}{60} < W \leq \tfrac{4}{60}\right] = \int_{2/60}^{4/60} 30 e^{-30w}\, dw = -e^{-30w}\Big|_{2/60}^{4/60} = e^{-1} - e^{-2} = 0.233$$

□

The exponential random variable is also a model for "time to failure" of a component or piece of equipment after it is put into service. In this context $1/\lambda$ is the average lifetime of the component, also called the "mean time to failure." Suppose that T represents the time to failure of a light bulb after it is installed. Then the probability that the bulb fails *after* some time t_0 is given by

$$\Pr[T > t_0] = \int_{t_0}^{\infty} \lambda e^{-\lambda t} dt = e^{-\lambda t_0}$$

The exponential random variable has what is called the *memoryless property*, which states that if the bulb has *not* failed at a time t_0 then the probability that it fails after an additional time t_1 (i.e., at time $t_0 + t_1$) is just $e^{-\lambda t_1}$. In other words, what happens up to time t_0 is "forgotten." The proof of this fact is rather straightforward:

$$\Pr[T > t_0 + t_1 | T > t_0] = \frac{\Pr\left[[T > t_0 + t_1] \cdot [T > t_0]\right]}{\Pr[T > t_0]} = \frac{\Pr[T > t_0 + t_1]}{\Pr[T > t_0]}$$

$$= \frac{e^{-\lambda(t_0 + t_1)}}{e^{-\lambda t_0}} = e^{-\lambda t_1} = \Pr[T > t_1]$$

3.4.3 Gaussian random variable

The last continuous random variable to be discussed here is the Gaussian or "normal" random variable. It arises naturally in all kinds of engineering and real-world problems and plays a central role in probability and statistics.

The Gaussian random variable has the PDF

$$f_X(x) = \frac{1}{\sqrt{2\pi\sigma^2}} e^{-(x-\mu)^2/2\sigma^2} \qquad -\infty < x < \infty \tag{3.19}$$

with the familiar "bell-shaped" curve shown in Fig. 3.16. The parameter μ represents

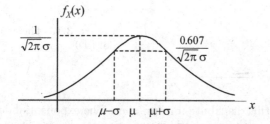

Figure 3.16 The Gaussian PDF.

the *mean* or average value of the random variable and the parameter σ^2 is called the *variance*. The square root of the variance (σ) is known as the *standard deviation* and is a measure of the spread of the distribution. It can be shown that the probability that any Gaussian random variable lies between $\mu - \sigma$ and $\mu + \sigma$ is approximately 0.683.

While the Gaussian density is an extremely important PDF in applications, integrals that appear in expressions like

$$\Pr[a < X \le b] = \frac{1}{\sqrt{2\pi\sigma^2}} \int_a^b e^{-(x-\mu)^2/2\sigma^2} dx \tag{3.20}$$

cannot be evaluated analytically. Instead, integrals of some standard functions such as

$$\Phi(x) = \frac{1}{\sqrt{2\pi}} \int_{-\infty}^x e^{-z^2/2} dz \tag{3.21}$$

can be computed to almost any degree of accuracy and are widely tabulated. $\Phi(x)$ can be recognized as the cumulative distribution function (CDF) of a Gaussian random variable with parameters $\mu = 0$ and $\sigma^2 = 1$. The CDF of a Gaussian random variable with arbitrary mean and variance (μ, σ^2) can therefore be written as

$$F_X(x) = \Pr[X \leq x] = \Phi\left(\frac{x - \mu}{\sigma}\right) \tag{3.22}$$

(see Prob. 3.18). The probability in (3.20) then can be computed as $F_X(b) - F_X(a)$.

In electrical engineering it is more common to use the Q function defined as

$$Q(x) = \frac{1}{\sqrt{2\pi}} \int_x^\infty e^{-z^2/2} dz = 1 - \Phi(x) \tag{3.23}$$

because it corresponds directly to the probabilities of detection and false alarm in a communication system. For a Gaussian random variable with parameters μ and σ^2 we have

$$\boxed{\Pr[X > x] = Q\left(\frac{x - \mu}{\sigma}\right)} \tag{3.24}$$

This result is depicted graphically in Fig. 3.17 where the area of the shaded region represents the probabililty of the event $X > x$. When x is to the right of the mean, the

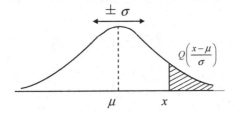

Figure 3.17 Probability of $X > x$ computed using the Q function (depicted for $x > \mu$).

region $X > x$ depicted in Fig. 3.17 is called a right "tail" of the distribution. Although the Q function is commonly thought of as a way to compute the probability when x is in the tail, the definition (3.23) and the relation (3.24) apply to *any* value of x. Therefore if x is chosen to the left of the mean as shown in Fig. 3.18, $Q((x - \mu)/\sigma)$

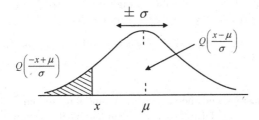

Figure 3.18 Probability of $X > x$ computed using the Q function (depicted for $x < \mu$).

represents the probability of the *unshaded* region depicted in the figure. In such cases, the argument of the Q function is *negative* and we can use the relation

$$\boxed{Q(-x) = 1 - Q(x)} \tag{3.25}$$

(see Prob. 3.19) to find the numerical value.

Let us consider a short numerical example.

Example 3.11: A Gaussian random variable has parameters $\mu = 1$ and $\sigma^2 = 4$. By using the tabulated values of the Q function on the inside cover of this book, find the probabilities (a) $\Pr[X > 3]$ and (b) $\Pr[X > -3]$.

(a) This corresponds to the case illustrated in Fig. 3.17. The Q function is used directly as in (3.24):

$$\Pr[X > 3] = Q\left(\frac{3-1}{2}\right) = Q(1) = 0.15866$$

where the value for $Q(1)$ is found in the table.

(b) Equation (3.24) can be applied in this case as well; however, the Q function turns out to have a negative argument:

$$\Pr[X > -3] = Q\left(\frac{-3-1}{2}\right) = Q(-2)$$

Although $Q(-2)$ does not appear in the table, (3.25) can be used to find

$$Q(-2) = 1 - Q(2) = 1 - 0.02275 = 0.97725$$

Therefore $\Pr[X > -3] = 0.97725$

□

The Q function can be used for other computations as well. Figure 3.18 shows the situation involving the left tail of the distribution defined by $-\infty < x < \mu$. The probability of the shaded region is

$$\Pr[X \leq x] = 1 - \Pr[X > x] = 1 - Q\left(\frac{x-\mu}{\sigma}\right) = Q\left(\frac{-x+\mu}{\sigma}\right)$$

where (3.25) was invoked in the last step. This result could also be obtained by exploiting the symmetry of the problem evident in Figs. 3.17 and 3.18.

Finally, the Q function can be used to compute the probability of a more general region for a Gaussian random variable. In particular,

$$\Pr[a < X \leq b] = Q\left(\frac{b-\mu}{\sigma}\right) - Q\left(\frac{a-\mu}{\sigma}\right) \tag{3.26}$$

which follows easily from (3.20) and (3.23).

The Gaussian distribution is familiar to engineers designing and analyzing electronic systems. The following example illustrates a typical such application.

Example 3.12: A radar receiver observes a voltage V which is equal to a signal (due to reflection from a target) plus noise:

$$V = s + N$$

Assume that the signal s is a constant and the noise N is a Gaussian random variable with mean $\mu = 0$ and variance $\sigma^2 = 4$ (mv^2). Assume further that the radar will detect the target if the voltage V is greater than 2 volts.

When the target is a certain distance from the radar, the signal s is equal to 3 volts. What is the probability that the radar detects the target?

The voltage V in this example is a Gaussian random variable with mean $\mu = s = 3$ and variance $\sigma^2 = 4$. (You can take this as a fact for now; it follows from the formula (3.33) discussed in Section 3.6.) The probability that the target is detected is given by

$$\Pr[V > 2] = Q\left(\frac{2-3}{2}\right) = Q(-0.5) = 1 - Q(0.5) = 1 - 0.31 = 0.69$$

where the Q function was found from the table and rounded.

□

With the parameters given above, the probability of detecting the target is not high. It is reasonable to turn the problem around and ask what conditions are necessary to achieve some desired (high) probability of detection.

Example 3.13:   As the target moves closer to the radar the signal becomes stronger. If the noise variance remains constant, what would be the signal strength s in volts required so that the probability of detecting the target is 0.98?

The probability of detection is given by $Q\left(\dfrac{2-s}{2}\right)$. Observing that the argument of the Q function is negative, we can write

$$1 - Q\left(\frac{-2+s}{2}\right) = 0.98 \quad \text{or} \quad Q\left(\frac{s-2}{2}\right) = 0.02$$

From the table it is found that the Q function is approximately 0.02 when its argument is 2.05. Therefore we require that $(s-2)/2 = 2.05$ or $s = 6.1$ volts.

□

3.4.4 Other types of continuous random variables

Several other types of continuous random variables and their distributions arise in later sections of this book. A number of these are listed on the inside front cover along with the sections in which they are described.

3.5 CDF and PDF for Discrete and Mixed Random Variables

3.5.1 Discrete random variables

The cumulative distribution function (CDF) is introduced in Section 3.3.1 as a probabilistic description for continuous random variables. In that section it is assumed that the CDF is continuous and differentiable; its derivative is the PDF.

The CDF can apply to discrete random variables as well, although the function as defined by (3.10) will exhibit discontinuities. For example, consider the Bernoulli random variable whose PMF is shown in Fig. 3.4. The CDF for this random variable is defined by

$$F_K(k) = \Pr[K \le k]$$

and is shown in Fig. 3.19.

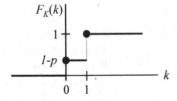

Figure 3.19 Cumulative distribution function for Bernoulli random variable.

Let us check out this plot according to the last equation, bearing in mind that K takes on only values of 0 and 1. Since the random variable K never takes on values less than 0, the function is 0 for $k < 0$. The probability that $K \le k$ when k is a number between 0 and 1 is just the probability that $K = 0$. Hence the value of the CDF is a constant equal to $1 - p$ in the interval $0 \leq k < 1$. Finally, the probability that $K \leq k$ for $k \geq 1$ is equal to 1. Therefore, the CDF is 1 in the interval $k \geq 1$.

This function clearly does not satisfy the conditions of continuity and differentiability previously assumed for *continuous* random variables. If we tolerate the discontinuities however, the CDF is a useful analysis tool since it allows for description of both discrete and continuous random variables.

A complete mathematical description of the CDF is beyond the scope of this text. An interesting discussion can be found in [3], however. For most engineering applications, we can make the following assumption:

> The CDF is continuous and differentiable everywhere except at a countable number of points, where it may have discontinuities.

At points of discontinuity, the "$\leq$" in the definition (3.10) implies that the CDF is continuous *from the right*. In other words, the function is determined by its value on the right rather than on the left (hence the dot in the illustration). The definition (3.10) also implies that the CDF must be *bounded between 0 and 1 and be monotonically increasing*.

With this broader definition of the CDF, the corresponding PDF cannot be defined unless a certain type of singularity is allowed. The needed singularity, however, is well known in engineering applications as the unit impulse, denoted by $\delta(x)$. A brief discussion of the unit impulse is provided in Appendix B.

The unit impulse $\delta(x)$ satisfies the following condition:

$$\int_{-\infty}^{x} \delta(\xi)d\xi = u(x) = \begin{cases} 1 & x \geq 0 \\ 0 & x < 0 \end{cases} \tag{3.27}$$

The function $u(x)$ defined on the right is the *unit step function* (see Appendix B) and can be used to account for any discontinuities that may appear in the CDF. *The impulse $\delta(x)$ can be regarded as the formal derivative of the step $u(x)$.* If the upper limit x in (3.27) becomes infinite, then the equation becomes

$$\int_{-\infty}^{\infty} \delta(\xi)d\xi = 1$$

which shows that the "area" of the impulse is equal to 1.

With this "function" available, the PDF for any random variable (continuous or discrete) can be defined as the derivative of the CDF. Fig. 3.20 shows the PDF for the

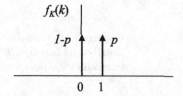

Figure 3.20 PDF for a Bernoulli random variable.

Bernoulli random variable whose CDF is depicted in 3.19. A comparison of this PDF to the PMF given in Fig. 3.3 shows that the discrete values have been replaced by impulses whos *areas* are equal to the corresponding probabilities. The areas are equal to the height of the discontinuities in the corresponding CDF. The general formula for expressing the PMF as a PDF with impulses is

$$f_K(k) = \sum_{\forall i} f_K[i]\delta(k - i) \tag{3.28}$$

Here the impulse occuring at $k = i$ is given the value (area) $f_K[i]$.

3.5.2 Mixed random variables

While we have so far only given examples of continuous and discrete random variables, the definition for the CDF and PDF allows for random variables that have aspects of both types. Such random variables are commonly referred to as *mixed* random variables. The CDF and PDF for a typical such random variable are illustrated in Fig. 3.21. For most values of x the probability $\Pr[X = x]$ is equal to 0; at the point $x = x_0$,

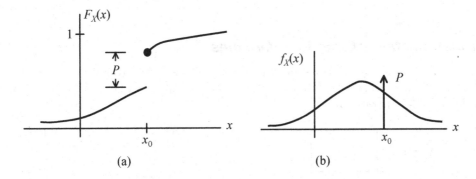

(a)   (b)

Figure 3.21 CDF and PDF for a random variable with a continuous and discrete component. (a) CDF. (b) PDF.

however, that probability is equal to a positive value P. (Note that the overall area under the density function, including the impulse, must integrate to 1.)

To see how such mixed random variables may occur, consider the following example.

Example 3.14: The input voltage X to the simple rectifier circuit shown below is a continuous Gaussian random variable with mean $m = 0$ and variance $\sigma^2 = 1$. The circuit contains an ideal diode which offers 0 resistance in the forward direction and infinite resistance in the backward direction. The circuit thus has the voltage input/output characteristic shown in the figure. It will be shown that the output voltage Y is a mixed random variable.

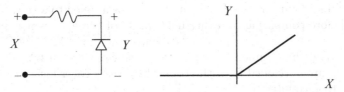

A more complete approach to problems of this type is developed in Section 3.6. For now, let us use the following argument. When the input voltage is positive, the output is equal to the input. Therefore, for positive values of X, the input and output PDF are identical. For negative values of the input, the output is 0. The probability of negative values is given by

$$\int_{-\infty}^{0} f_X(x)dx = 0.5$$

where $f_X(x)$ represents the Gaussian density of (3.19) with parameter $m = 0$. Since the probability of a negative input is 0.5, the probability of a 0 output is correspondingly 0.5. Thus the output PDF has an impulse with area of 0.5 at the location $y = 0$. The sketch below shows the input and output density functions. The output Y is a mixed random variable.

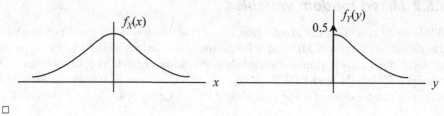

□

3.6 Transformation of Random Variables

In many important applications, one random variable is specified in terms of another. You can think of this as a *transformation* from a random variable X to a random variable Y as illustrated in Fig. 3.22.

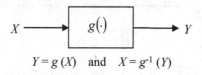

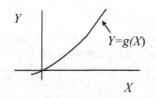

Figure 3.22 Transformation of a random variable.

Frequently this transformation is linear (or at least represented by a straight line). For instance, in Example 3.12, the output random variable V is given by $V = S + N$ where N is the random variable representing the noise and S is a constant. In some cases however, the transformation may be nonlinear. In the process of rectification, a voltage is applied to a diode or other nonlinear device to convert alternating current to direct current. In a neural network, a sum of inputs is applied to a sigmoidal or other nonlinearity before comparing to a threshold. In all of these cases it is important to be able to compute the output PDF ($f_Y(y)$) when you know the input PDF ($f_X(x)$).

In the following, three important cases are considered for the transformation and simple formulas are developed for each. These three cases cover most of the situations you are likely to encounter.

3.6.1 When the transformation is invertible

Figure 3.23 shows a portion of the transfer characteristic when g is an invertible

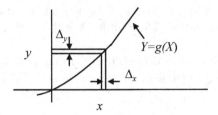

Figure 3.23 Section of invertible transfer characteristic.

function. Considering a point (x, y) on the diagram, we can write

$$\Pr[y < Y \le y + \Delta_y] = \Pr[x < X \le x + \Delta_x]$$

since both sides of the equation represent the same event. Thus

$$f_Y(y)\Delta_y = f_X(x)\Delta_x$$

Further, for any small Δ_x, we have to any desired degree of approximation

$$\Delta_y = \left|\frac{dy}{dx}\right|\Delta_x$$

(The absolute value is to allow for the fact that the slope of the curve might be negative.) Combining these last two equations and cancelling the common term Δ_x produces

$$\boxed{f_Y(y) = \frac{1}{|dy/dx|}f_X(x)\bigg|_{x=g^{-1}(y)}} \tag{3.29}$$

The entire expression on the right is evaluated at $x = g^{-1}(y)$ where g^{-1} represents the inverse transformation, because $f_Y(y)$ needs to be expressed in terms of y.

As an example of the use of this equation, suppose

$$Y = aX + b \tag{3.30}$$

Then $dy/dx = a$ and $g^{-1}(y) = (y - b)/a$. Therefore,

$$f_Y(y) = \frac{1}{|a|}f_X\left(\frac{y-b}{a}\right) \tag{3.31}$$

Notice that the absolute value of a appears outside of the function but not inside. This is important!

Although (3.31) is a very useful result, it is probably easier to memorize it and apply it as two separate special cases. These are summarized in the two boxes below.

For $b = 0$:

$$\boxed{\begin{aligned} Y &= aX \\ f_Y(y) &= \frac{1}{|a|}f_X\left(\frac{y}{a}\right) \end{aligned}} \tag{3.32}$$

while for $a = 1$:

$$\boxed{\begin{aligned} Y &= X + b \\ f_Y(y) &= f_X(y - b) \end{aligned}} \tag{3.33}$$

Example 3.15: Suppose random variables X and Y are related by $Y = 2X + 1$ and that X is uniformly distributed between -1 and $+1$. The PDF for X is shown in part (a) of the sketch below.

If the intermediate random variable $U = 2X$ is defined, then the PDF of U is given (from (3.32)) by $f_U(u) = \frac{1}{2} f_X\left(\frac{u}{2}\right)$. This PDF is shown in part (b) of the figure.

Now let $Y = U + 1$.

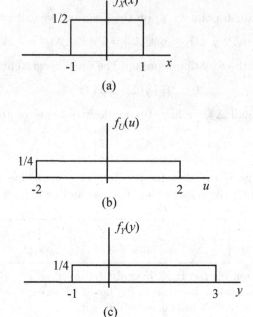

(a)

(b)

(c)

The PDF of Y is thus given (from (3.33)) by $f_Y(y) = f_U(y - 1) = \frac{1}{2} f_X\left(\frac{y-1}{2}\right)$. This final PDF is shown in part (c).

□

It is interesting to note in this example that both the original variable and the transformed variable turned out to be uniformly distributed. This is not a coincidence. Any time there is a straight-line transformation of the form (3.30), the *shape* of the PDF does not change. This is evident from (3.31) where it can be seen that f_Y is just a shifted and scaled version of f_X. Thus under a straight line transformation, a uniform random variable remains uniform (with different parameters), a Gaussian random variable remains Gaussian, and so on. Let us explore this point just a bit further for the Gaussian case.

Example 3.16: Random variables X and Y are related by $Y = aX + b$. X is a Gaussian random variable with mean μ and variance σ^2. Demonstrate that Y is also a Gaussian random variable. What are the parameters describing the random variable Y?

To answer this question, simply apply the transformation formula (3.31) using f_X as given by (3.19):

$$f_Y(y) = \frac{1}{|a|} \frac{1}{\sqrt{2\pi\sigma^2}} e^{-[((y-b)/a)-\mu]^2/2\sigma^2} = \frac{1}{\sqrt{2\pi a^2 \sigma^2}} e^{-[y-(a\mu+b)]^2/(2a^2\sigma^2)}$$

From this it is seen that Y has a Gaussian PDF with parameters $\mu' = a\mu + b$ and $\sigma'^2 = a^2\sigma^2$.

□

3.6.2 *When the transformation is not invertible*

First of all, let us mention that the formula (3.29) derived above works just fine when the transformation provides a one-to-one mapping of X to Y but Y does not take on all possible values on the real line. For example, the transformation $Y = e^X$ results in only positive values of Y. The inverse transformation therefore does not exist for negative values of Y. (Mathematicians would say that the transformation is one-to-one but not *onto*.) Nevertheless we can use (3.29) to write

$$f_Y(y) = y f_X(\ln y) \qquad 0 \le y < \infty$$

The main problem arises when the transformation is not one-to-one. This case is illustrated in Fig. 3.24. In this illustration, $g(x_1)$, $g(x_2)$ and $g(x_3)$ all produce the same

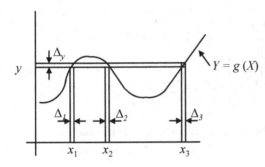

Figure 3.24 Section of noninvertible transfer characteristic.

value of y. In other words, the equation

$$g(x) = y$$

has three roots (x_1, x_2, x_3); there is no unique *inverse* for the function $g(x)$.

This case can be handled by a generalization of the technique applied in the previous subsection, however. The event "$y < Y \le y + \Delta_y$" is the sum of three disjoint events involving X. Therefore we can write

$$\Pr[y < Y \le y + \Delta_y] = \Pr[x_1 < X \le x_1 + \Delta_1] + \Pr[x_2 < X \le x_2 + \Delta_2] + \Pr[x_3 < X \le x_3 + \Delta_3]$$

or

$$f_Y(y)\Delta_y = f_X(x_1)\Delta_1 + f_X(x_2)\Delta_2 + f_X(x_3)\Delta_3$$

Note that the increments in x are generally different from each other because they must correspond to the *same* value for Δ_y and the slope of the function may be different at each point. Specifically, the increments satisfy the relation

$$\Delta_y = \left|\frac{dy}{dx}\right|_{x_i} \Delta_i \qquad i = 1, 2, 3$$

Then by combining the last two equations and cancelling the common term Δ_y we obtain the desired result,

$$\boxed{f_Y(y) = \sum_{i=1}^{r} \frac{1}{|dy/dx|} f_X(x)\Big|_{x=g_i^{-1}(y)}} \qquad (3.34)$$

which has been generalized to an arbitrary number of roots r. The notation $g_i^{-1}(y)$ is not a true inverse, but represents an algebraic expression that holds in the region of the root x_i. The application of the formula (3.34) is best illustrated through an example.

Example 3.17: The noise X in a certain electronic system is a Gaussian random variable with mean $\mu = 0$ and variance σ^2:

$$f_X(x) = \frac{1}{\sqrt{2\pi\sigma^2}}e^{-x^2/2\sigma^2} \qquad -\infty < x < \infty \qquad (1)$$

(see Fig. 3.16). The normalized instantaneous power of the noise is given by

$$Y = \frac{X^2}{\sigma^2}$$

where the parameter σ^2 can be interpreted as the average noise power.[4] The PDF of the noise power can be computed as follows.

The transformation given by the last equation is sketched below.

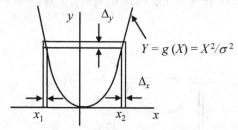

For a given value of $Y > 0$ the noise has two specific values, namely

$$x_1 = g_1^{-1}(y) = -\sigma\sqrt{y} \quad \text{and} \quad x_2 = g_2^{-1}(y) = +\sigma\sqrt{y}$$

The magnitude of the derivative is given by

$$\left|\frac{dy}{dx}\right| = |2x/\sigma^2| = 2\sqrt{y}/\sigma$$

The PDF of the noise power is then found by applying (3.34):

$$f_Y(y) = \frac{1}{2\sqrt{y}/\sigma}f_X(-\sigma\sqrt{y}) + \frac{1}{2\sqrt{y}/\sigma}f_X(+\sigma\sqrt{y})$$

Substituting the Gaussian density function (1) in this equation and simplifying yields

$$f_Y(y) = \frac{1}{\sqrt{2\pi y}}e^{-y/2} \qquad 0 \le x < \infty$$

(This PDF is known as a Chi-squared density with one degree of freedom.)
□

Using the CDF

An alternative approach for computing the density of a transformed random variable involves using the cumulative distribution function. This method, which works for both invertible and non-invertible transformations, can sometimes be algebraically simpler and thus is worth knowing about. The idea is to express the CDF of the new variable $F_Y(y)$ in terms of the CDF of the original variable $F_X(x)$ and take the derivative. The procedure is best illustrated by an example.

---

[4] This fact is discussed further in Chapter 9.

Example 3.18: To compute the PDF of the random variable Y in Example 3.17 we could first compute the CDF of Y and then take the derivative.

Begin by observing from the figure in Example 3.17 that

$$\Pr[Y \le y] = \Pr[x_2 \le X \le x_1] = \Pr\left[\frac{-\sqrt{y}}{\sigma} \le X \le \frac{\sqrt{y}}{\sigma}\right]$$

This equation can be written in terms of cumulative distribution functions as

$$F_Y(y) = F_X\left(\frac{\sqrt{y}}{\sigma}\right) - F_X\left(\frac{-\sqrt{y}}{\sigma}\right)$$

Taking the derivative with respect to y then yields

$$f_Y(y) = \frac{1}{2\sqrt{y}/\sigma} f_X(\sigma\sqrt{y}) + \frac{1}{2\sqrt{y}/\sigma} f_X(-\sigma\sqrt{y})$$

which leads to the same result as in Example 3.17.

□

3.6.3 When the transformation has flat regions or discontinuities

An extreme case of a non-invertible transformation occurs when the transformation has step discontinuities, or regions over which it is constant. The treatment of these cases is illustrated using the method involving the cumulative distribution function.

Transformation with flat region

Figure 3.25 shows a typical CDF and a transformation involving a flat region for

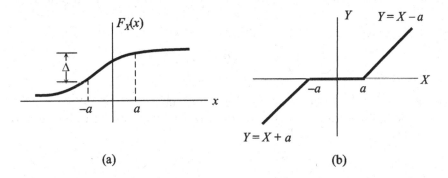

Figure 3.25 A typical CDF and a transformation. (a) CDF for X. (b) Transformation with flat region.

$-a < X \le a$. Since the transformation has three distinct regions let us consider three distinct cases.

(i) First, assume $y > 0$. This puts us on the rightmost branch of the curve so we can write

$$\Pr[Y \le y] = \Pr[X \le y + a]$$

or, in terms of the CDFs:

$$F_Y(y) = F_X(y + a); \quad y > 0 \tag{3.35}$$

Figure 3.26 shows the part of the transformation being dealt with (dark line) and

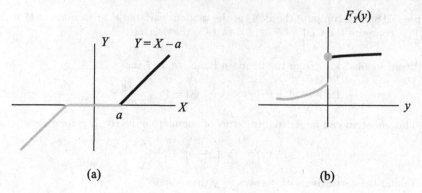

Figure 3.26 Development of CDF for Y. (a) Transformation. (b) CDF for Y.

the resulting portion of F_Y also shown with a dark line. Note that in this region F_Y is just equal to F_X shifted to the left.

(ii) Next consider $y = 0$. When y is 0, the event $Y \leq y$ corresponds to the event $X \leq a$. The probabilities of these two events can be stated in terms of the CDFs as

$$F_Y(y) = F_X(a); \qquad y = 0 \tag{3.36}$$

This represents a single point of the distribution shown as a grey dot in Fig. 3.26 (b).

(iii) Finally, consider $y < 0$. In this case $\Pr[Y \leq y] = \Pr[X \leq y - a]$; therefore

$$F_Y(y) = F_X(y - a); \qquad y < 0 \tag{3.37}$$

which corresponds to the last part of the distribution shown in grey in Fig. 3.26 (b).

These last three equations can be combined to write

$$F_Y(y) = \begin{cases} F_X(y + a) & y \geq 0 \\ F_X(y - a) & y < 0 \end{cases}$$

where the use of "$\geq$" takes care of the value at $y = 0$.

Before taking the derivative to obtain the PDF, let's be careful to notice that F_Y has a discontinuity of magnitude

$$\Delta = F_X(a) - F_X(-a)$$

at $y = 0$. This introduces an impulse of area Δ in the density function. The resulting PDF is

$$f_Y(y) = \begin{cases} f_X(y - a) & y \geq 0 \\ f_X(y + a) & y < 0 \end{cases} + \Delta\delta(y)$$

Figure 3.27 shows the form of the CDF and PDF for X and compares it to the CDF and PDF for Y.

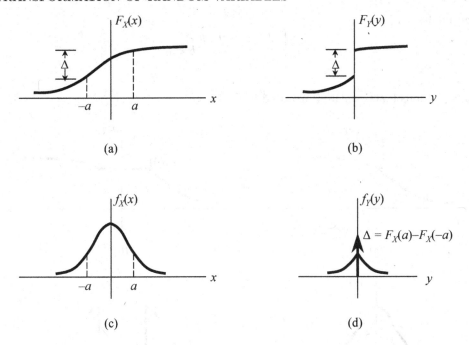

Figure 3.27 Distribution and density functions for the transformation of Fig. 3.26. (a) CDF of X. (b) CDF of Y. (c) PDF of X. (d) PDF of Y.

Transformation with discontinuity

A second case of transformations is one involving discontinuities, as in Fig. 3.28.

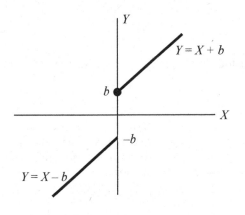

Figure 3.28 Transformation with discontinuity.

In this case we can write

$$\Pr[Y \le y] = \begin{cases} \Pr[x \le y - b] & y > b \\ \Pr[x \le 0] & -b < y \le b \\ \Pr[x \le y + b] & y \le -b \end{cases}$$

therefore,

$$F_Y(y) = \begin{cases} F_X(y - b) & y > b \\ F_X(0) & -b < y \le b \\ F_X(y + b) & y \le -b \end{cases}$$

and finally

$$f_Y(y) = \begin{cases} f_X(y-b) & y > b \\ 0 & -b < y \le b \\ f_X(y+b) & y \le -b \end{cases}$$

These various functions are illustrated in Fig. 3.29.

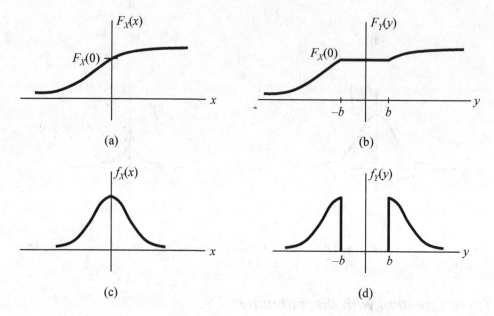

(a)

(b)

(c)

(d)

Figure 3.29 Distribution and density functions for the transformation of Fig. 3.28. (a) CDF of X. (b) CDF of Y. (c) PDF of X. (d) PDF of Y.

Transformation with both

As a final example, consider the transformation shown in Fig. 3.30, which has both

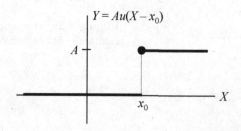

Figure 3.30 Transformation with both discontinuity and flat regions; u is the unit step function.

discontinuities and flat regions (and nothing else!) Proceeding as before, we write

$$\Pr[Y \le y] = \begin{cases} 0 & y < 0 \\ \Pr[X \le x_0] & 0 \le y < A \\ 1 & y \ge A \end{cases}$$

Writing the probabilities as CDFs then produces

$$F_Y(y) = \begin{cases} 0 & y < 0 \\ F_X(x_0) & 0 \le y < A \\ 1 & y \ge A \end{cases}$$

The function can be written compactly in terms of the unit step function.

$$F_Y(y) = F_X(x_0)u(y) + (1 - F_X(x_0)u(y - A)$$

The CDF is depicted in Fig. 3.31 (a).

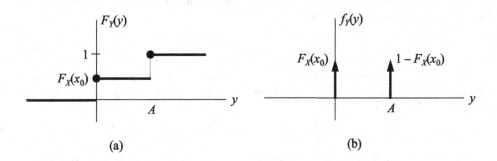

(a)   (b)

Figure 3.31 CDF and PDF corresponding to the transformation of Fig. 3.30.

The PDF is obtained by taking the derivative, thus producing two impulses.

$$f_Y(y) = F_X(x_0)\delta(y) + (1 - F_X(x_0)\delta(y - A)$$

The PDF is shown in 3.31 (b).

It is easy to generalize the preceding examples to other cases. In applying this method, it is important to observe whether or not each inequality is strict, especially if steps or impulses are involved in the original distribution.

3.7 Distributions Conditioned on an Event

Sometimes it is necessary to deal with a random variable in the context of a conditioning event. A good example of this is provided in the next section of this chapter.

For a discrete random variable, the *conditional* PMF is defined by

$$f_{K|A}[k|A] = \frac{\Pr[K = k, A]}{\Pr[A]} \tag{3.38}$$

where A is the conditioning event. In defining this quantity we have used a convenient shorthand notation on the right. In particular, the expression $\Pr[K = k, A]$ is interpreted as the joint probability of the event $\{s \in S : K(s) = k\}$ and the event A. The notation $f_{K|A}[k|A]$ although redundant, is quite conventional.

The conditional PMF has all of the usual properties of a probability mass function. In particular, it satisfies the two conditions

$$0 \le f_K[k] \le 1 \tag{3.39}$$

and

$$\sum_{k \in A} f_K[k] = 1 \tag{3.40}$$

where in the last expression the sum is taken over all values of k that the event A allows.

For a discrete *or* continuous (or mixed) random variable, the conditional cumulative distribution function is defined by

$$F_{X|A}[x|A] = \frac{\Pr[X \leq x, A]}{\Pr[A]} \tag{3.41}$$

This function has properties like any CDF; it is bounded between 0 and 1 and is a monotonically increasing function of x. The conditional PDF is then defined as the formal *derivative* of the conditional CDF,

$$f_{X|A}(x|A) = \frac{dF_{X|A}(x|A)}{dx} \tag{3.42}$$

where as before, the derivative is assumed to exist and can be represented by impulses at discontinuities. The conditional density is everywhere nonnegative and satisfies the condition

$$\int_{-\infty}^{\infty} f_{X|A}(x|A)dx = 1$$

For regions where there are no singularities, it is useful to think of the conditional PDF in terms of the following relation analogous to (3.16):

$$\Pr[x < X \leq x + \Delta_x | A] \approx f_{X|A}(x|A) \cdot \Delta_x \tag{3.43}$$

Equation 3.43 can be used to develop one more important result. Let us define the notation $\Pr[A|x]$ to mean

$$\Pr[A|x] = \lim_{\Delta_x \to 0} \Pr[A \mid x < X \leq x + \Delta_x] \tag{3.44}$$

This is the probability of the event A given that X is in an infinitesimally small increment near the value x. To show that this limit exists, we can use (2.17) of Chapter 2 and substitute (3.16) and (3.43) to write

$$\Pr[A|x < X \leq x + \Delta_x] = \frac{\Pr[x < X \leq x + \Delta_x \mid A] \cdot \Pr[A]}{\Pr[x < X \leq x + \Delta_x]} \approx \frac{f_{X|A}(x|A)\Delta_x \cdot \Pr[A]}{f_X(x)\Delta_x}$$

Since the Δ_x cancels, the limit in (3.44) exists, and these last two equations combine to yield

$$\Pr[A|x] = \frac{f_{X|A}(x|A)\Pr[A]}{f_X(x)} \tag{3.45}$$

The corresponding discrete form is easily shown to be

$$\Pr[A|k] = \frac{f_{K|A}[k|A]\Pr[A]}{f_K[k]} \tag{3.46}$$

Equations 3.45 and 3.46 are useful in situations where the probability of an event is influenced by the observation of a random variable. In statistics, problems of this type occur in what is known as statistical inference. In electrical engineering, the same type of problem comes about in applications of detection and classification. The latter applications are addressed in the next section.

Example 3.19: Consider a random variable X whose PDF is given below.

$$f_X(x) = \begin{cases} x & 0 \leq x < 1 \\ 2 - x & 1 \leq x \leq 2 \\ 0 & \text{otherwise} \end{cases}$$

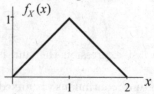

Let the event A be defined as the event "$X > 1$." Let us find the conditional PDF of the random variable.

The CDF of X can be found by integration. The result is sketched below. (The analytical expressions are not important in this example.)

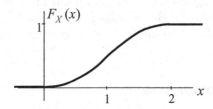

Next we find $F_{X|A}$ using (3.41). Observe that the numerator in (3.41) is just the original CDF restricted to the region where event A is true; that is,

$$\Pr[X \leq x, A] = \begin{cases} F_X(x) & x > 1 \\ 0 & \text{otherwise} \end{cases}$$

Additionally, it is easy to find that $\Pr[A] = 0.5$; thus from (3.41)

$$F_{X|A}[x|A] = \begin{cases} 2F_X(x) & x > 1 \\ 0 & \text{otherwise} \end{cases}$$

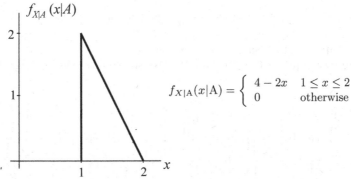

The conditional PDF is then found by taking the derivative (see (3.42)). The result is given below.

$$f_{X|A}(x|A) = \begin{cases} 4 - 2x & 1 \leq x \leq 2 \\ 0 & \text{otherwise} \end{cases}$$

Notice that both the conditional CDF and the conditional PDF are just scaled versions of the original (unconditional) CDF and PDF in the region where the condition applies, and are 0 elsewhere.

□

3.8 Applications Involving Random Variables

The tools developed in this chapter can be applied to a number of important engineering problems. Three different but related problems are described here. The first is digital communication, where a digital receiver must decide how to decode a received waveform into 0s and 1s. The second is the radar/sonar detection problem where the receiver must decide whether or not a target is present. The third is the classification problem where a digital computer with interface to a sensor must decide between (i.e., identify) two or more types of objects that are being observed. All of these applications involve the theory of *hypothesis testing* and result in *optimal decisions* for the particular problem area. Although the details of any of these applications can be complex, the associated principles of optimal decision can be developed with a basic knowledge of random variables.

3.8.1 Digital communication

Principles of optimal decision

In modern communication, which includes television, satellite radio, mobile telephoning, wireless networking and a host of other activities, the underlying technology is digital. In the simplest case shown in Fig. 3.32 a modulator maps a binary bit stream to a sequence of waveforms representing 1's or 0's and transmits these waveforms to a receiver. For example, in the method known as binary phase shift keying (BPSK)

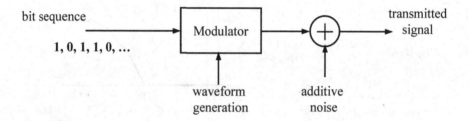

Figure 3.32 Digital signal transmission.

a 1 is mapped into a short sinusoidal burst while a 0 is mapped into a similar burst with opposite phase (see Fig. 3.33). The sinusoid allows the signal to propagate at

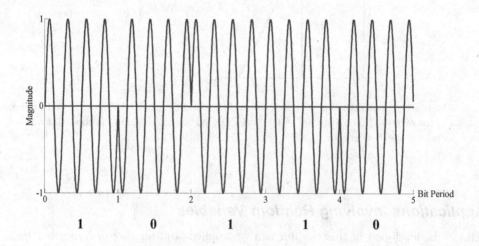

Figure 3.33 Waveforms for BPSK.

a suitably high frequency or in a band that is sufficiently free from interference by other users. In the various stages of transmission and propagation, noise is added to the signal. Since noise is a random phenomenon, the receiver cannot be certain about whether the received signal represents a 1 or a 0.

The receiver, depicted in Fig. 3.34, performs various operations of tuning and pro-

Figure 3.34 Receiver and processing prior to a binary decision (0 or 1).

cessing. Typically the received signal is multiplied by a local oscillator, filtered, and integrated to produce a single measurement represented by the random variable X. To complete the process, the receiver must then decide whether the value of X corresponds to logical 1 or logical 0.

Because the noise is additive and the processing by the receiver is linear, the random variable X will have a component due to the signal and a component due to the noise. Let us assume that the signal component is a constant s when a 1 is transmitted and $-s$ when a 0 is transmitted. If the noise is represented by a random variable N, then there are two distinct possibilities for X, referred to as *hypotheses* H_1 and H_0.

$$H_1: \quad X = s + N \quad \text{(the transmission is a 1)}$$
$$H_0: \quad X = -s + N \quad \text{(the transmission is a 0)}$$
(3.47)

Given an observation x for the random variable X, the receiver needs to decide in an optimal way, between H_1 and H_0.

The hypotheses H_1 and H_0 can be regarded as *events*. Therefore, using the ideas developed in Section 3.7, the following quantities can be defined for $i = 0, 1$:

term	name
$f_{X|H_i}(x|H_i)$	Conditional density function
$\Pr[H_i]$	Prior probability
$\Pr[H_i|x]$	Posterior probability

The *conditional density function* is the PDF for the observation X, which depends on whether a 1 or a 0 was transmitted. The *prior probabilities* reflect our knowledge of the likelihood of 1s and 0s in the bit stream and are assumed to be known. (Usually they are the same.) Finally, the *posterior probabilities* represent the likelihood that a 1 or a 0 was transmitted after observing the random variable X. According to (3.45), these quantities are related as

$$\Pr[H_i|x] = \frac{f_{X|H_i}(x|H_i)\Pr[H_i]}{f_X(x)}; \qquad i = 0, 1$$
(3.48)

Now, let us turn to the decision procedure and suppose that it is desired to minimize the probability of making an error. Consider the following logical argument. To minimize $\Pr[\text{error}]$ it is necessary to minimize $\Pr[\text{error}|x]$ for each possible value of x. This follows from a form of the principle of total probability written for random variables as

$$\Pr[\text{error}] = \int_{-\infty}^{\infty} \Pr[\text{error}|x]\, f_X(x)\,dx \qquad (3.49)$$

(Compare this formula with (2.15) of Chapter 2.) Now, let x_o denote an observed value of X and suppose the posterior probability curves are as shown in Fig. 3.35. If

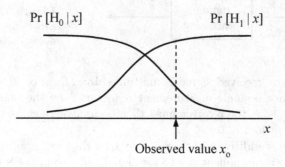

Pr $[H_0 \mid x]$ Pr $[H_1 \mid x]$

Observed value x_o

Figure 3.35 Typical plots of posterior probability.

we decide that x_o corresponds to H_0, we will be right with probability $\Pr[H_0|x_o]$ and wrong with probability $\Pr[H_1|x_o]$. Of course, if we decide that x_o corresponds to H_1, we will be right with probability $\Pr[H_1|x_o]$ and wrong with probability $\Pr[H_0|x_o]$. To minimize the probability of a wrong decision (error) we should choose the hypothesis with the highest posterior probability. In the case illustrated, $\Pr[H_1|x_o] > \Pr[H_0|x_o]$ so we should choose H_1.

A generalization of this argument shows that H_1 should be chosen *whenever* $\Pr[H_1|x] > \Pr[H_0|x]$ and H_0 should be chosen otherwise. The optimal decision rule, that minimizes the probability of error, is written as

$$\Pr[H_1|x] \underset{H_0}{\overset{H_1}{\gtrless}} \Pr[H_0|x] \qquad (3.50)$$

This result is intuitively satisfying; the optimal decision rule is to *choose the hypothesis that is most likely* when conditioned on the observation.

To put the decision rule in a more workable form, we can substitute (3.48) in (3.50) and cancel the common denominator term to obtain

$$f_{X|H_1}(x|H_1)\Pr[H_1] \underset{H_0}{\overset{H_1}{\gtrless}} f_{X|H_0}(x|H_0)\Pr[H_0]$$

Finally, if we define

$$P_i = \Pr[H_i]; \qquad i = 0,1$$

then the decision rule can be written in the form

$$\boxed{\dfrac{f_{X|H_1}(x|H_1)}{f_{X|H_0}(x|H_0)} \underset{H_0}{\overset{H_1}{\gtrless}} \lambda} \qquad (3.51)$$

where $\lambda = P_0/P_1$. The term on the left side of (3.51) is a fundamental quantity in statistical decision theory called the *likelihood ratio* while the variable λ is referred to as the *threshold*. The decision rule itself is known as a *likelihood ratio test*. When the threshold is set equal to P_0/P_1, this test minimizes the probability of error.

The case of Gaussian noise

In many cases in the real world the noise has a Gaussian PDF with 0 mean:

$$f_N(n) = \frac{1}{\sqrt{2\pi\sigma^2}} e^{-n^2/2\sigma^2} \tag{3.52}$$

The conditional densities for the two hypotheses in (3.47) are then given by

$$f_{X|H_1}(x|H_1) = f_N(x-s) = \frac{1}{\sqrt{2\pi\sigma^2}} e^{-(x-s)^2/2\sigma^2}$$

and

$$f_{X|H_0}(x|H_0) = f_N(x+s) = \frac{1}{\sqrt{2\pi\sigma^2}} e^{-(x+s)^2/2\sigma^2}$$

In this Gaussian case, substituting these equations into (3.51), taking the logarithm and simplifying, leads to the following decision rule (see Prob. 3.62).

$$x \underset{H_0}{\overset{H_1}{\gtrless}} \tau; \qquad \text{where} \quad \tau = \frac{\sigma^2}{2s} \ln \frac{P_0}{P_1} \tag{3.53}$$

Here τ is an optimal threshold (for x) that minimizes the probability of error. Notice that when $P_1 = P_0$, the threshold τ is equal to 0.

Computing the error

The probability of error, which was chosen as the criterion for optimal decision, serves also as a measure of performance of the communication system. This performance measure can be computed as follows.

Two types of errors can be made in the communication system; a 0 can be mistaken for a 1 or a 1 can be mistaken for a 0. We will characterize these two types of errors by their *conditional* probabilities and define

$$\varepsilon_1 \overset{\text{def}}{=} \Pr[\text{error}|H_1]; \qquad \varepsilon_0 \overset{\text{def}}{=} \Pr[\text{error}|H_0] \tag{3.54}$$

These errors are illustrated graphically in Figs. 3.36 and 3.37.

Figure 3.36 shows the situation when a 1 is transmitted. The probability for the

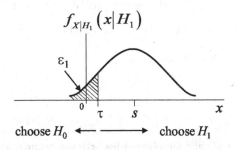

Figure 3.36 Illlustration of a type 1 error.

observation is the conditional density $f_{X|H_1}$. The decision rule (3.53) says to choose H_1 when the observation lies above the threshold and to choose H_0 otherwise. Therefore an error is made when the observation lies below the threshold; the probability ε_1 of that error is the shaded region under the density.

Figure 3.37 shows the corresponding situation when a 0 is transmitted. In this case an error is made when the observation lies *above* the threshold. The probability ε_0 is the shaded area under the density.

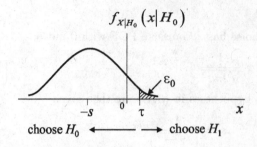

$f_{X|H_0}(x|H_0)$

ε_0

$-s$ 0 τ x

choose H_0 ⟵⟶ choose H_1

Figure 3.37 Illlustration of a type 0 error.

The figures illustrate that

$$\varepsilon_1 = \Pr[X \le \tau|H_1]$$
$$\varepsilon_0 = \Pr[X > \tau|H_0]$$

The (unconditional) probability of error can then be computed by using the principle of total probability. In particular,

$$\Pr[\text{error}] = \Pr[\text{error}|H_0]\Pr[H_0] + \Pr[\text{error}|H_1]\Pr[H_1]$$

or

$$\Pr[\text{error}] = \varepsilon_0\,P_0 + \varepsilon_1\,P_1 \tag{3.55}$$

If the noise is Gaussian, then the conditional error probabilities can be computed using the Q function:

$$\varepsilon_1 = Q\left(\frac{-\tau + s}{\sigma}\right) ; \qquad \varepsilon_0 = Q\left(\frac{\tau + s}{\sigma}\right) \tag{3.56}$$

(see Section 3.4.3). In the usual case where $P_1 = P_0 = 1/2$ the threshold τ is equal to zero and (3.55) reduces to

$$\Pr[\text{error}] = Q\left(\frac{s}{\sigma}\right) \tag{3.57}$$

The error probability depends completely on the ratio s/σ or equivalently on the quantity

$$\text{SNR} = 20\log_{10} s/\sigma \tag{3.58}$$

which is known as the *signal-to-noise ratio* in decibels (dB). Performance curves for the optimal receiver can be generated by plotting the errror probability versus SNR.

Example 3.20: Consider a communication system where the signal s in (3.47) is equal to 2 mv and the noise is a zero-mean Gaussian random variable with variance $\sigma^2 = 4\,\text{mv}^2$. The conditional densities for the two hypotheses are then given by

$$H_1\,(X = s + N): \quad f_{X|H_1}(x|H_1) = f_N(x - 2) = \frac{1}{\sqrt{2\pi\cdot4}}e^{-(x-2)^2/(2\cdot4)}$$

$$H_0\,(X = -s + N): \quad f_{X|H_0}(x|H_0) = f_N(x + 2) = \frac{1}{\sqrt{2\pi\cdot4}}e^{-(x+2)^2/(2\cdot4)}$$

Substituting these equations into (3.51) and simplifying leads to the optimal decision rule (3.53):

$$x \underset{H_0}{\overset{H_1}{\gtrless}} \tau ; \qquad \tau = \ln\frac{P_0}{P_1}$$

If the prior probabilities are equal ($P_0 = P_1 = 0.5$), then the threshold τ is equal to 0 and the conditional error probabilities are given by (see (3.56))

$$\varepsilon_1 = \varepsilon_0 = Q\left(\frac{2}{2}\right) = Q(1) = 0.1587$$

The probability of error for the binary communication system from (3.55) is then

$$\Pr[\text{error}] = \varepsilon_0 \, P_0 + \varepsilon_1 \, P_1 = 0.1587$$

It is instructive to compare the performance of the decision rule (3.53) for an arbitrary threshold τ. If we let the threshold τ take on a range of values, the probability of error (3.55) can be plotted for each value of the threshold as shown below.

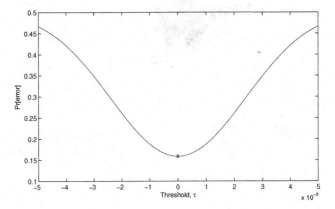

Since the prior probabilities are equal, the minimum error of 0.1587 occurs, as expected, at $\tau = 0$.

The error probabilities computed above are quite high and represent rather poor performance. This relates to the fact that the signal-to-noise ratio (3.58) is quite low (0 dB). If the level of the signal is increased, say to 4 mv, the SNR increases to about 6 dB and the probability of error decreases to approximately 0.0228. A performance curve for the system is provided below which shows probability of error versus SNR and the 0 and 6 dB points shown as an open square and open circle on the curve. From the plot it is also seen that to achieve a probability of error less than 10^{-2} the SNR needs to be greater than about 7.4 dB (open diamond with dotted lines).

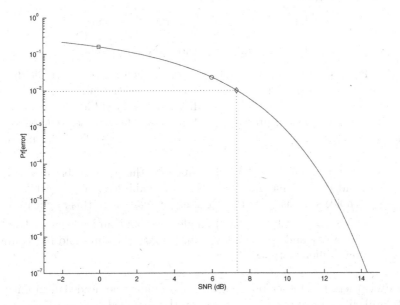

□

3.8.2 Radar/sonar target detection

A related problem involving optimal decision is the radar or sonar detection problem. In the radar case depicted in Fig. 3.38, an electromagnetic pulse[5] is sent out which

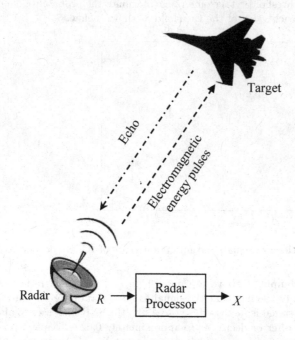

Target

Echo

Electromagnetic energy pulses

Radar R →

Radar Processor → X

Figure 3.38 Depiction of radar target detection.

reflects from an object (if present) called the *target*. The radar receives the reflected pulse and after processing, produces a random variable X which has a component due to the reflection from the target and a component due to noise. The following hypotheses can be defined

$$H_1 : \quad X = s + N \quad \text{(a target is present)}$$

$$H_0 : \quad X = N \quad\quad \text{(no target is present)}$$

(3.59)

where s is the component due to the target and N is a random variable due to noise.

    Since the problem is similar to the binary communication problem developed above, a similar analysis to minimize the probability of error would lead to a likelihood ratio test as the optimal decision procedure. Minimizing the error as defined by (3.55) is not the best criterion for this problem, however, for at least the following reasons:

- The threshold in (3.51) is defined by the prior probabilities P_1 and P_0. We may not know these probabilities, and an arbitrary choice such as $P_1 = P_0 = 1/2$ is not realistic in this application.

- The test involves a fixed single threshold and results in fixed errors ε_1 and ε_0. We may want to vary the threshold to "trade off" different types of errors.

Fortunately, there is another procedure that provides the needed flexibility and also leads to a likelihood ratio test. It is based on the so-called Neyman-Pearson criterion.

---

[5] In the sonar case, a sound pressure pulse is used.

To describe the procedure, let us first identify some relevant terms. In the context of target detection, the error term ε_0 is called the probability of *false alarm*. This is the probability that the system decides that a target is present, while in reality no target is present. The error term ε_1 is called the probability of a *missed detection*. This is the probability that the system decides that there is no target, when a target is in fact present. A related term is the quantity $1 - \varepsilon_1$ which is called the probability of *detection*. In the radar problem it is conventional to use the terms p_{fa} and p_d, defined as

$$
\begin{aligned}
p_{fa} &= \varepsilon_0 \\
p_d &= 1 - \varepsilon_1
\end{aligned}
$$

In the Neyman-Pearson procedure, the false alarm probability is fixed (say $p_{fa} = \alpha$) and the detection probability p_d is maximized subject to this constraint. This leads to a likelihood ratio test of the form

$$
\frac{f_{X|H_1}(x|H_1)}{f_{X|H_0}(x|H_0)} \underset{H_0}{\overset{H_1}{\gtrless}} \lambda \tag{3.60}
$$

where λ is a threshold that relates directly to the chosen value α for the probability of false alarm. The threshold can be varied to trade off higher detection rates for increased false alarms.

Unfortunately, deriving the Neyman-Pearson test is an exercise in the calculus of variations and would be an unnecessary diversion at this point. (Details can be found in [4].) Knowing that the result has the form (3.60), however, is sufficient motivation to explore the problem further.

The detection-false alarm trade-off

Let us now consider the specific case where the noise is Gaussian with PDF (3.52). The conditional density functions under each of the two hypotheses are then given by

$$
f_{X|H_1}(x|H_1) = f_N(x - s) = \frac{1}{\sqrt{2\pi\sigma^2}} e^{-(x-s)^2/2\sigma^2}
$$

and

$$
f_{X|H_0}(x|H_0) = f_N(x) = \frac{1}{\sqrt{2\pi\sigma^2}} e^{-(x)^2/2\sigma^2}
$$

Substituting these equations in the likelihood ratio (3.60) and taking logarithms leads to the decision rule (see Prob. 3.62)

$$
x \underset{H_0}{\overset{H_1}{\gtrless}} \tau \tag{3.61}
$$

where $\tau = 1 + 2\ln\lambda$ is a threshold (for x) that relates to the desired probability of false alarm.

Figure 3.39 shows the two conditional density functions on the same plot and an arbitrary threshold τ. The probability of false alarm p_{fa} is the shaded area to the right of the threshold which is under the density function for H_0. The probability of detection p_d is the (unshaded) area to the right of the threshold which is under the density function for H_1. The shaded area to the *left* of the threshold is the probability of missed detection $1 - p_d$. Observe that the threshold τ for any desired false alarm probability α can be found by solving

$$
\alpha = \int_\tau^\infty f_{X|H_0}(x|H_0)dx
$$

The result can be expressed as an inverse Q function.

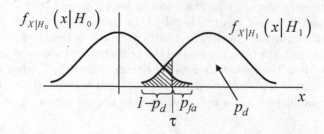

Figure 3.39 Conditional density functions showing probability of detection (p_d) and probability of false alarm (p_{fa}).

It is easy to see from Fig. 3.39 that as the threshold moves to the left both p_d and p_{fa} increase, while if it moves to the right these two probabilities decrease. Thus in order to increase detection probability we have to accept higher probability of false alarm. In radar and sonar detection problems it is common to plot p_d versus p_{fa} as the threshold τ is varied. Such a plot is called a receiver operating characteristic (ROC) and shows in a quantitative sense how high probability of detection can be traded off for low probability of false alarm. This is illustrated in the example below.

Example 3.21:   Consider a detection problem where the signal s in (3.59) is equal to 2 mv and the noise is a zero-mean Gaussian random variable with variance $\sigma^2 = 4\,\mathrm{mv}^2$. The conditional densities for the two hypotheses are then given by

$$H_0: \quad f_{X|H_0}(x|H_0) = f_N(x) = \frac{1}{\sqrt{2\pi \cdot 4}} e^{-x^2/(2 \cdot 4)}$$

$$H_1: \quad f_{X|H_1}(x|H_1) = f_N(x-2) = \frac{1}{\sqrt{2\pi \cdot 4}} e^{-(x-2)^2/(2 \cdot 4)}$$

Substituting these equations into (3.60) and simplifying leads to the decision rule (3.61):

$$x \underset{H_0}{\overset{H_1}{\gtrless}} \tau$$

The probabilities of detection and false alarm are given by

$$p_d = \Pr[X > \tau|H_1] \quad = \quad \int_\tau^\infty f_{X|H_1}(x|H_1)\,dx$$

$$p_{fa} = \Pr[X > \tau|H_0] \quad = \quad \int_\tau^\infty f_{X|H_0}(x|H_0)\,dx$$

As an example, for $\tau = 1$, these probabilities are given by

$$p_d = Q\left(\frac{1-2}{2}\right) = 0.6915 \qquad p_{fa} = Q\left(\frac{1-0}{2}\right) = 0.3085$$

If the threshold τ is now allowed to take on a range of values, we can plot the receiver operating characteristic (p_d versus p_{fa}). Two curves are shown below. The solid curve corresponds to the parameters specified above, which result in a 0 dB signal-to-noise ratio. The point on the curve for $\tau = 1$ is marked with a small square and corresponds to the computed values of $p_d = 0.6915$ and $p_{fa} = 0.3085$.

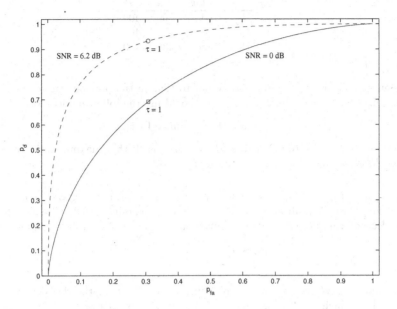

If the received signal s doubles in value (as it would as the target gets closer) then the SNR increases to 6.2 dB and the ROC changes to the curve shown with the dashed line. Here a threshold setting of $\tau = 1$ results in $p_d = 0.9332$ and $p_{fa} = 0.3085$, which represents better performance. Note that when the SNR is increased, but the noise power and the threshold τ are held fixed, the false alarm probability does not change. The detection probability increases, however, and the entire curve shifts toward the upper left corner of the graph.

□

3.8.3 Object classification

In a large number of problems, it is desired to provide automatic classification of objects based on some type of measurement. Examples of this include optical text recognition, military target identification, sorting of recyclable materials, and others. This is typically studied under the topic of Pattern Recognition [5, 6] and the statistical form of pattern recognition can be thought of as an expansion of the communication and detection problems described above. The approach can be illustrated through the simple example presented below.

Example 3.22: In a certain manufacturing plant, employees are required to wear colored badges. The color of the badge determines the level of access to any particular area of the plant; employees have to present their badges to a badge reader that can recognize the color.

Assume the badges are nominally red, green or blue. The color of the badge can vary, however, due to age, lighting conditions, inconsistency in printing or other factors. Let the actual color be described by its wavelength X in microns and assume that X is a Gaussian random variable with mean μ and variance σ^2 given in the table below. A hypothesis H_i is also associated with each color for the purpose of making decisions.

color	μ	σ^2
H_1: red	700	400
H_2: green	546	400
H_3: blue	436	400

It can be shown by an argument similar to that in the previous subsection, that to minimize the error in classification, one should use the following decision rule:

$$\text{choose } H_k = \text{argmax } \Pr[H_i|x]$$

The equation is read to mean H_k is the hypothesis H_i that maximizes the probability. This decision rule is the generalization of (3.50).

By applying (3.48) the decision rule can be written in terms of the conditional densities and the prior probabilities. Further, if the prior probabilities $\Pr[H_i]$ are all equal, then the decision rule becomes just a comparison of the conditional density functions:

$$\text{choose } H_k = \text{argmax } f_{X|H_i}(x|H_i)$$

For this case, where the prior probabilities are all equal, the decision regions are defined by whichever density function is larger. This is illustrated in the figure below.

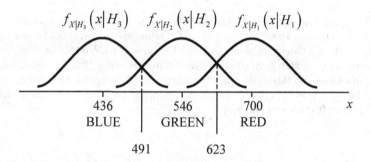

For example, if the color X is measured to be $x = 625$ microns, the density function $f_{X|H_1}$ evaluated at this point is larger than $f_{X|H_2}$ or $f_{X|H_3}$ (see figure). Therefore, the object would be classified as red. The decision region for "red" is thus the region where the red density function $f_{X|H_1}$ is highest.

The decision region boundaries shown here are centered between the means of the densities. This occurs because the priors are equal and because the variance σ^2 under each hypothesis is the same. If the variances were different (as they well could be) a more complex set of decision regions would result.

The probabilities of misclassification for the three colored badges are computed as follows.

The probability that a blue badge is misclassified is given by

$$\Pr[X > 491|H_3] = Q\left(\frac{491 - 436}{20}\right) = Q(2.75) \approx 2.98 \times 10^{-3}$$

The probability that a red badge is misclassified is given by

$$\Pr[X < 623|H_1] = 1 - Q\left(\frac{623 - 700}{20}\right) = Q\left(\frac{-623 + 700}{20}\right) = Q(3.85) \approx 5.91 \times 10^{-5}$$

Finally, the probability that a green badge is misclassified is given by the sum of two terms:

$$\Pr[X > 623|H_2] + \Pr[X < 491|H_2] \quad = \quad Q\left(\frac{623 - 546}{20}\right) + Q\left(\frac{-491 + 546}{20}\right)$$

$$= \quad Q(3.85) + Q(2.75) \approx 3.04 \times 10^{-3}$$

□

In problems involving classification, it is typical to have more than one type of measurement. The principles of classification are the same, however, but involve the joint density functions of the combined set of measurements.

3.9 Summary

Many probabilistic experiments result in measurements or outcomes that take on numerical values. Such measurements are modeled as random variables, which are a mapping from event space to the real line.

Random variables may be discrete, continuous, or a combination of both. Purely discrete random variables take on only a countable number of values with nonzero probability. Such random variables can be characterized by the probability mass function (PMF). Continuous random variables can take on an uncountably infinite number of possible values and are characterized by the cumulative distribution function (CDF) and its derivative called the probability density function (PDF). For a continuous random variable, the PDF $f_X(x)$ has the following interpretation. The probability that X is in the interval $x < X \le x + \Delta_x$ for an arbitrarily small increment Δ_x is given by $f_X(x)\Delta_x$. This probability goes to 0 as $\Delta_x \to 0$.

Discrete and mixed random variables can be also be characterized by the CDF; this function then has discontinuities at discrete points where the probability of the random variable is non-zero. If impulses are allowed, then the PDF can be considered as the formal derivative of the CDF. Thus PDFs containing impulses can represent both continuous and discrete random variables and are especially useful when the random variable has both continuous and discrete components.

A number of useful discrete and continuous distributions for random variables are described in this chapter. One of the most important and ubiquitous is the Gaussian. Since integrals of the Gaussian density function cannot be expressed in closed form, several related normalized quantities have been computed and listed in tables. The Q function is used in this text because of its prevelance in communications and electrical engineering. Another function, more common in mathematics and statistics is the "error function," described in Appendix C.

When one random variable is expressed in terms of another, i.e., $Y = g(X)$, it may be necessary to compute the PDF of the new random variable (Y) from the PDF of the original (X). Some basic methods for dealing with transformations of random variables are discussed in this chapter. An especially important case is when the transformation can be represented on a graph as a straight line. In this case, the form of the density does not change; only a shift and scaling occurs.

Distributions may be conditioned on an *event* and the probability of an event can be conditioned on the observance of a random variable. This concept is introduced and developed in preparation for the last section that presents some important applications.

The last section discusses three related problems, namely binary communication, radar/sonar target detection, and finally object classification. While the application areas are diverse, principles of optimal decision bind these three problems together. The problems are developed in a basic approach involving a single random variable; however, principles of optimal decision are developed in a manner that requires no more knowledge than that presented in this chapter.

References

[1] George B. Thomas, Jr., Maurice D. Weir, and Joel Hass. *Thomas' Calculus*. Pearson, New York, twelfth edition, 2010.

[2] Alberto Leon-Garcia. *Probability, Statistics, and Random Processes for Electrical Engineering*. Pearson, New York, third edition, 2008.

[3] Wilbur B. Davenport, Jr. *Probability and Random Processes*. McGraw-Hill, New York, 1970.

[4] Harry L. Van Trees. *Detection, Estimation, and Modulation Theory - Part I*. John Wiley & Sons, New York, 1968.

[5] Keinosuke Fukunaga. *Introduction to Statistical Pattern Recognition*. Academic Press, New York, second edition, 1990.

[6] Richard O. Duda, Peter E. Hart, and David G. Stork. *Pattern Classification*. John Wiley & Sons, New York, second edition, 2001.

Problems

Discrete random variables

3.1 In a set of independent trials, a certain discrete random variable K takes on the values

$$1, 0, 0, 0, 1, 1, 0, 1, 1, 1, 0, 1$$

(a) Plot a histogram for the random variable.

(b) Normalize the histogram to provide an estimate of the PMF of K.

3.2 Repeat Problem 3.1 for the discrete random variable J that takes on the values

$$1, 3, 0, 2, 1, 2, 0, 3, 1, 1, 3, 1, 0, 1, 2, 3, 0, 2, 1, 3, 3, 2$$

3.3 A random variable that takes on values of a sequence of 4QAM symbols, I, is transmitted over a communication channel:

$$3, 0, 1, 3, 2, 0, 2, 1, 3, 0, 0, 2, 3, 1, 3, 0, 2, 1, 0, 3$$

The channel introduces additive noise, J, equivalent to adding the following values to the above symbol sequence:

$$0, 0, 1, 0, 0, 1, 1, 0, 0, 0, 0, 1, 0, 1, 0, 0, 0, 1, 0, 1$$

The addition is performed in modulo form.

(a) Form the resulting sequence: $I+J$ modulo 4. Let K be the random variable that takes on these values.

(b) Plot histograms for random variables I and K.

    (c) Normalize the histograms to provide an estimate of the respective probability mass functions.

3.4 Consider transmission of ASCII characters, 8 bits long. Define a random variable as the number of 0s in a given ASCII character.

    (a) Determine the range of values that this random variable takes.

    (b) Determine and plot the probability mass function.

3.5 Consider transmission of packets with a fixed size of 3 bits. The probability of a binary 1 is 0.56. Let the number of binary 1s in a packet be the random variable I.

    (a) Determine and plot the probability mass function (PMF) of this random variable.

    (b) What is the probability that a given packet contains all binary 1s or all binary 0s?

Common discrete probability distributions

3.6 Sketch the PMF $f_K[k]$ for the following random variables and label the plots.

    (a) Uniform. $m = 0, n = 9$

    (b) Bernoulli. $p = 1/10$

    (c) Binomial. $n = 10, p = 1/10$

    (d) Geometric. $p = 1/10$ (plot $0 \leq k \leq 9$ only)

For each of these distributions, compute the probability of the event $1 \leq k \leq 4$.

3.7 In the transmission of a packet of 128 bytes, byte errors occur independently with probability $1/9$. What is the probability that less than five byte errors occur in the packet of 128 bytes?

3.8 In transmission of a very long (you may read "infinite") set of characters, byte errors occur independently with probability $1/9$. What is the probability that the first error occurs within the first five bytes transmitted?

3.9 The Poisson PMF has the form

$$f_K[k] = \frac{\alpha^k}{k!} e^{-\alpha} \qquad k = 0, 1, 2, \ldots$$

where α is a parameter.

    (a) Show that this PMF sums to 1.

    (b) What are the restrictions on the parameter α, that is, what are the allowable values?

3.10 For each of the following random variables:

    (i) geometric, $p = 1/8$

    (ii) uniform, $m = 0, n = 7$

    (iii) Poisson, $\alpha = 2$

    (a) Sketch the PMF.

    (b) Compute the probability that the random variable is greater than 5.

    (c) Find the value of the parameters such that $\Pr[K > 5] \leq 0.5$.

3.11 The probability mass function of a geometric random variable I is given to be

$$f_I[i] = 0.2(0.8)^i, \quad i \geq 0$$

(a) Find the probability that the random variable is greater than 1 and less than 5.

(b) If it is desired that $\Pr[I \geq l] \approx 0.6$, what should be the value of l?

Continuous random variables

3.12 A random variable X has the probability density function

$$f_X(x) = \begin{cases} Cx^2, & 0 \leq x \leq 1 \\ C(2-x), & 1 < x < 2 \\ 0 & \text{otherwise} \end{cases}$$

(a) Find the constant C.

(b) Find $\Pr[0.75 < X \leq 1.5]$.

(c) Find the value of α such that $\Pr[X \leq \alpha] = 0.9$.

3.13 The probability density function of a random variable X is given by

$$f_X(x) = \begin{cases} \frac{1}{2}x, & 0 \leq x \leq 1 \\ \frac{1}{6}(4-x), & 1 < x < c \\ 0 & \text{otherwise} \end{cases}$$

(a) Find the constant c. Plot the probability density function.

(b) Determine the probability $\Pr[0.35 < X \leq 6]$.

3.14 The PDF for a random variable X is given by

$$f_X(x) = \begin{cases} 2x/9 & 0 \leq x \leq 3 \\ 0 & \text{otherwise} \end{cases}$$

(a) Sketch $f_X(x)$ versus x.

(b) What is the probability of each of the following events?

    (i) $X \leq 1$

    (ii) $X > 2$

    (iii) $1 < X \leq 2$

(c) Find and sketch the CDF $F_X(x)$.

(d) Use the CDF to check your answers to part (b).

3.15 The PDF of a random variable X is given by

$$f_X(x) = \begin{cases} \alpha x, & 0 \leq x \leq 2 \\ \frac{\alpha}{2}(6-x), & 2 < x \leq 6 \\ 0 & \text{otherwise} \end{cases}$$

(a) Find the constant α. Plot the PDF.

(b) Determine the CDF of the random variable for the following ranges of values:

(i) $2 < x \leq 6$
(ii) $8 < x \leq 10$.

Complete expressions and numerical values are required.

3.16 The PDF of a random variable X is given by:

$$f_X(x) = \begin{cases} 0.4 + Cx, & 0 \leq x \leq 5 \\ 0 & \text{otherwise.} \end{cases}$$

(a) Find the value C that makes f_X a valid PDF.
(b) Determine: $\Pr[X > 3]$, $\Pr[1 < X \leq 4]$.
(c) Find the CDF, $F_X(x)$.

3.17 A PDF for a continuous random varaiable X is defined by

$$f_X(x) = \begin{cases} C & 0 < x \leq 2 \\ 2C & 4 < x \leq 6 \\ C & 7 < x \leq 9 \\ 0 & \text{otherwise} \end{cases}$$

where C is a constant.

(a) Find the numerical value of C.
(b) Compute $\Pr[1 < X \leq 8]$.
(c) Find the value of M for which

$$\int_{-\infty}^{M} f_X(x)dx = \int_{M}^{\infty} f_X(x)dx = \tfrac{1}{2}$$

M is known as the *median* of the random variable.

3.18 Show that the CDF of a Gaussian random variable with arbitrary mean and variance can be written as in (3.22) where Φ is defined by (3.21).

3.19 Show that the Q function satisfies the identity (3.25).

3.20 Is the following expression a valid probability density function? Support your answer.

$$f_X(x) = \begin{cases} \tfrac{3}{2} - x, & 0 \leq x \leq 2 \\ 0 & \text{otherwise} \end{cases}$$

3.21 For what values or ranges of values of the parameters would the following PDFs be valid? If there are no possible values that would make the PDF valid, state

(a)
$$f_X(x) = \begin{cases} B + A\sin x & 0 \le x \le 2\pi \\ 0 & \text{otherwise.} \end{cases}$$

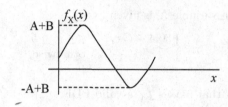

(b)

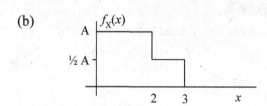

(c)

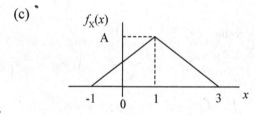

that.

3.22 The probability density function of a random variable X is shown below.

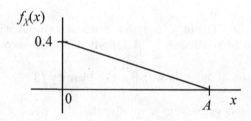

(a) Find the constant A. Write a mathematical expression for the PDF.

(b) Find the CDF for the case: $0 \le x \le A$.

3.23 The CDF for a random variable X is given by

$$F_X(x) = \begin{cases} A\arctan(x) & x > 0 \\ 0 & x \le 0 \end{cases}$$

(a) Find the constant A.

(b) Find and sketch the PDF.

3.24 The probability density function of a random variable X is given by

$$f_X(x) = \begin{cases} 0.5\sin x, & 0 \le x \le c \\ 0.5, & x = 1 \\ 0 & \text{otherwise} \end{cases}$$

(a) Find the constant c. Plot the probability density function.
(b) Determine and plot the cumulative distribution function (CDF). Mark appropriate values along both axes.

3.25 A random variable X has the CDF specified below.

$$F_X(x) = \begin{cases} 0 & x \le -\dfrac{\pi}{2\alpha} \\[2mm] \dfrac{1}{2}(1+\sin \alpha x) & -\dfrac{\pi}{2\alpha} < x \le \dfrac{\pi}{2\alpha} \\[2mm] 1 & x \ge \dfrac{\pi}{2\alpha} \end{cases}$$

(a) Sketch $F_X(x)$.
(b) For what values of the real parameter α is this expression valid?
(c) What is the PDF $f_X(x)$? Write a formula and sketch it.
(d) Determine the probability of the following events:

(i) $X \le 0$
(ii) $X \le -1$
(iii) $|X| \le .001$ (approximate)

3.26 The PDF of a random variable X is shown below:

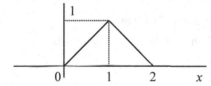

(a) Find the CDF.
(b) Find the following probabilities

(i) $\Pr[0.5 \le X \le 1.5]$
(ii) $\Pr[1 \le X \le 5]$
(iv) $\Pr[X < 0.5]$
(v) $\Pr[0.75 \le X \le 0.7501]$

3.27 Given the CDF,

$$F_X(x) = \begin{cases} 0 & x < 0 \\ Cx^2 & 0 \le x \le 2 \\ 1 & x > 2 \end{cases}$$

(a) Find C.
(b) Sketch this function.
(c) Find the PDF and sketch it.
(d) Find $\Pr[0.5 \le X \le 1.5]$.

3.28 The cumulative distribution function (CDF) of a certain random variable X is

$$F_X(x) = \begin{cases} 0 & x \le 0 \\ x & 0 < x \le \frac{1}{2} \\ 1 & \frac{1}{2} \le x \end{cases}$$

(a) Sketch $F_X(x)$.

(b) Determine and sketch the PDF of X.

3.29 The PDF of a random variable X is of the form

$$f_X(x) = \begin{cases} Cxe^{-x} & x > 0 \\ 0 & x \le 0 \end{cases}$$

Where C is a normalization constant.

(a) Find the constant C and sketch $f_X(x)$.

(b) Compute the probability of the following events:

  (i) $X \le 1$
  (ii) $X > 2$

(c) Find the probability of the event $x_0 < X \le x_0 + 0.001$ for the following values of x_0:

  (i) $x_0 = 1/2$
  (ii) $x_0 = 2$

  What value of x_0 is "most likely," i.e., what value of x_0 would give the *largest* probability for the event $x_0 < X \le x_0 + 0.001$?

(d) Find and sketch the CDF.

Common probability density functions

3.30 Sketch the PDFs of the following random variables:

(a) Uniform:

  (i) $a = 2$, $b = 6$;
  (ii) $a = -4$, $b = -1$;
  (iii) $a = -7$, $b = 1$;
  (iv) $a = 0$, $b = a + \delta$, where $\delta \to 0$.

(b) Gaussian:

  (i) $\mu = 0$, $\sigma = 1$;
  (ii) $\mu = 1.5$, $\sigma = 1$;
  (iii) $\mu = 0$, $\sigma = 2$;
  (iv) $\mu = 1.5$, $\sigma = 2$;
  (v) $\mu = -2$, $\sigma = 1$;
  (vi) $\mu = -2$, $\sigma = 2$;
  (vii) $\mu = 0$, $\sigma \to 0$.

(c) Exponential:

  (i) $\lambda = 1$;
  (ii) $\lambda = 2$;
  (iii) $\lambda \to 0$;

(iv) $\lambda \to \infty$;

(v) Can λ be negative? Explain.

3.31 The interarrival time T between "hits" on a certain web page is an exponential random variable with parameter $\lambda = 0.5$ hits/minute. The average interarrival time is given by $1/\lambda$.

(a) What is the probability that the interarrival time between hits is greater than twice the average interarrival time?

(b) What is the probability that the interarrival time between hits is less than half the average interarrival time?

(c) What are the answers to parts (b) and (c) if λ changes to 10 hits/min?

3.32 The *median M* of a random variable X is defined by the condition

$$\int_{-\infty}^{M} f_X(u)du = \int_{M}^{\infty} f_X(u)du$$

Find the median for:

(a) The uniform PDF.

(b) The exponential PDF.

(c) The Gaussian PDF.

3.33 A certain random variable X is described by a Gaussian PDF with parameters $\mu = -1$ and $\sigma^2 = 2$. Use the Q function to compute the probability of the following events:

(a) $X \le 0$

(b) $X > +1$

(c) $-2 < X \le +2$

3.34 Given a Gaussian random variable with mean $\mu = 1$ and variance $\sigma^2 = 3$, find the following probabilities:

(a) $\Pr[X > 1]$ and $\Pr\left[X > \sqrt{3}\right]$.

(b) $\Pr\left[|X - 1| > \sqrt{12}\right]$ and $\Pr\left[|X - 1| < 6\right]$.

(c) $\Pr\left[X > 1 + \sqrt{3}\right]$ and $\Pr\left[X > 1 + 3\sqrt{3}\right]$.

CDF and PDF for discrete and mixed random variables

3.35 Consider the random variable described in Prob. 3.3.

(a) Sketch the CDF for I and K.

(b) Using the CDF, compute the probability that the transmitted signal is between 1 and 2 (inclusive).

(c) Using the CDF, compute the probability that the received signal is between 1 and 2 (inclusive).

3.36 Determine and plot the CDF for the random variable described in Prob. 3.4.

3.37 The PDF of a random variable X is of the form

$$f_X(x) = \begin{cases} Ce^{-x} + \frac{1}{3}\delta(x-2) & x > 0 \\ 0 & x \le 0 \end{cases}$$

Where C is a constant.

(a) Find the constant C and sketch $f_X(x)$.

(b) Find and sketch the CDF.

(c) What is the probability of the event $1.5 < X \leq 2.5$?

3.38 Given the PDF of a random variable X:

$$f_x(x) = \begin{cases} Cx + 0.5\delta(x-1), & 0 \leq x \leq 3 \\ 0 & \text{otherwise.} \end{cases}$$

(a) Find C and sketch the PDF.

(b) Determine CDF of X.

(c) What are the probabilities for the following events?

   (i) $0.5 \leq X \leq 2$
   (ii) $X > 2$.

3.39 The CDF of a random variable X is

$$F_X(x) = \begin{cases} 0 & x \leq 0 \\ x & 0 < x \leq \frac{1}{2} \\ 1 & \frac{1}{2} \leq x \end{cases}$$

(a) Sketch $F_X(x)$.

(b) Determine and sketch the PDF of X.

3.40 Given the cumulative distribution function of random variable X shown in the figure (not drawn to scale and all slant lines have the same slope) below, determine the following.

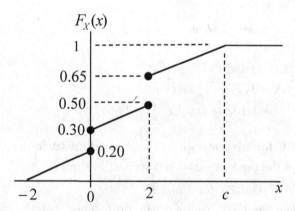

(a) The constant c that makes F_X a valid CDF.

(b) The probability density function of X (an accurate plot of the PDF will do). Mark all axes clearly and indicate all numerical values.

(c) The probabilities of the events $-2 < X < 0$ and $0 < X \leq 2.5$.

Transformation of random variables

3.41 A voltage X is applied to the circuit shown below and the output voltage Y is measured.

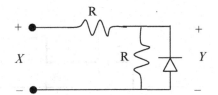

The diode is assumed to be ideal, acting like a short circuit in the forward direction and an open circuit in the backward direction. If the voltage X is a Gaussian random variable with parameters $m = 0$ and $\sigma^2 = 1$, find and sketch the PDF of the output voltage Y.

3.42 A random variable X is *uniformly distributed* (i.e., it has a uniform PDF) between -1 and $+1$. A random variable Y is defined by $Y = 3X + 2$.

   (a) Sketch the density function for X and the function describing the transformation from X to Y.

   (b) Find the PDF and CDF for the random variable Y and sketch these functions.

3.43 The Laplace PDF is defined by

$$f_X(x) = \frac{\alpha}{2} e^{-\alpha|x-\mu|} \qquad -\infty < x < \infty$$

Consider the transformation $Y = aX + b$. Demonstrate that if X is a Laplace random variable, then Y is also a Laplace random variable. What are the parameters α' and μ' corresponding to the PDF of Y?

3.44 Consider a Laplace random variable with PDF

$$f_X(x) = e^{-2|x|}, \qquad -\infty < x < \infty.$$

This random variable is passed through a 4-level uniform quantizer with output levels $\{-1.5d, -0.5d, 0.5d, 1.5d\}$.

   (a) Determine the PMF of the quantizer output.

   (b) Find $\Pr[X > 2d]$ and $\Pr[X < -2d]$.

3.45 Given the following transfer characteristic, write down the input-output mapping. The input takes values in the range of $-\infty$ to $+\infty$.

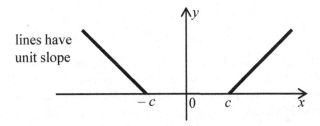

3.46 Let $Y = e^X$, where X is a Gaussian random variable with zero mean and variance 2. Find the PDF of Y in terms of the PDF of X.

3.47 The transfer characteristic of a quantizer is shown below. The input random variable X is uniformly distributed in the range of -8 to $+5$.

   (a) Determine and plot the probability mass function of Y.

   (b) Determine and plot the cumulative distribution function (CDF) of Y.

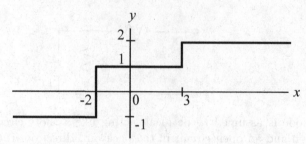

3.48 The transfer characteristic of a memoryless system is given to be:

$$y = \begin{cases} 0, & x < 3 \\ 2x - 6, & 3 \le x \le 8 \\ 10, & x > 8 \end{cases}$$

(a) If the range of the input values is from 2 to 6, what is the range of the output values?

(b) If the input is a uniform random variable in the range of 1 to 6, determine and plot the output PDF.

3.49 Consider a half-wave rectifier with a Gaussian input having mean 0 and standard deviation 2.

(a) Find the CDF of the output.

(b) Find the PDF of the output by differentiating the CDF.

3.50 Consider a clipper with the following transfer characteristic:

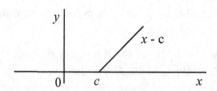

The random variable X is exponential with $\lambda = 2$.

(a) Determine the CDF of the output.

(b) Find the PDF of output by differentiating the CDF.

3.51 Consider the transfer characteristic of a limiter as shown below. The input random variable X is exponentially distributed with $\lambda = 0.5$. Determine and plot the probability density function of Y. Clearly mark all the axes and the numerical values.

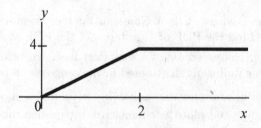

3.52 The transfer characteristic of a clipper is given by

$$y = \begin{cases} 0, & x < 2 \\ 2x - 4, & x > 2 \end{cases}$$

The probability density function (PDF) of the input random variable X is given as

$$f_X(x) = \begin{cases} \frac{1}{3}, & 3 \le x \le 6 \\ 0 & \text{otherwise} \end{cases}$$

(a) Draw the system transfer function.

(b) Determine and plot the PDF of the output random variable $f_Y(y)$.

3.53 Consider the transfer characteristic of a limiter, given by

$$y = \begin{cases} -2, & x < -2 \\ x, & -2 \le x \le 2 \\ 2, & x > 2 \end{cases}$$

The probability density function (PDF) of the input random variable X is given as

$$f_X(x) = \begin{cases} \frac{1}{6}, & -3 \le x \le 3 \\ 0 & \text{otherwise} \end{cases}$$

(a) Draw the system transfer characteristic.

(b) Determine and plot the PDF of the output random variable $f_Y(y)$. All axes must be clearly marked with appropriate values.

3.54 Consider the transfer characteristic of a memoryless system as given by

$$y = \begin{cases} 0, & x < -1 \\ x + 1, & x \ge -1 \end{cases}$$

The probability density function (PDF) of the input random variable X is uniform in the range of -3 to 5.

(a) Draw the system transfer characteristic. (All values must be clearly indicated.)

(b) Determine and plot the PDF of the output random variable $f_Y(y)$.

Distributions conditioned on an event

3.55 For the PDF given in Example 3.19, find and sketch the conditional density $f_{X|A}(x|A)$ using the following choices for the event A:

(a) $X < 1$

(b) $\frac{1}{3} < X < \frac{5}{3}$

(c) $|\sin(X)| < \frac{1}{2}$

3.56 Consider the exponential density function

$$f_X(x) = \begin{cases} \lambda e^{-\lambda x} & x \ge 0 \\ 0 & \text{otherwise} \end{cases}$$

Find and sketch the conditional density function $f_{X|A}(x|A)$ where A is the event $X \ge 2$.

3.57 Let K be a discrete uniform random variable with parameters $n = -2$, $m = 2$. Find the conditional PMF $f_{K|A}[k|A]$ for the following choices of the event A:

(a) $K < 0$

(b) $K \geq 0$

Applications

3.58 In a certain communication problem, the received signal X is a random variable given by

$$X = 2 + W$$

where W is a random variable representing the noise in the system. The noise W has a PDF which is sketched below:

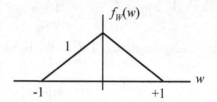

(a) Find and sketch the PDF of X.

(b) Using your answer to Part (a), find $\Pr[X > 2]$.

3.59 In a certain radar system, the noise W is a random variable with pdf as sketched below.

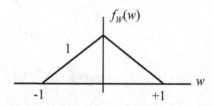

When a certain type of target is present the received signal X is given by $X = 1.5 + W$. When no target is present, the signal is just equal to the noise. We are concerned with just these two cases.

(a) Plot the pdf $f_X(x)$ for each of the two cases.

(b) The radar has been programmed to declare a detection if $X > 1$. What is the probability of detection (p_d) and the probability of false alarm (p_{fa})?

3.60 A signal is sent over a channel where there is additive noise. The noise N is uniformly distributed between -10 and 10 as shown below:

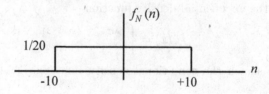

The receiver is to decide between two possible events based on the value of a random variable X. These two possible events are:

$$H_0 : \quad X = N \qquad \text{(no signal sent, only noise received)}$$
$$H_1 : \quad X = S + N \quad \text{(signal was sent, signal plus noise received)}$$

Here S is a fixed constant set equal to 18; it is *not* a random variable.

(a) Draw the density functions $f_{X|H_0}(x|H_0)$ and $f_{X|H_1}(x|H_1)$ in the same picture. Be sure to label your plot carefully.

(b) What decision rule, based on the value of X, would guarantee perfect detection of the signal (i.e., $\Pr[\text{Error}|H_1] = 0$), but minimize the probability of false alarm (i.e., $\Pr[\text{Error}|H_0]$)? What is this probability of false alarm?

(c) What decision rule would make the probability of a false alarm go to 0 but maximize the probability of detection? What is the probability of detection for this decision rule?

(d) If we could choose another value of S rather than S=18, what is the *minimum* value of S that would make it possible to have perfect detection and 0 false alarm probability?

3.61 The noise W in the Byzantic Ocean south of Antartica has the PDF sketched below:

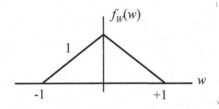

The navy of Lower Slobbovia conducts reconnaissance operations under the ice and uses the following detection rule:

$$X > \tau \quad \text{decide target present}$$
$$X < \tau \quad \text{decide no target present}$$

where X is a random variable representing the sonar output.

The officers on board know that the sonar output X is given by

$$H_1 : \quad X = W + A \quad \text{when a target is present}$$
$$H_0 : \quad X = W \qquad \text{when no target is present}$$

where A is an unknown constant (not a random variable). The crew has decided to set their threshold in the decision rule to $\tau = 0.75$.

(a) For the given threshold setting, what is the probability of a false alarm?

(b) Write an expression for the false alarm as a function of τ in the range of values $0 \le \tau \le 1$.

(c) If a giant seal results in a value of $A = 1.5$, what is the probability that the crew will *not* detect it with a threshold set at $\tau = 0.75$?

3.62 Carry out the algebraic steps involved to show that the decision rule of (3.51) reduces to the simpler form given by (3.53). Show the differences that arise to obtain (3.61).

3.63 In a simple detection scheme, the received signal is $X = A+W$, where it is given that A is the desired signal of magnitude 9 Volts, and the noise is uniformly distributed in the range -5 to 5. A detection probability, p_d, of 0.99 is desired.

(a) Find the value of threshold, x_T, required to obtain this probability of detection.

(b) Determine the corresponding probability of false alarm, p_{fa}.

3.64 In a simple detection scheme, the received signal sample is $X = A + W$, where A is the desired signal sample (a constant value) of magnitude 6 Volts, and the noise W is Laplacian distributed whose PDF is given by

$$f_X(x) = e^{-2|x|}, \quad -\infty < x < \infty$$

The desired false alarm probability $p_{fa} = 0.001$.

(a) Find the value of threshold, x_T, required to obtain this probability of false alarm.

(b) Determine the corresponding probability of detection, p_d.

3.65 In a detection scheme, the received signal value is $X = A + W$, where A is the desired signal value (a constant) of magnitude 6 Volts, and the PDF of noise W is given by

$$f_X(x) = \frac{1}{2\sqrt{2\pi}} e^{-x^2/8}, \quad -\infty < x < \infty$$

A false alarm probability, p_{fa}, of 0.01 is desired.

(a) Find the value of threshold, x_T, required to obtain this probability of false alarm.

(b) Determine the corresponding probability of detection, p_d.

3.66 In a binary communication system, a binary 0 is transmitted using an electrical pulse of magnitude $+2.5$ volts and a binary 1 is transmitted as a pulse of magnitude 0 volts. The additive noise in the channel is known to be Gaussian with a mean of zero and a variance of 1.2. The communication channel is modeled as a binary symmetric channel.

(a) What is the threshold level for this system?

(b) Calculate the four probabilities that characterize this channel model.

(c) Calculate the average signal-to-noise (SNR) in decibels (dB). Assume equal numbers of binary 0s and binary 1s.

Computer Projects

Project 3.1

This project requires writing computer code to generate random variables of various kinds and measure their probability distributions.

1. Generate a sequence of each of the following types of random variables; each sequence should be at least 10,000 points long.

(a) A binomial random variable. Let the number of Bernoulli trials be $n = 12$. Recall that the binomial random variable is defined as the number of 1s in n trials for a Bernoulli (binary) random variable. Let the parameter p in the Bernoulli trials be $p = 0.5109$.

    (b) A Poisson random variable as a limiting case of the binomial random variable with $p = 0.0125$ or less and $n = 80$ or more while maintaining $\alpha = np = 1$.

    (c) A type 1 geometric random variable with parameter $p = 0.09$.

    (d) A (continuous) uniform random variable in the range $[-2, 5]$.

    (e) A Gaussian random variable with mean $\mu = 1.3172$ and variance $\sigma^2 = 1.9236$.

    (f) An exponential random variable with parameter $\lambda = 1.37$.

2. Estimate the CDF and PDF or PMF as appropriate and plot these functions next to their theoretical counterparts. Compare and comment on the results obtained.

MATLAB programming notes

To generate uniform and Gaussian random variables in Step 1(d) and 1(e), use the MATLAB functions "rand" and "randn" respectively with an appropriate transformation.

To generate an exponential random variable, you may use the function "expon" from the software package for this book. This function uses the "transformation" method to generate the random variable.

In Step 2 you may use the functions "pmfcdf" and "pdfcdf" from the software package.

Project 3.2

In this project you will explore the problem of detection of a simple signal in Gaussian noise.

1. Consider a binary communication scheme where a logical 1 is represented by a constant voltage of value $A = 5$ volts, and a logical 0 is represented by zero volts. Additive Gaussian noise is present, so during each bit interval the received signal is a random variable X described by

$$X = A + N \quad \text{if a 1 was sent}$$
$$X = N \quad \text{if a 0 was sent}$$

The noise N is a Gaussian random variable with mean zero and variance σ^2. The signal-to-noise ratio in decibels is defined by

$$\text{SNR} = 10 \log_{10} \frac{A^2}{\sigma^2} = 20 \log_{10} \frac{A}{\sigma}$$

For this study the SNR will be taken to be 10 dB. This formula allows you to compute a value for σ^2.

    (a) Let H_0 be the event that a logical 0 was sent and H_1 be the event that a 1 was sent. Show that the decision rule for the receiver to minimize the probability of error is of the form

$$x \underset{H_0}{\overset{H_1}{\gtrless}} \tau$$

Evaluate the threshold for the case where $P_0 = P_1 = 0.5$ where P_0 and P_1 are the prior probabilities of a 0 and 1, i.e., $P_0 = \Pr[H_0]$ and $P_1 = \Pr[H_1]$.

(b) Now consider a decision rule of the above form where τ is any arbitrary threshold. Evaluate ε_0 and ε_1 and plot the probability of error (3.55) as a function of τ for $1 \leq \tau \leq 4$. (Continue to use $P_0 = P_1 = 0.5$ in this formula.) Find the minimum error, and verify that it occurs for the value of τ computed in Step 1(a).

(c) Repeat Step 1(b) using $P_0 = 1/3$ and $P_1 = 2/3$. Does the minimum error occur where you would expect it to occur? Justify your answer.

2. The detection of targets (by a radar, for example) is an analogous problem. The signal, which is observed in additive noise, will have a mean value of A if a target is present (event H_1) , and mean 0 if only noise (no target) is present (event H_0). Let p_d represent the probability of detecting the target and p_{fa} represent the probability of a false alarm.

(a) Plot the ROC for this problem for the range of values of τ used in Step 1 (above). Indicate the direction of increasing τ to show how the probability of detection can be traded off for the probability of false alarm by adjusting the threshold.

(b) Show where the two threshold values used in Step 1(b) and 1(c) lie on the ROC curve, and find values of p_d and p_{fa} corresponding to each threshold value.

3. In this last part of the project you will simulate data to see how experimental results compare to the theory.

(a) i. Generate 1000 samples of a Gaussian random variable with mean 0 and variance σ^2 as determined in Step 1 for the 10 dB SNR. Generate 1000 samples of another random variable with mean 5 and variance σ^2. These two sets of random variables represent the outcomes in 1000 trials of the events H_0 and H_1. For each of these 2000 random variables make a decision according to the decision rule in Step 1(a) using the values of τ that you found in Steps 1(a) and 1(b). Form estimates for the errors ε_1 and ε_0 by computing the relative frequency, that is:

$$\hat{\varepsilon}_i = \frac{\#\text{misclassified samples for } H_i}{1000} \quad \text{for } i = 0, 1$$

ii. Also compute p_d, p_{fa}, and the probability of error (3.55) using these estimates. Plot the values on the graphs generated in Steps 1 and 2 for comparison.

(b) Repeat Step (a)(i) above using two other threshold values of your choice for τ. You can use the *same* 2000 random variables generated in Step (i); just make your decision according to the new thresholds. Plot these points on the ROC of Step 2.

MATLAB programming notes

The MATLAB function "erf" is available to help evaluate the integral of the Gaussian density.

To generate uniform and Gaussian random variables in Step 1(d) and 1(e), use the MATLAB functions "rand" and "randn" respectively with an appropriate transformation.

4 Expectation, Moments, and Generating Functions

One of the cornerstones in the fields of probability and statistics is the use of averages. The most important type of average for random variables involves a weighting by the PDF or PMF and is known as *expectation*. This chapter discusses expectation and how it applies to the analysis of random variables. It is shown that some of the most important descriptors of the random variable X are not just its expectation, but also the expectation of powers of the random variable such as X^2, X^3, and so on. These averages are called *moments*; knowing the complete set of moments for a random variable is equivalent to knowing the density function itself. This last idea is manifested through what are known as "generating functions," which are also discussed in this chapter. Generating functions are (Laplace, Fourier, z) transformations of the density function and thus provide a different but equivalent probabilistic description of a random variable. The name "generating function" comes from the fact that moments are easily derived from these functions by taking derivatives.

As an application involving expectation, this chapter discusses the concept of entropy. Entropy is the expected value of information in an experiment, where "information" is defined as in Chapter 2. A theorem due to Shannon is cited that provides a lower bound on the average number of bits needed to code a message in terms of the entropy of the code words. This result, although developed in the first half of the twentieth century, has clear relevance to modern problems of digital communication, data compression and coding of digital information on computers and computer networks.

4.1 Expectation of a Random Variable

In the previous chapter, a random variable is defined as a function $X(s)$ that assigns a real number to each outcome s in the sample space of a random experiment. The *expectation* or *expected value* of a random variable can be thought of as the "average value" of the random variable, where the average is computed weighted by the probabilities. More specifically, if X is a random variable defined on a discrete sample space S, then the expectation $\mathcal{E}\{X\}$ is given by

$$\mathcal{E}\{X\} = \sum_{S} X(s)\Pr[s] \tag{4.1}$$

where the sum is over all elements of the sample space. The following example, based on Example 3.1 of the previous chapter, illustrates the idea of expectation of a random variable.

Example 4.1: Recall from Example 3.1 that the cost of buying 3 CDs from Simon depends on the number of good and bad CDs that result (good CDs cost $0.36 while bad CDs cost the customer $0.50). What is the expected value of the cost C of buying three CDs at Simon's?

The events of the sample space and the cost associated with each event are listed below:

Events:	BBB	BBG	BGB	BGG	GBB	GBG	GGB	GGG
Cost:	\$1.50	\$1.36	\$1.36	\$1.22	\$1.36	\$1.22	\$1.22	\$1.08
Probability:	1/8	1/8	1/8	1/8	1/8	1/8	1/8	1/8

All of the events in the sample space are equally likely and have probability 1/8. The expected value of the cost is the weighted sum:

$$\mathcal{E}\{C\} = (1/8)1.50 \; + \; (1/8)1.36 + (1/8)1.36 + (1/8)1.22$$
$$+(1/8)1.36 + (1/8)1.22 + (1/8)1.22 + (1/8)1.08 = \$1.29$$

□

4.1.1 Discrete random variables

Expectation for a random variable can be computed in a more convenient way. The expectation of a discrete random variable is *defined* as

$$\mathcal{E}\{K\} = \sum_k k\, f_K[k] \tag{4.2}$$

Thus the expected value is the sum of all the values of the random variable weighted by their probabilities. Since the sum may be infinite, the expectation is defined only if the sum converges absolutely[1] [1]. Let us compute the expectation of the previous example using this formula.

Example 4.2: The PMF for the cost C of buying Simon's CDs was computed in Example 3.1 and is shown below.

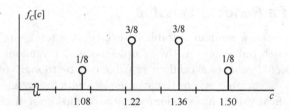

Using (4.2), the expected value of the cost is given by

$$\mathcal{E}\{C\} = 1.08 \cdot f_C[1.08] + 1.22 \cdot f_C[1.22] + 1.36 \cdot f_C[1.36] + 1.50 \cdot f_C[1.50]$$

Then substituting the values of $f_C[c]$ from the figure yields

$$\mathcal{E}\{C\} = 1.08 \cdot (1/8) + 1.22 \cdot (3/8) + 1.36 \cdot (3/8) + 1.50 \cdot (1/8) = 1.29$$

□

The fact that both examples produce the same result shouldn't be surprising. After all, we are just slicing up the probability in different ways.

---

[1] Absolute convergence is defined by: $\sum_k |k|\, f_K[k] < \infty$.

4.1.2 Continuous random variables

The expectation of a continuous random variable can also be defined in terms of the sample space. Conceptually the procedure is similar to (4.1); however, the result is written as

$$\mathcal{E}\{X\} = \int_S X(s)dP(s) \qquad (4.3)$$

where the expression on the right is a Lebesgue-Stieltjes integral and $P(s)$ is an appropriately defined probability measure for the sample space.[2]

A more convenient definition however, is given in terms of the probability density function (PDF) when it exists, as we have assumed. For a continuous random variable the expectation is defined to be

$$\boxed{\mathcal{E}\{X\} = \int_{-\infty}^{\infty} x f_X(x)dx} \qquad (4.4)$$

whenever the infinite integral converges absolutely [5].[3]

This formula can also be applied to discrete and mixed random variables as long as impulses are allowed in the PDF. In particular, let the PDF for a discrete random variable be represented as

$$f_K(k) = \sum_i f_K[i]\,\delta(k-i) \qquad (4.5)$$

where $f_K[i]$ denotes the PMF and δ is the unit impulse. In other words, the PDF is represented as a series of impulses with areas equal to the probability of the random variable. Then substitution of this expression into (4.4) leads directly to (4.2). Alternatively, if a continuous random variable X is related to a discrete random variable K through quantization as in Section 3.3.3, then (4.4) can be thought of as the limiting version of (4.2).

In light of the preceding discussion, (4.4) will henceforth be considered to be the definition of expectation. For most analyses, it is convenient to regard $\mathcal{E}\{\cdot\}$ as a linear *operator*. The definition of this operator will have to be expanded slightly for the treatment of multiple random variables in Chapter 5.

4.1.3 Invariance of expectation

An extremely important property of expectation is *invariance* under a transformation. This property is really grounded in the interpretation of expectation in the sample space as provided by (4.1) and (4.3). It can be explained, however, using the definition (4.4).

Let X be a random variable and let $Y = g(X)$. Then (4.4) states that

$$\mathcal{E}\{Y\} = \int_{-\infty}^{\infty} y f_Y(y)dy \qquad (4.6)$$

What is *not* so obvious is that $\mathcal{E}\{Y\}$ can also be computed from

$$\mathcal{E}\{Y\} = \int_{-\infty}^{\infty} g(x) f_X(x)dx \qquad (4.7)$$

---

[2] Some further discussion of this idea can be found in [2], [3], and especially in [4], which provides an excellent treatment for engineers.

[3] Absolute convergence is defined as: $\int_{-\infty}^{\infty} |x| f_X(x)dx < \infty$.

In other words, to compute the expectation of $Y = g(X)$ it is not necessary to transform and find the PDF $f_Y(y)$. We can simply place the function $g(x)$ under the integral along with the probability density f_X for X and integrate! Thus for any function $g(X)$, expectation of that quantity can be defined to mean:

$$\mathcal{E}\{g(X)\} = \int_{-\infty}^{\infty} g(x) f_X(x) dx \tag{4.8}$$

A simple argument can be given here to convince you of this result.

Let (4.6) be represented by the discrete approximation

$$\mathcal{E}\{Y\} \approx \sum_{\forall k} y_k f_Y(y_k) \Delta_y$$

The term $f_Y(y_k)\Delta_y$ in the summation is the probability that Y is in a small interval $y_k < y \le y_k + \Delta_y$. Now refer to Fig 4.1. The probability that Y is in that small interval

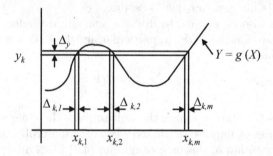

Figure 4.1 Transformation $Y = g(X)$.

is the same as the sum of the probabilities that X is in each of the corresponding small intervals shown in the figure. In other words,

$$f_Y(y_k)\Delta_y = f_X(x_{k,1})\Delta_{k,1} + f_X(x_{k,2})\Delta_{k,2} + \cdots + f_X(x_{k,m})\Delta_{k,m}$$

where m is the number of intervals in x that map into the interval $(y_k, y_k + \Delta_y]$ in y. Substituting this expression in the previous equation yields

$$\mathcal{E}\{Y\} \approx \sum_{k}\sum_{i=1}^{m} y_k f_X(x_{k,i})\Delta_{k,i} = \sum_{k}\sum_{i=1}^{m} g(x_{k,i}) f_X(x_{k,i})\Delta_{k,i}$$

Since the double sum merely covers all possible regions of x, we can reindex the variables and replace it with the single summation

$$\mathcal{E}\{Y\} = \sum_{l} g(x_l) f_X(x_l)\Delta_l$$

which in the limit becomes the integral (4.7).

Although Fig. 4.1 depicts a continuous function, the result (4.8) is actually quite general and applies for almost any function $g(X)$ likely to be encountered.

Our argument was based on the usual Riemann integral; however, in advanced treatments of random variables, it is possible to state results in terms of Lebesgue integration for which invariance of expectation is almost a direct consequence. In most cases, the Lebesgue and the Riemann integrals produce identical results, so operations of integration are carried out in the usual way.

Invariance of expectation can be expressed for discrete random variables as

$$E\{g(K)\} = \sum_k g(k)\, f_K[k] \tag{4.9}$$

Although this equation is implied by (4.8) whenever the PDF contains impulses, it is sometimes useful to have an explicit formula for the discrete case.

Let us close this section with one further example on the use of expectation.

Example 4.3:   The continuous random variable Θ is uniformly distributed between $-\pi$ and π. The expected value of this random variable is found using (4.4) as

$$E\{\Theta\} = \int_{-\infty}^{\infty} \theta f_\Theta(\theta)d\theta = \int_{-\pi}^{\pi} \theta \frac{1}{2\pi}d\theta = \left.\frac{\theta^2}{4\pi}\right|_{-\pi}^{\pi} = 0$$

Now consider the random variable $Y = A\cos^2(\Theta)$ where A is assumed to be a constant (not a random variable). Applying (4.8) yields

$$\begin{aligned}
E\{Y\} &= E\{A\cos^2(\Theta)\} = \int_{-\pi}^{\pi} A\cos^2(\theta)\frac{1}{2\pi}d\theta \\
&= \frac{A}{2\pi}\int_{-\pi}^{\pi} \tfrac{1}{2}(1+\cos 2\theta)d\theta = \left.\frac{A}{4\pi}(\theta + \tfrac{1}{2}\sin 2\theta)\right|_{-\pi}^{\pi} = \frac{A}{2}
\end{aligned}$$

□

The same result for $E\{Y\}$ can be obtained by finding $f_Y(y)$ and using (4.6). This would require much more work, however.

4.1.4 Properties of expectation

As we proceed in the study of random variables, you will find that we often use a number of properties of expectation, without ever needing to carry out the integral or summation involved in the definition of expectation. Some of the most important properties result from viewing the expectation $E\{\cdot\}$ as a linear operator.

Consider any two functions of X, say $g(X)$ and $h(X)$ and form a linear combination $ag(X) + bh(X)$ where a and b are any constants. Using 4.8, we can write

$$\begin{aligned}
E\{ag(X) + bh(X)\} &= \int_{-\infty}^{\infty} (ag(x) + bh(x))\, f_X(x)dx \\
&= a\int_{-\infty}^{\infty} g(x)f_X(x)dx + b\int_{-\infty}^{\infty} h(x)f_X(x)dx \\
&= aE\{g(X)\} + bE\{h(X)\}
\end{aligned}$$

If we let $Y = g(X)$ and $Z = h(X)$, then

$$E\{aY + bZ\} = aE\{Y\} + bE\{Z\} \tag{4.10}$$

which defines expectation as a *linear operator*.

A number of properties of expectation are now apparent which are listed in Table 4.1. The first property follows from the fact that the PDF integrates to 1. The second and third properties follow from the linearity (4.10), and the last follows from the first, and the observance that $E\{X\} = m_X$ is a constant.

$$\mathcal{E}\{c\} = c \quad (c \text{ is a constant})$$

$$\mathcal{E}\{X + c\} = \mathcal{E}\{X\} + c$$

$$\mathcal{E}\{cX\} = c\mathcal{E}\{X\}$$

$$\mathcal{E}\{\mathcal{E}\{X\}\} = \mathcal{E}\{X\}$$

Table 4.1 Some simple properties of expectation.

4.1.5 Expectation conditioned on an event

Occasionally there is a need to compute expectation conditioned on an event. If A is an event, then the expectation conditioned on the event A is defined by

$$\mathcal{E}\{X|A\} = \int_{-\infty}^{\infty} x f_{X|A}(x|A) dx \qquad (4.11)$$

Here $f_{X|A}$ denotes a density function for X conditioned on the event A, which usually controls one or more of the parameters.

Many applications and variations of this idea are possible. For example, since A and A^c are mutually exclusive and collectively exhaustive events, it is possible to write

$$f_X(x) = f_{X|A}(x|A) \Pr[A] + f_{X|A^c}(x|A^c) \Pr[A^c]$$

This is just a form of the principle of total probability. It follows from this last expression that

$$\mathcal{E}\{X\} = \mathcal{E}\{X|A\} \Pr[A] + \mathcal{E}\{X|A^c\} \Pr[A^c]$$

4.2 Moments of a Distribution

Expectation can be used to describe properties of the PDF or PMF of random variables. If we let $g(X) = X^n$ for $n = 1, 2, 3, \ldots$ in (4.8) and (4.9), the quantity

$$\mathcal{E}\{X^n\} = \int_{-\infty}^{\infty} x^n f_X(x) dx \qquad (4.12)$$

for a continuous random variable or

$$\mathcal{E}\{X^n\} = \sum_{\forall k} k^n f_K[k] \qquad (4.13)$$

for a discrete random variable is called the nth *moment* of the distribution. The first moment ($n = 1$) is called the *mean*, and represents the average value of the random variable for the given PDF. The mean will be denoted by the variable m with an appropriate subscript, i.e.,

$$m_X = \mathcal{E}\{X\} \quad \text{or} \quad m_K = \mathcal{E}\{K\} \qquad (4.14)$$

Second and higher order moments further characterize the PDF in a way that is explained below. In the special case where X is a random electrical signal (voltage), X^2 represents the "power" associated with the signal; therefore, the second moment $\mathcal{E}\{X^2\}$ represents *average signal power*.

4.2.1 Central moments

When the mean of a random variable is not 0, then the moments centered about the mean are generally more useful than the moments defined above. These *central moments* are defined by

$$
\begin{aligned}
\mathcal{E}\{(X - m_X)^n)\} &= \int_{-\infty}^{\infty} (x - m_X)^n f_X(x)dx \quad (a) \text{ CONTINUOUS RV} \\
\mathcal{E}\{(K - m_K)^n\} &= \sum_{\forall k}(k - m_K)^n \, f_K[k] \qquad\qquad (b) \text{ DISCRETE RV}
\end{aligned}
\tag{4.15}
$$

The second central moment is called the *variance* and is given by[4]

$$
\sigma_X^2 = \text{Var}\,[X] = \mathcal{E}\{(X - m_X)^2\}
\tag{4.16}
$$

(Both the symbol σ_X^2 and the notation Var$\,[X]$ are commonly used to denote the variance of X.) The square root of the variance (σ_X) is called the *standard deviation* and has the same units as the random variable X. The quantity σ_X is important because it is a measure of the *width* or *spread* of the distribution.

For the Gaussian PDF, the variance appears explicitly in the formula (i.e., $\sigma_X^2 = \sigma^2$ in (3.19)), while for other PDFs the variance and standard deviation must be expressed in terms of the other parameters. For example, the variance of a uniform random variable (see Prob. 4.35) is given by

$$
\sigma_X^2 = \frac{(b-a)^2}{12}
$$

and thus the standard deviation σ_X is approximately $0.289(b-a)$. The integral

$$
\int_{m_X-\sigma_X}^{m_X+\sigma_X} f_X(x)dx
$$

of the PDF from one standard deviation below the mean to one standard deviation above the mean provides a rough indication of how the probability is distributed. For the Gaussian PDF this integral is approximately 0.683 while for the uniform PDF it is $1/\sqrt{3}$ or 0.577. This indicates that for the Gaussian random variable, the probability is distributed more densely in the region closer to the mean.

The following example shows how to compute the mean, variance, and standard deviation for an exponential random variable.

Example 4.4: Suppose X is an exponential random variable:

$$
f_X(x) = \begin{cases} \lambda e^{-\lambda x} & x \geq 0 \\ 0 & x < 0 \end{cases}
$$

To help in the evaluation of moments, we can make use of the formula

$$
\int_0^{\infty} u^n e^{-u} du = n! \qquad n = 0, 1, 2, 3, \ldots
$$

(See e.g., [6].)

---

[4] To avoid unnecessary redundancy, we assume in the following that the random variable X can be either continuous or discrete. The expectation is then carried out using whichever formula (i.e., (4.15) (a) or (b)) applies.

The mean is given by

$$m_X = \int_{-\infty}^{\infty} x f_X(x) dx = \int_0^{\infty} x \lambda e^{-\lambda x} dx$$

Making the change of variables $u = \lambda x$ yields

$$m_X = \frac{1}{\lambda} \int_0^{\infty} u e^{-u} du = \frac{1}{\lambda} \cdot 1! = \frac{1}{\lambda}$$

The variance is computed from

$$
\begin{aligned}
\sigma_X^2 &= \int_{-\infty}^{\infty} (x - m_X)^2 f_X(x) dx = \int_0^{\infty} (x - \frac{1}{\lambda})^2 \lambda e^{-\lambda x} dx \\
&= \int_0^{\infty} x^2 \lambda e^{-\lambda x} dx - \frac{2}{\lambda} \int_0^{\infty} x \lambda e^{-\lambda x} dx + \frac{1}{\lambda^2} \int_0^{\infty} \lambda e^{-\lambda x} dx \\
&= \frac{1}{\lambda^2} \int_0^{\infty} u^2 e^{-u} du - \frac{2}{\lambda^2} \int_0^{\infty} u e^{-u} du + \frac{1}{\lambda^2} \\
&= \frac{1}{\lambda^2} \cdot 2! - \frac{2}{\lambda^2} \cdot 1! + \frac{1}{\lambda^2} = \frac{1}{\lambda^2}
\end{aligned}
$$

From this it is seen that the mean and standard deviation (σ_X) of the exponential random variable are the same; both are equal to $1/\lambda$.

□

4.2.2 Properties of variance

The variance is such an important descriptor for a random variable, that it is appropriate to list some of its properties, as in Table 4.2. Although the properties are stated

$$
\boxed{
\begin{aligned}
&\text{Var}\,[c] = 0 \quad\; (c \text{ is a constant}) \\
&\text{Var}\,[X + c] = \text{Var}\,[X] \\
&\text{Var}\,[cX] = c^2\, \text{Var}\,[X]
\end{aligned}
}
$$

Table 4.2 Some properties of variance.

without proof, they are easy to show. Note especially the last property and the fact that the constant in front of the variance on the right-hand side is *squared*. This occurs because the variance is the expectation of a *squared* quantity. Check yourself out on the use of these properties. Suppose the variance of X is equal to 1 and $Y = 2X + 3$. Find the variance of Y.[5]

The last formula involving variance that will be mentioned is given below:

$$\boxed{\text{Var}\,[X] = \mathcal{E}\{X^2\} - m_X^2} \tag{4.17}$$

This states that the variance can be found by computing the second moment and subtracting the mean squared. Frequently, this method for computing the variance is much easier than using the formula (4.16) directly.

[5] Answer: Var $[Y] = 4$.

The proof of (4.17) is a good example of using $\mathcal{E}\{\cdot\}$ as a linear *operator* and applying the properties of expectation listed in Table 4.1. Begin with (4.16) and expand the argument algebraically:

$$\mathcal{E}\{(X - m_X)^2\} = \mathcal{E}\{X^2 - 2m_X X + m_X^2\}$$

Since the expectation is a linear operator, we can distribute the expectation over the three terms and write

$$\mathcal{E}\{(X - m_X)^2\} = \mathcal{E}\{X^2\} - \mathcal{E}\{2m_X X\} + \mathcal{E}\{m_X^2\} = \mathcal{E}\{X^2\} - 2m_X\mathcal{E}\{X\} + m_X^2$$

where in the last step properties in Table 4.1 have been used. Finally, since the expectation $\mathcal{E}\{X\}$ is equal to m_X, by collecting terms, (4.17) follows.

An example, involving a discrete random variable, illustrates the use of (4.17).

Example 4.5: A Bernoulli random variable has the PMF shown in Fig. 3.3 of Chapter 3.

The mean is computed as follows:

$$m_K = \mathcal{E}\{K\} = \sum_{k=0}^{1} k\, f_K[k] = (1 - p) \cdot 0 + p \cdot 1 = p$$

The variance is computed by first computing the second moment:

$$\mathcal{E}\{K^2\} = \sum_{k=0}^{1} k^2\, f_K[k] = (1 - p) \cdot 0^2 + p \cdot 1^2 = p$$

Then (4.17) is applied to compute the variance:

$$\sigma_K^2 = \mathcal{E}\{K^2\} - m_K^2 = p - p^2 = p(1 - p)$$

(You can check this result by computing σ_K^2 from (4.16) directly.)

$\square$

4.2.3 Some higher-order moments

Some normalized quantities related to the third and fourth order moments are sometimes used in the analysis of non-Gaussian random variables. These quantities are called the *skewness* and *kurtosis* and are defined by

$$\alpha_3 = \frac{\mathcal{E}\{(X - m_X)^3\}}{\sigma_X^3} \qquad \text{(a) SKEWNESS}$$

$$\alpha_4 = \frac{\mathcal{E}\{(X - m_X)^4\}}{\sigma_X^4} - 3 \quad \text{(b) KURTOSIS}$$

(4.18)

The skewness and kurtosis for a Gaussian random variable are identically 0.

The skewness can be used to measure the asymmetry of a distribution. PDFs or PMFs which are symmetric about their mean have *zero* skewness. The kurtosis is sometimes used to measure the deviation of a distribution from the Gaussian PDF. In other words distributions that are in this sense "closer to Gaussian" have lower values of kurtosis. The two quantities α_3 and α_4 belong to a more general family of higher-order statistics known as *cumulants* [7, 8]. These quantities have found use in various areas of signal processing (see e.g., [9]). Some problems on computing skewness and kurtosis are included at the end of this chapter (see Prob. 4.45).

4.3 Generating Functions

In the analysis of signals and linear systems, Fourier, Laplace, and z-transforms are essential to show properties of the signal or system that are not readily apparent in the time domain. For similar reasons formal transforms applied to the PDF or PMF can be useful tools in problems involving probability and random variables. One especially important use of the transforms is in generating *moments* of the distribution; hence the name "generating functions." This section provides a short introduction to these generating functions for continuous and discrete random variables.

4.3.1 The moment generating function

The moment generating function (MGF) corresponding to a random variable X is defined by

$$M_X(s) = \mathcal{E}\left\{e^{sX}\right\} = \int_{-\infty}^{\infty} f_X(x)e^{sx}dx \tag{4.19}$$

where s is a complex variable taking on values such that the integral converges.[6] These values of s define a region of the complex plane called the "region of convergence." The definition (4.19) provides two interpretations, *both of which are useful*. The MGF can be thought of as either: (1) the expected value of e^{sX}, *or* (2) as the Laplace transform of the PDF.[7] The MGF is most commonly used for continuous random variables; however, it can be applied to mixed or discrete random variables as long as impulses are allowed in the PDF.

When the MGF is evaluated at $s = j\omega$ (i.e., on the imaginary axis of the complex plane), the result is the Fourier tranform:

$$M_X(j\omega) = \mathcal{E}\left\{e^{j\omega X}\right\} = \int_{-\infty}^{\infty} f_X(x)e^{j\omega x}dx$$

This quantity, which is called the characteristic function, is often used instead of the MGF to avoid discussion of region of convergence and some subtle mathematical difficulties that can arise when we stray off the $s = j\omega$ axis. Notice that on the $j\omega$ axis, there is absolute convergence for the integral:

$$\int_{-\infty}^{\infty} |f_X(x)||e^{j\omega x}|dx = \int_{-\infty}^{\infty} f_X(x)dx = 1 < \infty$$

Therefore the characteristic function always exists and the MGF always converges *at least* on the $s = j\omega$ axis.

The ability to derive moments from the MGF is apparent if we expand the term e^{sX} in a power series and take the expectation[8]

$$M_X(s) = \mathcal{E}\left\{e^{sX}\right\} = 1 + \mathcal{E}\{sX\} + \mathcal{E}\left\{\tfrac{1}{2}s^2X^2\right\} + \mathcal{E}\left\{\tfrac{1}{3!}s^3X^3\right\} + \cdots$$

$$= 1 + \mathcal{E}\{X\}s + \tfrac{1}{2}\mathcal{E}\left\{X^2\right\}s^2 + \tfrac{1}{3!}\mathcal{E}\left\{X^3\right\}s^3 + \cdots$$

The moments $\mathcal{E}\{X^n\}$ appear as coefficients of the expansion in s.

Given the moment generating function $M_X(s)$, the moments can be derived by

---

[6] Lately it is common to define the MGF for only *real* values of s (e.g., [10]). Although this simplifies the mathematical discussion, it leads to cases where the MGF may not exist.

[7] Notice that the transform definition (4.19) uses s instead of $-s$ which is common in electrical engineering. Hence any arguments involving the left and right half planes have to be reversed.

[8] This expansion is possible in the region of convergence since there are no poles at $s = 0$. [11]

taking derivatives with respect to s and evaluating the result at $s = 0$.[9] To see how this works, we can use the last equation to write

$$\frac{dM_X(s)}{ds}\bigg|_{s=0} = \mathcal{E}\{X\} + 2 \cdot \tfrac{1}{2}\mathcal{E}\{X^2\}s + 3 \cdot \tfrac{1}{3!}\mathcal{E}\{X^3\}s^2 + \cdots \bigg|_{s=0} = \mathcal{E}\{X\}$$

Then by taking the derivative once again:

$$\frac{d^2M_X(s)}{ds^2}\bigg|_{s=0} = \mathcal{E}\{X^2\} + 3 \cdot 2 \cdot \tfrac{1}{3!}\mathcal{E}\{X^3\}s + \cdots \bigg|_{s=0} = \mathcal{E}\{X^2\}$$

Generalizing this result produces the formula

$$\boxed{\mathcal{E}\{X^n\} = \frac{d^n M_X(s)}{ds^n}\bigg|_{s=0}} \tag{4.20}$$

The following examples illustrate the calculation of the MGF and the use of (4.20) to compute the mean and variance.

Example 4.6:   Refer to the exponential random variable defined in Example 4.4. The PDF is given by

$$f_X(x) = \begin{cases} \lambda e^{-\lambda x} & x \geq 0 \\ 0 & x < 0 \end{cases}$$

The MGF is computed by applying (4.19)

$$\begin{aligned} M_X(s) &= \int_{-\infty}^{\infty} f_X(x)e^{sx}dx = \int_0^{\infty} \lambda e^{-\lambda x}e^{sx}dx \\ &= \lambda \int_0^{\infty} e^{-(\lambda - s)x}dx = \frac{\lambda}{-(\lambda - s)}e^{-(\lambda - s)x}\bigg|_0^{\infty} = \frac{\lambda}{\lambda - s} \end{aligned}$$

The integral exists as long as $\mathrm{Re}\,[s] < \lambda$. This inequality defines the region of convergence.

□

Example 4.7:   To compute the mean of the exponential random variable from its MGF, apply (4.20) with $n = 1$:

$$m_X = \mathcal{E}\{X\} = \frac{dM_X(s)}{ds}\bigg|_{s=0} = \frac{d}{ds}\left(\frac{\lambda}{\lambda - s}\right) = \frac{\lambda}{(\lambda - s)^2}\bigg|_{s=0} = \frac{1}{\lambda}$$

To compute the variance, first compute the second moment using (4.20):

$$\mathcal{E}\{X^2\} = \frac{d^2 M_X(s)}{ds^2}\bigg|_{s=0} = \frac{2\lambda}{(\lambda - s)^3}\bigg|_{s=0} = \frac{2}{\lambda^2}$$

Then use (4.17) to write

$$\sigma_X^2 = \mathcal{E}\{X^2\} - m_X^2 = \frac{2}{\lambda^2} - \left(\frac{1}{\lambda}\right)^2 = \frac{1}{\lambda^2}$$

The results agree with the results of Example 4.4.

□

---

[9] While the derivative of a function of a complex variable may not always exist, we assume here that they exist in our region of interest.

4.3.2 *The probability generating function*

The moments for a discrete integer-valued random variable K are most easily dealt with by using the probability generating function (PGF) defined by

$$G_K(z) = \mathcal{E}\{z^K\} = \sum_{k=-\infty}^{\infty} f_K[k]z^k \tag{4.21}$$

where z is a complex variable in the *region of convergence* (i.e., the region where the infinite sum converges). Again, two interpretations are equally valid; the PGF can be thought of as either the expectation of z^K or the z-transform of the PMF.[10] The name probability generating function comes from the fact that if $f_K[k] = 0$ for $k < 0$ then

$$G_K(z) = f_K[0] + zf_K[1] + z^2f_K[2] + z^3f_K[3] + \cdots$$

From this expansion it is easy to show that

$$\frac{1}{k!}\frac{d^k G_K(z)}{dz^k}\bigg|_{z=0} = f_K[k] = \Pr[K = k] \tag{4.22}$$

Our interest in the PGF, however, is more in generating moments than it is in generating probabilities. For this, it is not necessary to require that $f_K[k] = 0$ for $k < 0$. Rather we can deal with the full two-sided transform defined in (4.21).

The method for generating moments can be seen clearly by using the first form of the definition in (4.21), i.e.,

$$G_K(z) = \mathcal{E}\{z^K\}$$

The derivative of this expression is[11]

$$\frac{dG_K(z)}{dz} = \mathcal{E}\{Kz^{K-1}\}$$

If this is evaluated at $z = 1$, the term z^{K-1} goes away and leaves the formula

$$\mathcal{E}\{K\} = \frac{dG_K(z)}{dz}\bigg|_{z=1} \tag{4.23}$$

This is the mean of the discrete random variable. To generate higher order moments, we repeat the process. For example,

$$\frac{d^2 G_K(z)}{dz^2}\bigg|_{z=1} = \mathcal{E}\{K(K-1)z^{K-2}\}\big|_{z=1} = \mathcal{E}\{K^2\} - \mathcal{E}\{K\}$$

While this result is not as "clean" as the corresponding result for the MGF, we can use the last two equations to express the second moment as

$$\mathcal{E}\{K^2\} = \frac{d^2 G_K(z)}{dz^2} + \frac{dG_K(z)}{dz}\bigg|_{z=1} \tag{4.24}$$

Table 4.3 summarizes the results for computing the first four moments of a discrete random variable using the PGF. The primes in the table denote derivatives.

---

[10] Once again, there is a difference with electrical engineering, where the convention is to define the z-transform in terms of negative powers of z.

[11] It is assumed valid to interchange the operations of differentiation and expectation. This is true in most cases.

$$\mathcal{E}\{K\} = G'(z)|_{z=1}$$

$$\mathcal{E}\{K^2\} = G''(z) + G'(z)|_{z=1}$$

$$\mathcal{E}\{K^3\} = G'''(z) + 3G''(z) + G'(z)|_{z=1}$$

$$\mathcal{E}\{K^4\} = G''''(z) + 6G'''(z) + 7G''(z) + G'(z)|_{z=1}$$

Table 4.3 Formulas for moments computed from the PGF. G', G'' etc. denote derivatives of $G(z)$.

Before closing this section, let us just mention a relationship between the MGF and the PGF. If K is an integer-valued discrete random variable, then a PDF for K can be defined as in (4.5). Substituting (4.5) in (4.19) then yields

$$M_K(s) = \sum_{k=-\infty}^{\infty} f_K[k]e^{sk}$$

This is the same as (4.21) evaluated at $z = e^s$. Hence for an integer-valued discrete random variable, we have the relationship

$$M_K(s) = G_K(e^s) \tag{4.25}$$

Let us end this section with an example of using the PGF to compute the mean and variance of a discrete random variable.

Example 4.8: Consider a Poisson random variable whose PMF is given by

$$f_K[k] = \begin{cases} \dfrac{\alpha^k}{k!}e^{-\alpha} & k \geq 0 \\ 0 & k < 0 \end{cases}$$

The PGF for this random variable can be computed from the second form in (4.21), namely

$$G_K(z) = \sum_{k=-\infty}^{\infty} f_K[k]z^k = \sum_{k=0}^{\infty} \frac{\alpha^k}{k!}e^{-\alpha}z^k = e^{-\alpha}\sum_{k=0}^{\infty} \frac{(\alpha z)^k}{k!} = e^{-\alpha}e^{\alpha z} = e^{\alpha(z-1)}$$

Given the above PMF, the mean can then be computed using (4.23):

$$m_K = \mathcal{E}\{K\} = \frac{dG_K(z)}{dz}\bigg|_{z=1} = e^{-\alpha}\alpha e^{\alpha z}\big|_{z=1} = \alpha$$

The second moment can be computed from (4.24):

$$\mathcal{E}\{K^2\} = \frac{d^2 G_K(z)}{dz^2} + \frac{dG_K(z)}{dz}\bigg|_{z=1} = e^{-\alpha}\alpha^2 e^{\alpha z} + e^{-\alpha}\alpha e^{\alpha z}\big|_{z=1} = \alpha^2 + \alpha$$

Finally, the variance is computed from

$$\sigma_K^2 = \mathcal{E}\{K^2\} - m_K^2 = \alpha^2 + \alpha - (\alpha)^2 = \alpha$$

□

4.4 Application: Entropy and Source Coding

In Chapter 2 we discussed the topics of information theory and coding. Recall that the *information* associated with an event A_i is defined as

$$I(A_i) = -\log \Pr[A_i]$$

where the logarithm is taken with base 2 and the information is measured in "bits." Now reconsider the problem where the "events" represent the transmission of symbols over a communication channel. As such, the events A_i are mutually exclusive and collectively exhaustive.

With the concept of expectation developed in this chapter, we are in a position to formally define *entropy H* for the set of symbols as the average information, that is,

$$H = \mathcal{E}\{I\} = -\sum_i \Pr[A_i] \log \Pr[A_i] \tag{4.26}$$

Now suppose that a binary code is associated with each symbol to be transmitted. Let the discrete random variable L represent the length of a code word and define the average length of the code as

$$m_L = \mathcal{E}\{L\} = \sum_l l\, f_L[l] \tag{4.27}$$

Shannon's source coding theorem can be stated as follows [12]:

Theorem. *Given any discrete memoryless source with entropy H, the average length m_L of any lossless code for encoding the source satisfies*

$$m_L \geq H \tag{4.28}$$

Here the term "lossless" refers to the property that the output of the source can be completely reconstructed without any errors or substitutions. The theorem can be illustrated by a continuation of Example 2.14 of Chapter 2.

Example 4.9: Recall that in the coding of the message "ELECTRICAL ENGINEERING" the probability of the letters (excluding the space) are represented by their relative frequency of occurrence and thus are assigned the probabilities listed in the first table in Example 2.14. The codeword and length of the codeword associated with each letter by the Shannon-Fano algorithm are also shown in the table.

It can be seen that the random variable L representing the length of the codeword takes on only three possible values. The PMF describing the random variable can be contructed by adding up the probabilities of the various length codewords. This PMF is shown below:

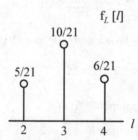

The average length of the codewords is computed from (4.27) and is given by

$$m_L = \mathcal{E}\{L\} = 2 \cdot \tfrac{5}{21} + 3 \cdot \tfrac{10}{21} + 4 \cdot \tfrac{6}{21} = 3.047 \text{ (bits)}$$

Now, let us compute the source entropy. From (4.26), the entropy is given by

$$H = -\tfrac{5}{21}\log\left(\tfrac{5}{21}\right) + 3\left(-\tfrac{3}{21}\log\left(\tfrac{3}{21}\right)\right) + 4\left(-\tfrac{2}{21}\log\left(\tfrac{2}{21}\right)\right) + 2\left(-\tfrac{1}{21}\log\left(\tfrac{1}{21}\right)\right)$$

$$= 3.006 \text{ bits}$$

The result shows that the bound (4.28) in Shannon's theorem is satisfied. The average code length produced by the Shannon-Fano procedure is very close to the bound, however. This shows that the Shannon-Fano code is very close to an optimal code for this message.

□

The original definition of entropy is in terms of source and channel coding. The basic definition extends to any random variables, however (not just codewords). The entropy for any discrete random variable K is defined as

$$H_K = -\mathcal{E}\left\{\log(f_K[K])\right\} = -\sum_{\forall k} f_K[k]\log(f_K[k]) \qquad (4.29)$$

while the entropy for a continuous random variable X is given by

$$H_X = -\mathcal{E}\left\{\log(f_X(X))\right\} = -\int_{-\infty}^{\infty} f_X(x)\log(f_X(x))dx \qquad (4.30)$$

These formulas have many uses in engineering problems and a number of optimal design procedures are based on minimizing or maximizing entropy.

4.5 Summary

The expectation of a random variable $\mathcal{E}\{X\}$ is a sum or integral over all possible values of the random variable weighted by their associated probability. Further, for any function $g(X)$, the expected value $\mathcal{E}\{g(X)\}$ can be computed by summing or integrating $g(X)$ with the PMF or PDF of X.

The moments of a random variable are expected values of powers of X. Thus $m_X = \mathcal{E}\{X\}$ is the first moment (also called the *mean*), $\mathcal{E}\{X^2\}$ is the second moment, and so on. *Central* moments are expectations of powers of the term $(X - m_X)$. The quantity $\sigma_X^2 = \mathcal{E}\{(X - m_X)^2\}$ is known as the *variance* of the distribution and measures how the random variable is spread about the mean.

The moment generating function is defined as $\mathcal{E}\{e^{sX}\}$ where s is a complex-valued parameter. The MGF can also be interpreted as the Laplace transform of the PDF. The power series expansion of the MGF reveals the moments of the distribution as coefficients. Thus, knowing all the moments of a distribution is equivalent to knowing the distribution. Some results and properties for random variables are discovered more easily from the MGF than by using the PDF.

For discrete random variables, a number of operations are carried out more conveniently using the probability generating function. The PGF is defined as $\mathcal{E}\{z^K\}$ and is the z-transform of the PMF.

The last section of this chapter discusses the concept of entropy as average information. Shannon's theorem states that entropy provides a lower bound on the average number of bits needed to code a message without loss.

References

[1] William Feller. *An Introduction to Probability Theory and Its Applications - Volume I.* John Wiley & Sons, New York, second edition, 1957.

[2] Athanasios Papoulis and S. Unnikrishna Pillai. *Probability, Random Variables, and Stochastic Processes*. McGraw-Hill, New York, fourth edition, 2002.

[3] Henry Stark and John W. Woods. *Probability, Random Processes, and Estimation Theory for Engineers*. Prentice Hall, Inc., Upper Saddle River, New Jersey, third edition, 2002.

[4] Wilbur B. Davenport, Jr. *Probability and Random Processes*. McGraw-Hill, New York, 1970.

[5] William Feller. *An Introduction to Probability Theory and Its Applications - Volume II*. John Wiley & Sons, New York, second edition, 1971.

[6] Daniel Zwillinger. *CRC Standard Mathematical Tables and Formulae*. CRC Press, Boca Raton, 2003.

[7] David R. Brillinger. *Time Series: Data Analysis and Theory*. Holden-Day, Oakland, California, expanded edition, 1981.

[8] Murray Rosenblatt. *Stationary Sequences and Random Fields*. Birkhauser, Boston, 1985.

[9] Chrysostomos L. Nikias and Athina P. Petropulu. *Higher-Order Spectra Analysis: A Nonlinear Signal Processing Framework*. Prentice Hall, Inc., Upper Saddle River, New Jersey, 1993.

[10] Sheldon M. Ross. *A First Course in Probability*. Prentice Hall, Inc., Upper Saddle River, New Jersey, sixth edition, 2002.

[11] Ruel V. Churchill and James Ward Brown. *Complex Variables and Applications*. McGraw-Hill Book Company, New York, fourth edition, 1984.

[12] Claude E. Shannon and Warren Weaver. *The Mathematical Theory of Communication*. University of Illinois Press, Urbana, IL, 1963.

[13] Alvin W. Drake. *Fundamentals of Applied Probability Theory*. McGraw-Hill, New York, 1967.

Problems

Expectation of a random variable

4.1 Rolf is an engineering student at the Technical University. Rolf plans to ask one of two women students for a Saturday night date: Claudia Schönstück or Ursula von Doppeldoof. Rolf is more attracted to Claudia but Ursula is more available. Further, Claudia likes the finer things in life and a night with her is more expensive. All things considered, Rolf estimates that landing a date with Claudia is a "50/50 chance" and Rolf will end up spending 200 euros. On the other hand, Ursula is about twice as likely to accept a date with Rolf as not, and Ursula is happy with beer and pretzels (40 euros).

Rolf has the following procedure to determine whom he will ask first. He tosses a 1-euro coin. If it comes up "heads" he asks Claudia. If it comes up "tails," he flips it again (just to be sure). If it comes up "heads" this time he asks Claudia; otherwise he asks Ursula.

(a) What are the probabilities that Rolf will initially ask Claudia or Ursula?

According to the outcome of his toss(es) of the coin, Rolf asks either Claudia or Ursula. If he asks Claudia and she turns him down, then he asks Ursula. If he asks Ursula (first) and she turns him down, he is so despondent that he does not have the courage to ask Claudia. If he does not have a date, he spends the night drinking beer and eating pretzels by himself, so the evening still costs 40 euros.

(b) What is the probability that Rolf has a Saturday night date with Claudia?

(c) What is the expected value of the money that Rolf spends on a Saturday night?

(d) If C is a random variable representing Rolf's expenditure on a Saturday night, sketch the PMF for C. Compute $\mathcal{E}\{C\}$ using the PMF.

4.2 Consider the rolling of a pair of fair dice. The PMF for the number K that is rolled is given in Chapter 3. What is $\mathcal{E}\{K\}$?

4.3 A *discrete* uniform random variable K is described by the PMF $f_K[k] = 1/5$ for $k = 0, 1, 2, 3, 4$. Find the mean m_K and compute the fourth central moment $\mathcal{E}\{(K - m_K)^4\}$.

4.4 Consider transmission of packets with a fixed size of 3 bits. The probability of a binary 1 is 0.56. If the number of binary 1s in a packet is the random variable I, calculate its mean.

4.5 Consider a random experiment in which you first roll a die with outcomes $\{i = 1, 2, 3, 4, 5, 6\}$ and then toss a coin with outcomes $\{j = 0, 5\}$. An outcome of the experiment is a pair: (i, j). All outcomes of the experiment are equally likely. Define the random variable K to be the difference: $k = i - j$.

(a) Determine and plot the probability mass function of this random variable $f_K[k]$. The axes of the plot must be clearly marked with appropriate values.

(b) Find the mean value of this random variable, $\mathcal{E}\{K\}$.

4.6 Eduardo de Raton, known to his business partners as *Ed the Rat*, has invited you to play a "friendly" game of dice, which you dare not refuse. Ed always uses tetrahedral dice and plays on a glass coffee table with a mirror mounted below.[12] This table is useful to Ed in some of his other games. This situation is useful to you as well in that there are fewer cases you need to consider than with six-sided dice.

The basic dice game is as follows. Both you and Ed have one die; he rolls first then you roll. If you match him, you win; if you don't match him, you lose. Ed's die happens to be loaded. The probability that Ed rolls a 1, 2, or 3 is each $1/5$ while the probability that he rolls a 4 is $2/5$. Your die is a fair one; the probability that you roll a 1, 2, 3, or 4 is each equal to $1/4$. If you win, Ed pays you what was rolled on the dice (Example: Ed rolls 3 and you roll 3. Ed pays you $6.) If you lose you pay Ed $3 (i.e., you "win" $-3).

(a) Draw the sample space for the game and label each elementary event with its probability.

(b) What is the expected value of your winnings? (This value may be positive or negative.)

(c) At your option, you can play another game with Ed. This game is called the "high stakes" game. This game is the same as the basic game except that if "doubles" are rolled you win twice the amount of money shown on the dice. (Example: Ed rolls 3 and you roll 3. You win $2 \times 6 = \$12$.) Except

---

[12] Tetrahedral dice have four sides with 1, 2, 3, and 4 dots; they were apparently invented by Alvin Drake [13]. The outcome of the roll is determined by which side is facing *down*.

if a double 4 is rolled, *you pay Ed.* (Ed rolls 4 and you roll 4, you *lose* $16.) Which game should you play (the basic game or the high stakes game) if you want to maximize the money you win or minimize the money you lose (expected value)?

4.7   Let K be a discrete uniform random variable with parameters $n = -2$ and $m = 3$. Define the random variable $I = 5K + 2$. Determine the following expectations:

(a) $\mathcal{E}\{K\}$
(b) $\mathcal{E}\{I\}$
(c) $\mathcal{E}\{K + I\}$
(d) $\mathcal{E}\{K \cdot I\}$

Hint: Don't forget to use $\mathcal{E}\{\cdot\}$ as a linear operator.

4.8   Consider the continuous uniform random variable with PDF

$$f_X(x) = \begin{cases} \dfrac{1}{2b} & -b \le x \le b \\[2mm] 0 & \text{otherwise} \end{cases}$$

(a) What is $\mathcal{E}\{X\}$?
(b) What is $\mathcal{E}\{X|X > 0\}$?

4.9   Given the cumulative distribution function of a certain random variable X,

$$F_X(x) = \begin{cases} 0 & x \le 0 \\ x & 0 < x \le \frac{1}{2} \\ 1 & \frac{1}{2} \le x \end{cases}$$

find the mean $\mathcal{E}\{X\}$.

4.10   Consider the discrete random variable I, with PMF shown below.

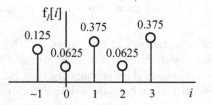

(a) What is $\mathcal{E}\{I|I > 0\}$?
(b) What is $\mathcal{E}\{I|I \le 0\}$?

4.11   Claudia is a student at the Technical University majoring in statistics. On any particular day of the week, her friend Rolf may ask her for a Saturday night date with probability 3/4. Claudia, however, is more attracted to Roberto de la Dolce, who is very handsome, drives an expensive Italian car, and really knows how to treat women! Roberto has other women he likes however, so the probability that he asks Claudia out in any particular week is only 2/3. Roberto is also very self-confident and does not plan his activities early in the week. Let D represent the event that Roberto asks Claudia for a Saturday night date. Then the day of the week on which he asks her (beginning on Monday) is a random variable K with PMF $f_{K|D}[k|D] = k/15$ for $1 \le k \le 5$ and 0 otherwise. Claudia is aware of this formula and the probabilities.

Claudia needs to plan whether or not to accept if Rolf asks her for a date before Roberto; thus she decides to rate her emotional state (α) for the week on a scale of 0 to 10. A date with Roberto is actually way off the scale but she assigns it a 10. She further determines that a date with with Rolf is worth 4 points, and a Saturday night without a date is worth −5 points. She decides that if Rolf asks her out on the kth day of the week she will compute the expected value of α given that she accepts and the expected value of α given that she does not accept. Then she will make a choice according to which expected value is larger.

(a) Make a plot of the probability that Roberto does *not* ask Claudia for a date given that he has not asked her by the end of the lth day $1 \le l \le 5$.

(b) Sketch the conditional PMF for K given that Roberto asks Claudia out but has not done so by the end of the second day. Given this situation, what is the probability that Roberto asks her out

  (i) on the third day ($k = 3$)?
  (ii) on the fifth day ($k = 5$)?

(c) By the middle of the third day of the week Roberto has not asked Claudia for a date; but Rolf decides to ask her. Will she accept Rolf or not?

(d) Rolf has not studied statistics (not even probability), and thinks that his chances for a date with Claudia will be better if he asks her earlier in the week. Is he right or wrong?

(e) What is the optimal strategy for Rolf (i.e., when should he ask Claudia) in order to maximize the probability that Claudia will accept if he asks her for a Saturday night date?

Hint: It is useful to draw a sample space for this problem.

4.12 The probability density function of a random variable X is given by

$$f_X(x) = \begin{cases} \frac{1}{2}x, & 0 \le x \le 1 \\ \frac{1}{6}(4 - x), & 1 < x < c \\ 0 & \text{otherwise} \end{cases}$$

(a) Find the constant c.
(b) Calculate the expectation: $E[3 + 2X]$.

4.13 The probability density function of a random variable X is given by

$$f_X(x) = \begin{cases} Ae^{-x}, & 0 \le x \le 3 \\ 0 & \text{otherwise.} \end{cases}$$

(a) Find the constant A. Plot the probability density function.
(b) Calculate the first moment of this random variable $E[X]$.

Moments of a distribution

4.14 Let K be a random variable representing the number shown at the roll of a die. The PMF is given by

$$f_K[k] = \begin{cases} 1/6 & 1 \le k \le 6 \\ 0 & \text{otherwise} \end{cases}$$

Determine the mean and variance of the random variable K.

4.15 If a discrete random variable takes on three values $\{0, 1, 2\}$ with equal probabilities, what is its standard deviation?

4.16 In a set of independent trials, a random variable K takes on the following values: 0, 0, 8, 8, 8, 0, 5, 8, 5, 0, 8, 0, 8, 0, 8, 8.

(a) From the given data, obtain an estimate of the probability mass function of the random variable.

(b) Find the mean of K.

(c) Find the variance of K.

4.17 In a set of independent trials, a discrete random variable J takes on the following values: 1, 1, 9, 6, 1, 6, 1, 9, 6, 9, 6, 9, 1, 6, 6.

(a) Obtain an estimate and plot the probability mass function of this random variable.

(b) Calculate the mean $\mathcal{E}\{J\}$ and the variance $\text{Var}\,[J]$.

4.18 In a set of independent trials, a certain discrete random variable I takes on the values: 5, 7, 2, 8, 2, 7, 2, 7, 7, 8, 7, 2, 7, 2, 5, 7.

(a) Estimate the probability mass function $f_I[i]$ using the given sequence of values.

(b) Using the PMF obtained in (a), determine the second moment of I.

4.19 A discrete random variable K has the PMF shown below.

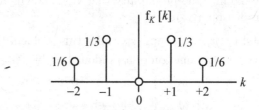

Find the mean and variance of this random variable.

4.20 In a certain digital control system a 0 is represented by a negative 1 volt level and a 1 is represented by a positive 1 volt level. The voltage level is modeled by a discrete random variable K with PMF

$$
f_K[k] = \begin{cases} p & k = +1 \\ 1 - p & k = -1 \\ 0 & \text{otherwise} \end{cases}
$$

(a) Sketch the PMF.

(b) Find $\mathcal{E}\{K\}$, $\mathcal{E}\{K^2\}$, and $\text{Var}\,[K]$.

(c) For what value of p is the variance maximized and what is this maximum value?

4.21 Consider a nonsymmetric binary communication channel. The input symbols occur with probability $\Pr[0_S] = 2/5$ and $\Pr[1_S] = 3/5$. The channel probabilities are given as $\Pr[1_R|1_S] = 0.92$ and $\Pr[1_R|0_S] = 0.13$. Let random variable S taking on values $0, 1$ represent the input and random variable R taking on values $0, 1$ represent the output.

(a) Draw the probability mass function for S. Compute the mean and the variance of S.

(b) Determine and draw the PMF of the output R. Find the mean and the variance of R.

4.22 Find the mean and variance of a discrete uniform random variable in terms of the parameters m and n. <u>Hint:</u> To find the variance, use (4.17).

4.23 Consider the discrete random variable I, defined in Prob. 4.10.

(a) Find the expected value of random variable I.

(b) Determine the second moment of I.

(c) Compute the variance of I

 (i) starting with the definition of the second central moment,
 (ii) using the relationship among mean, second moment, and variance.

 Compare the results.

(d) What is the $\Pr[I > m_I]$ where m_I is the mean?

4.24 A type 0 geometric random variable takes on the values 0, 1, 2, 3, ...

(a) What is the mean of this random variable?

(b) Determine the second moment and variance.

(c) Given that the parameter $p = 1/3$, compute the mean and the variance.

4.25 Find the mean and the variance for the following two distributions:

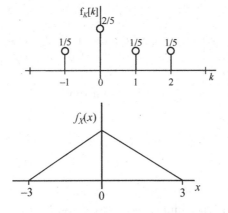

4.26 The PDF of a random variable X is given in the figure below:

(a) Find the mean of X.

(b) Calculate the variance of X.

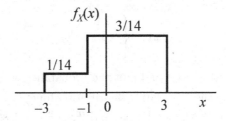

4.27 Consider the transfer characteristic of a limiter shown below. The input random variable X is exponentially distributed with $\lambda = 0.5$. Find the second moment of Y.

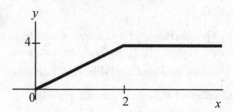

4.28 Given the cumulative distribution function of random variable X shown in the figure (not drawn to scale and all slant lines have the same slope) below, find its mean and the variance.

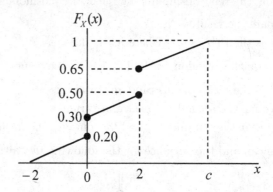

4.29 A certain *discrete* uniform random variable K is described by $f_K[k] = 1/5$ for $k = -2, -1, 0, 1, 2$ and zero otherwise. Compute the third and fourth moments of this random variable.

4.30 The probability density function of a random variable X is given by

$$f_X(x) = \begin{cases} A, & 1 \le x \le 4 \\ 0.2\delta(x-2), & x = 2 \\ 0.3\delta(x-5), & x = 5 \\ 0, & \text{otherwise.} \end{cases}$$

(a) Find the constant A. Plot the probability density function.

(b) Determine the cumulative distribution function.

(c) Calculate $\Pr[1.5 \le X \le 2.5]$ and $\Pr[3 \le X < 6]$.

(d) Find the mean and variance of X.

4.31 The probability density function of a random variable X is given by

$$f_X(x) = \begin{cases} 0.4e^{-0.4x}, & 0 \le x \le 3 \\ c\delta(x-1), & x = 1 \\ 0, & \text{otherwise.} \end{cases}$$

(a) Find the constant c and plot the probability density function.

(b) Determine the cumulative distribution function. And plot it.

(c) Given that $\mathcal{E}\{X\} = 1.1446$ and $\mathcal{E}\{X^2\} = 1.8076$, determine the mean and variance of $Y = 3X$.

4.32 Consider a random variable with the following probability density function

$$f_X(x) = \begin{cases} \frac{1}{\sqrt{2\pi}\sigma} \exp -\frac{(x-m)^2}{2\sigma^2} & -A \le x \le A \\ 0.1\delta(x-1), & x = 1 \\ 0 & \text{otherwise.} \end{cases}$$

where $m = 0$ and $\sigma = 2$.

(a) Find the constant A to make it a valid probability density function and plot it.

(b) Calculate these probabilities: $\Pr[X \le 0]$, $\Pr[X \ge 1]$ and $\Pr[X > 3]$.

(c) Determine the cumulative distribution function and plot it.

(d) Find the mean value of the random variable.

4.33 A random variable X has the PDF shown below.

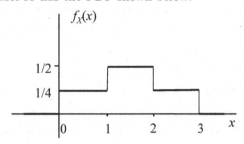

(a) Sketch the cumulative distribution function (CDF).

(b) Find the mean $E\{X\}$.

(c) Determine the variance of X.

4.34 The probability density function of a random variable X is shown below.

(a) Find the value of A that makes it a valid probability density function.

(b) Determine the first moment $E\{X\}$.

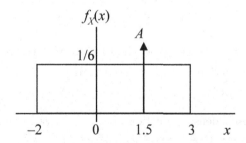

4.35 Find the mean and variance of a continuous uniform random variable in terms of the parameters a and b. Hint: To find the variance, use (4.17).

4.36 A random variable X has a mean of 2 and a variance of 2. A new random variable Y is formed as follows: $Y = \sqrt{2}X$. Find the mean and the standard deviation of the random variable Y.

4.37 A random variable X has a variance of $1/6$. The probability density function of this random variable is shown below. What is its second moment?

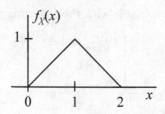

4.38 Consider a continuous uniform random variable X in the range of 0 to 5.

(a) Determine $\mathcal{E}\{X\}$.

(b) Find $\mathcal{E}\{X|1 \leq X \leq 3.5\}$.

4.39 A random variable Y is defined in terms of another random variable X as

$$Y = 2X + 3$$

X is known to be a Gaussian random variable with mean $\mu = 1$ and variance $\sigma^2 = 1$.

(a) What is the mean of Y?

(b) What is the variance of Y?

(c) Is Y a Gaussian random variable?

4.40 The officers of the Lower Slobbovian Navy have decided to increase the gain on their receiver. What they do not realize is that increasing the gain also increases the noise. Therefore, in the case that a giant seal is present, the receiver output is given by

$$X = 2W + 3$$

The noise PDF is shown below.

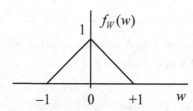

(a) Sketch and carefully label the PDF for the receiver output X.

(b) Find the variance of the noise, Var $[W]$.

(c) Find the variance of the sonar output, Var $[X]$.

4.41 The probability density function of a random variable X is shown below:

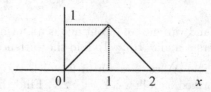

(a) Write a set of algebraic expressions that describe the density function.

(b) Compute the mean and variance of this random variable.

4.42 The PDF of a random variable X is shown below.

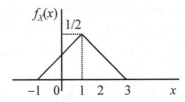

(a) Find the mean m_X and the second moment of this random variable.

(b) Determine the variance σ_X^2.

(c) Determine $\Pr[m_X - \sigma_X \leq X \leq m_X + \sigma_X]$.

4.43 The probability density function of a Laplace random variable is given by

$$f_X(x) = \frac{\alpha}{2} e^{-\alpha|x-\mu|}, \quad -\infty < x < \infty.$$

(a) Find the mean m_X, second moment, and variance σ_X^2.

(b) Determine the following probabilities for $\alpha = 4$ and $\mu = 1$.

   (i) $\Pr[|X - m_X| > 2\sigma_X]$

  (ii) $\Pr[|X - m_X| < \sigma_X]$

4.44 Find the mean and variance of the random variable in Prob. 3.12 of Chapter 3.

4.45 Find the *skewness* and *kurtosis* of a continuous uniform random variable with the PDF defined in Prob. 4.8. How would the results change if the PDF were uniform with a lower limit $a \neq -b$?

4.46 Show that the mean and variance of a Gaussian random variable are given by μ and σ^2 respectively, where these are the parameters that appear in the PDF. <u>Hint:</u> Make a change of variables and integrate "by parts."

Generating functions

4.47 The probability density function of a random variable X is shown below:

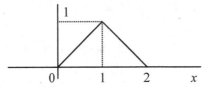

(a) Write a set of algebraic expressions that describe the density function.

(b) Compute the Moment Generating Function.

(c) Show how you would use the MGF to compute the mean.

4.48 A Gaussian random variable has mean μ and variance σ^2.

(a) Show that the moment generating function (MGF) for the Gaussian random variable is given by

$$M_X(s) = e^{\left(\mu s + \frac{1}{2}\sigma^2 s^2\right)}$$

<u>Hint:</u> Use the technique of "completing the square."

(b) Assume that $\mu = 0$ and use the MGF to compute the first four moments of X as well as the variance, skewness, and kurtosis.

(c) What are the mean, variance, skewness, and kurtosis for $\mu \neq 0$?

4.49 The moment generating function of a continuous random variable X is given to be

$$M_X(s) = \frac{3}{1 - 2s}.$$

Find the second moment.

4.50 A random variable X is described by the Laplace density:

$$f_X(x) = \tfrac{1}{2} e^{-|x|}$$

(a) Find the MGF $M_X(s) = \mathcal{E}\{e^{sX}\}$. Assume that $-1 < \mathrm{Re}\,[s] < +1$ in order to evaluate the integral.

(b) What is the mean of X?

(c) What is the variance of X?

(d) Find a transformation such that the new random variable Y has a mean of $+1$. Find and sketch the density function $f_Y(y)$.

(e) Find a transformation such that the new random variable Y has a variance equal to 4. What is the density function $f_Y(y)$?

4.51 Find the moment generating function of random variable X whose probability density function is given by

$$f_X(x) = \lambda e^{-\lambda(x-1)}, \quad x \geq 1.$$

4.52 Find the PGF for the random variable described in Prob. 4.20. Use it to compute the first four moments of the random variable.

4.53 The probability generating function of a discrete random variable I is given to be

$$G_I(z) = \frac{2}{5 - 3z}.$$

Find the second moment.

4.54 Consider the geometric random variable as defined in Prob. 4.24.

(a) Determine the PGF.

(b) From the PGF, find the mean and the variance.

Application: entropy

4.55 Let K be a Bernoulli random variable with parameter p.

(a) Find an expression for the entropy H_K of the Bernoulli random variable.

(b) Sketch H_K as a function of p for $0 < p < 1$. For what value of p is the entropy maximized?

4.56 The message "SHE HAS A KEEN EYE" is to be transmitted.

(a) Using the Morse code chart provided below, represent this message in the code.

(b) Change the message to a binary string of 1s and 0s by representing a dash as binary 1 and a dot as binary 0.

(c) Calculate the average codeword length in bits per letter.

(d) If you consider the whole set of 26 letters, what is the average codeword length in bits? Assume the letters are equally likely.

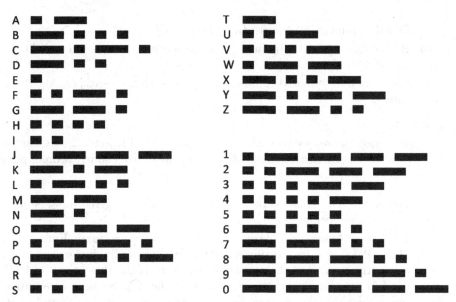

International Morse code: A dash is equal to three dots, and the space between parts of the same letter is equal to one dot. The space between two letters is equal to three dots, and the space between two words is equal to seven dots

4.57 Let X be a Gaussian random variable with mean μ and variance σ^2. The entropy in units of "nats" is given by (4.30) using the natural (Naperian) logarithm instead of logarithm to the base 2. Find a simple expression for the entropy of the Gaussian random variable. <u>Hint:</u> Don't get hung up on the integration. Use the first form of the expression in (4.30) and recognize that the expectation of certain terms produces the variance σ^2.

4.58 Consider the situation described in Prob. 4.6.

(a) Is the entropy for the first (basic) game greater than, less than, or equal to the entropy for the second (high stakes) game?

(b) If the second game were played with a fair set of dice (i.e., both pairs of dice are fair) is the entropy for the first game greater than, less than, or equal to the entropy for the second game?

Computer Projects

Project 4.1

The objective of this project is to estimate the mean and variance of random variables of various kinds. This project follows Project 3.1 of Chapter 3.

1. For each type of random variable described in Project 3.1 Step 1, calculate the theoretical mean and variance.

2. Generate the random sequences as specified in Project 3.1 Step 1.

3. Estimate the mean and the variance of these random variables using the following formulas:

$$\mathcal{E}\{X\} \approx M_n = \frac{1}{n}\sum_{i=1}^{n}X_i \qquad \mathrm{Var}\,[X] \approx \Sigma_n^2 = \frac{1}{n}\sum_{i=1}^{n}(X_i - M_n)^2$$

(These formulas are discussed in Chapter 6.)

4. Repeat Steps 2 and 3 by increasing the sequence length n to (a) 20,000, and (b) 40,000.

5. Compare the estimated values in each case with the corresponding theoretical values.

Project 4.2

This project serves as an extension to Computer Project 2.2 of Chapter 2 to explore the formal definition of entropy and the source coding theorem presented in this chapter.

1. Follow Steps 1, 2, and 3 of Project 2.2 to code the given message using the Huffman algorithm.

2. Let L be a random variable representing the codeword length. Find an estimate of the PMF of L by plotting a histogram of the individual codeword lengths and normalizing it so the probabilities sum to 1. Use this estimated PMF $\hat{f}_L[l]$ in (4.27) to compute the mean length of the codewords, $\hat{m}_L$. Compare this mean length to the average codeword length computed in Step 4 of Project 2.2. (They should be the same.)

3. Consider the average information for the set of symbols computed in Step 2 of Project 2.2. This is an estimate of the source entropy H. Compare this estimate of the entropy to the mean codeword length $\hat{m}_L$ computed above, and verify that Shannon's source coding theorem (4.28) is satisfied experimentally.

5

Two and More Random Variables

This chapter expands the scope of discussion to two and more random variables. Among the topics introduced is that of dependency among random variables; specifically, where one random variable is *conditioned* upon another. As part of this we define what it means for random variables to be *independent*.

The chapter begins with the case of two discrete random variables to motivate the concept of dependency among random variables, and develops the concepts of joint and conditional probability mass functions. It moves on to continuous random variables and carries on a parallel development there. The chapter then discusses expectation for two random variables and defines the notion of "correlation."

Several properties and formulas involving sums of multiple random variables are then discussed. These properties and formulas are needed sooner or later in any course in probability and statistics, and they fit well with the developments in this chapter.

5.1 Two Discrete Random Variables

Recall that a random variable is a mapping from the sample space for an experiment to the set of real numbers. Many problems involve two or more random variables. If these random variables pertain to the same experiment, then these random variables may be interrelated; thus it is necessary to have a *joint* characterization of these random variables. To make the discussion more specific, consider the following example involving two discrete random variables.

Example 5.1: In the secure communication link provided by British Cryptologic Telecommunication Ltd (popularly known as "BRITTEL"), the overall probability of bit errors is 0.2. But errors tend to occur in bursts, so once an error occurs, the probability that the next bit has an error is 0.6, while if an error does not occur, the probability that the next bit is in error is only 0.1.

We start observing the transmission at some random time and observe two successive bits. An event tree and sample space corresponding to this experiment is shown in the figure on the next page, where "E" represents the occurrence of an error. Recall that the probabilities listed for the the four outcomes in the sample space are found by multiplying the probabilities along the branches of the tree leading to the outcome. Let the random variables K_i represent the total number of errors that have occured after the ith bit. The values of random variables K_1 and K_2 after observing two bits are also shown in the figure. These random variables are clearly interrelated.

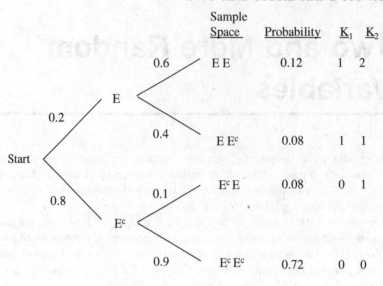

This example serves to motivate the need for joint treatment of two random variables. The remainder of this section continues to discuss various aspects of the analysis of discrete random variables while continuing to use this example for illustration.

5.1.1 The joint PMF

For problems involving two discrete random variables, the probability that the random variables take on any given pair of values is specified by the *joint* probability mass function (PMF), defined by[1]

$$f_{K_1 K_2}[k_1, k_2] \overset{\text{def}}{=} \Pr[K_1 = k_1, K_2 = k_2] \tag{5.1}$$

The joint PMF for the random variables described in the Example 5.1 is illustrated in Fig 5.1. The joint PMF is represented as a three-dimensional plot with values of (k_1, k_2) represented in the horizontal plane and values of their probabilities represented in the vertical direction. (Zero values of probability are not explicitly shown in Fig. 5.1.) In general the joint PMF could have non-zero values for $-\infty < k_1 < \infty$ and $-\infty < k_2 < \infty$. Since the set of all the possible values of k_1 and k_2 represent a mapping of all outcomes in the sample space, however, it follows that

$$\sum_{k_1=-\infty}^{\infty} \sum_{k_2=-\infty}^{\infty} f_{K_1 K_2}[k_1, k_2] = 1 \tag{5.2}$$

Needless to say, the individual values of $f_{K_1 K_2}[k_1, k_2]$, which represent probabilities, must all lie between 0 and 1.

Since K_1 and K_2 are both random variables, each can be described by its own PMF

---

[1] A more precise notation involving the sample space could be given (see (3.1) and the equation preceding in Chapter 3). The notation of the right of (5.1) is common, however, and less cumbersome.

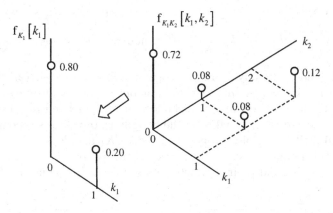

Figure 5.1 Joint PMF for the random variables K_1 and K_2 in Example 5.1 and marginal PMF for K_1.

$f_{K_1}[k_1]$ and $f_{K_2}[k_2]$. These so-called *marginal* PMFs are obtained from the equations

$$f_{K_1}[k_1] = \sum_{k_2=-\infty}^{\infty} f_{K_1 K_2}[k_1, k_2] \quad \text{(a)}$$

$$f_{K_2}[k_2] = \sum_{k_1=-\infty}^{\infty} f_{K_1 K_2}[k_1, k_2] \quad \text{(b)}$$

(5.3)

In other words the marginal PMFs are obtained by summing over the other variable. As such, they can be considered to be the *projections* of the joint density on the margins of the area over which they are defined. Fig. 5.1 shows the marginal PMF for K_1, which is formed by projecting (summing) over values of K_2. The marginal PMF for the variable K_2 is formed in an analogous way and is shown in Fig. 5.2.

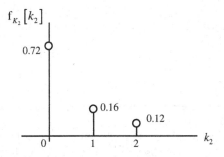

Figure 5.2 Marginal PMF for random variable K_2.

Equations 5.3 follow directly from the *principle of total probability* [(2.11) of Chapter 2]. To see this, let B be the event "$K_1 = k_1$" and A_i be the events "$K_2 = i$." Using these events in (2.11) produces (5.3)(a). Equation 5.3(b) can be obtained in a similar manner.

5.1.2 Independent random variables

Two discrete random variables are defined to be *independent* if their PMFs satisfy the condition

$$f_{K_1 K_2}[k_1, k_2] = f_{K_1}[k_1] \cdot f_{K_2}[k_2] \qquad \text{(for independent random variables)} \qquad (5.4)$$

The definition follows directly from the corresponding definition (2.12) for events. Notice that (5.4) is a *special condition* for random variables and does not apply in general! In particular, if two random variables are *not* independent, there is *no way* that the joint PMF can be inferred from the marginals. In that case the marginals are insufficient to describe any joint properties between K_1 and K_2.

To examine this point further, let us check out independence for our example. The two marginal densities and their product are represented in the matrix below on the left

	0.72	.0.16	0.12			0.720	0.080	0
0.80	0.576	0.128	0.096			0.720	0.080	0
0.20	0.144	0.032	0.024			0	0.080	0.120

The values along the left and top edges of the matrix represent values of $f_{K_1}[k_1]$ and $f_{K_2}[k_2]$ computed earlier, while the other values in the matrix represent the product of these values $f_{K_1}[k_1] \cdot f_{K_2}[k_2]$. The matrix on the right represents the values of the joint PMF from Fig. 5.1. Clearly the values in the two matrices are not the same; hence the random variables fail to satisfy (5.4) and are *not* independent. This conclusion satisfies intuition because from the tree diagram in Example 5.1, the value of K_2 definitely depends upon the value of K_1.

5.1.3 Conditional PMFs for discrete random variables

When random variables are not independent, they are conditionally *dependent* upon one another. This leads to the need to define a conditional PMF for random variables. For discrete random variables, the conditional PMF is defined as

$$f_{K_1|K_2}[k_1|k_2] \stackrel{\text{def}}{=} \Pr[K_1 = k_1 | K_2 = k_2] \qquad (5.5)$$

(Recall that the vertical bar '|' is read as "given.") Since all of these PMFs (joint, marginal, conditional) represent probabilities directly, we can use the definition of conditional probability for events (2.13) to write

$$\boxed{f_{K_1|K_2}[k_1|k_2] = \frac{f_{K_1 K_2}[k_1, k_2]}{f_{K_2}[k_2]}} \qquad (5.6)$$

Let us illustrate this computation in terms of our example.

The observed values of K_1 and K_2 are clearly related as has been seen (i.e., they are not independent). Suppose we had somehow missed observing the value of K_1 but had observed the value of K_2. Then we might be interested in knowing the probabilities for K_1 *given* the value of K_2. That is, we would want to know the conditional PMF (5.6).

The values of the joint PMF are shown in Fig. 5.1 and the marginal PMF for K_2 is sketched in Fig 5.2. The conditional PMF is represented by a *set* of plots, shown in Fig. 5.3, corresponding to each possible value of k_2. The plots of Fig. 5.3 represent normalized *slices* of the joint PMF in Fig. 5.1. That is, for each value of k_2 with non-zero probability, $f_{K_1|K_2}[k_1|k_2]$ is computed from (5.6) by taking a slice of the joint PMF in the k_1 direction and dividing the values by $f_{K_2}[k_2]$ at that location. For example, to compute $f_{K_1|K_2}[k_1|0]$ shown in Fig. 5.3 (a), we would fix $k_2 = 0$ and divide the joint

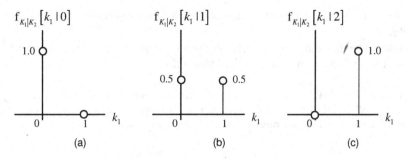

Figure 5.3 Conditional PMF $f_{K_1|K_2}[k_1|k_2]$. (a) $k_2 = 0$. (b) $k_2 = 1$. (c) $k_2 = 2$.

PMF along the line $k_2 = 0$ in Fig. 5.1 by the value $f_{K_2}[0] = 0.72$. A similar procedure is followed to compute the plots shown in Fig. 5.3 (b) and (c).

Observe from Fig. 5.3 that the conditional PMF is represented by a function of a single variable (k_1) and *parameterized* by the other variable (k_2). It is indeed a probability mass function in the first variable in that

$$\sum_{k_1=-\infty}^{\infty} f_{K_1|K_2}[k_1|k_2] = 1 \qquad (5.7)$$

for any value of k_2 (see Prob. 5.6). The summation

$$\sum_{k_2=-\infty}^{\infty} f_{K_1|K_2}[k_1|k_2]$$

however, over the conditioning variable, has absolutely no meaning.

5.1.4 Bayes' rule for discrete random variables

A final result to be mentioned here is a form of Bayes' rule for discrete variables. Writing (5.6) in the form

$$f_{K_1|K_2}[k_1|k_2]f_{K_2}[k_2] = f_{K_1K_2}[k_1, k_2] = f_{K_2|K_1}[k_2|k_1]\,f_{K_1}[k_1]$$

and then rearranging yields

$$f_{K_1|K_2}[k_1|k_2] = \frac{f_{K_2|K_1}[k_2|k_1]\,f_{K_1}[k_1]}{f_{K_2}[k_2]} \qquad (5.8)$$

Then since the denominator term can be computed as

$$f_{K_2}[k_2] = \sum_{k_1=-\infty}^{\infty} f_{K_1K_2}[k_1, k_2] = \sum_{k_1=-\infty}^{\infty} f_{K_2|K_1}[k_2|k_1]\,f_{K_1}[k_1]$$

substituting this result in (5.8) produces the formula

$$f_{K_1|K_2}[k_1|k_2] = \frac{f_{K_2|K_1}[k_2|k_1]\,f_{K_1}[k_1]}{\sum_{k_1=-\infty}^{\infty} f_{K_2|K_1}[k_2|k_1]\,f_{K_1}[k_1]} \qquad (5.9)$$

This form of Bayes' rule is useful when it is desired to "work backward" and infer the probability of some fundamental (but usually unobservable) random variable from another related random variable that can be measured or observed directly. Equations 5.8 and 5.9 are not really new results if you consider that the various PMFs represent probabilities of events. In this case they are direct consequences of (2.18) and (2.19) of Chapter 2.

Let us close the discussion of discrete random variables with an example that further illustrates some of the ideas presented in this subsection.

Example 5.2:   The Navy's new underwater digital communication system is not perfect. In any sufficiently long period of operation, the number of communication errors can be modeled as a Poisson random variable K_1 with PMF

$$f_{K_1}[k_1] = e^{-\alpha}\frac{\alpha^{k_1}}{k_1!} \qquad 0 \leq k_1 < \infty$$

where the parameter α represents the average number of errors occuring in the time period.

The contractor responsible for developing this system, having realized that the communication system is not perfect, has implemented an error detection system at the receiver. Unfortunately, the error detection system is not perfect either. Given that K_1 errors occur, only K_2 of those errors are detected. K_2 is a random variable with a binomial distribution conditioned on K_1:

$$f_{K_2|K_1}[k_2|k_1] = \binom{k_1}{k_2}p^{k_2}(1-p)^{(k_1-k_2)} \qquad 0 \leq k_2 \leq k_1$$

where p is the probability of detecting a single error. To make matters worse, the contractor, Poseidon Systems Incorporated, who had never heard of the Poisson distribution, thought that their system was being dubbed by the Navy as a "Poseidon random variable," and threatened to sue.

The admiral in charge of development has called for a performance analysis of the overall system and has required the contractor to compute $f_{K_2}[k_2]$, the marginal PMF for the number of errors detected, and the conditional PMF $f_{K_1|K_2}[k_1|k_2]$ for the number of errors occuring given the number of errors corrected. The admiral is a graduate of the Naval Postgraduate School with a master's degree in operations research, and claims that both of these distributions are Poisson. Is he correct?

To find the marginal PMF, let us first write the joint PMF as

$$\begin{aligned} f_{K_1K_2}[k_1, k_2] &= f_{K_2|K_1}[k_2|k_1]\,f_{K_1}[k_1] \\ &= \binom{k_1}{k_2}p^{k_2}(1-p)^{(k_1-k_2)} \cdot e^{-\alpha}\frac{\alpha^{k_1}}{k_1!} \qquad 0 \leq k_2 \leq k_1 < \infty \end{aligned}$$

This joint PMF is non-zero only in the region of the k_1, k_2 plane illustrated below.

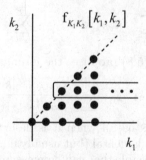

Then to find the marginal from the joint PMF, we expand the binomial coefficient in the expression for $f_{K_1K_2}[k_1,k_2]$ and sum over k_1. (The region of summation is also shown in the figure.) The algebraic steps are as follows:

$$f_{K_2}[k_2] = e^{-\alpha}p^{k_2}\sum_{k_1=k_2}^{\infty}\frac{\alpha^{k_1}}{k_1!}\frac{k_1!}{(k_1-k_2)!\,k_2!}(1-p)^{(k_1-k_2)}$$

$$= \frac{e^{-\alpha}p^{k_2}\alpha^{k_2}}{k_2!}\sum_{k_1=k_2}^{\infty}\frac{[\alpha(1-p)]^{(k_1-k_2)}}{(k_1-k_2)!} = \frac{e^{-\alpha}(\alpha p)^{k_2}}{k_2!}\sum_{k=0}^{\infty}\frac{[\alpha(1-p)]^{k}}{k!}$$

$$= \frac{e^{-\alpha}(\alpha p)^{k_2}}{k_2!}e^{\alpha(1-p)} = \frac{e^{-\alpha p}(\alpha p)^{k_2}}{k_2!}$$

Thus the marginal PMF for the number of errors detected is in fact also a Poisson PMF, but with parameter αp.

The conditional PMF $f_{K_1|K_2}$ can now be found from (5.8) or (5.9). From (5.8), we have

$$f_{K_1|K_2}[k_1|k_2] = \frac{f_{K_2|K_1}[k_2|k_1]\,f_{K_1}[k_1]}{f_{K_2}[k_2]}$$

$$= \frac{\binom{k_1}{k_2}p^{k_2}(1-p)^{(k_1-k_2)}\cdot e^{-\alpha}\frac{\alpha^{k_1}}{k_1!}}{\frac{e^{-\alpha p}(\alpha p)^{k_2}}{k_2!}} = e^{-\alpha(1-p)}\alpha^{(k_1-k_2)}(1-p)^{(k_1-k_2)}$$

$$= e^{-\alpha(1-p)}[\alpha(1-p)]^{(k_1-k_2)} \qquad 0 \le k_2 \le k_1 < \infty$$

Therefore this conditional PMF is Poisson as well, with parameter $\alpha(1-p)$; but the distribution is shifted to the right. This shift is understandable since the number of errors detected cannot be greater than the number of errors occuring in the data, i.e., the probability that $K_1 < K_2$ must be 0.

□

5.2 Two Continuous Random Variables

The probabilistic characterization of two continuous random variables involves the joint probability density function (PDF). As in the case of a single random variable, the PDF will be approached via its relationship to the (joint) cumulative distribution function.

5.2.1 Joint distributions

The joint cumulative distribution function (CDF) for two random variables X_1 and X_2 is defined as

$$F_{X_1X_2}(x_1,x_2) \stackrel{\text{def}}{=} \Pr[X_1 \le x_1, X_2 \le x_2] \qquad (5.10)$$

A typical CDF for two continuous random variables is plotted in Fig. 5.4(a) The function $F_{X_1X_2}(x_1,x_2)$ goes to 0 as either variable x_1 or x_2 approaches minus infinity, satisfying our intuition that the probability of any variable being less than $-\infty$ should be 0. The CDF is then monotonically increasing in both variables and equals or approaches 1 as both variables approach infinity, again satisfying intuition. Although

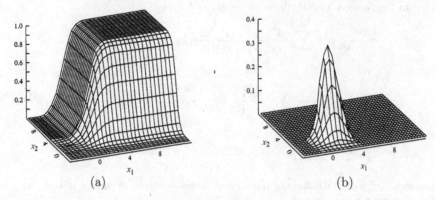

Figure 5.4 Typical joint CDF and PDF for continuous random variables. (a) Cumulative distribution function. (b) Probability density function.

the CDF shown in Fig. 5.4(a) for two *continuous* random variables is continuous, the definition (5.10) can apply to discrete or mixed random variables. In this case the CDF although monotonic, may have discontinuities. At such discontinuities, the function is *continuous from the right* in each variable.

The joint PDF is defined as the following second derivative of the CDF:

$$f_{X_1 X_2}(x_1, x_2) = \frac{\partial^2 F_{X_1 X_2}(x_1, x_2)}{\partial x_1 \partial x_2} \tag{5.11}$$

The PDF corresponding to the joint CDF of Fig. 5.4(a) is shown in Fig. 5.4(b) and resembles a small pile of sand placed on the x_1, x_2 plane. It can be seen from this picture, however, that the corresponding CDF, which is related to the PDF by integrating:

$$F_{X_1 X_2}(x_1, x_2) = \int_{-\infty}^{x_1} \int_{-\infty}^{x_2} f_{X_1 X_2}(u_1, u_2) du_2 du_1 \tag{5.12}$$

has the properties discussed above.

If impulses, or even line singularities[2] are allowed in the density, then discrete or mixed random variables can also be treated in terms of the PDF. The procedure is similar to what has been discussed in Chapter 3 for single random variables.

The interpretation of the joint PDF for continuous random variables in terms of probability is also extremely important. If a_1 and a_2 are two particular values of x_1 and x_2, and Δ_1 and Δ_2 are two small increments, then

$$\boxed{\Pr[a_1 < X_1 \le a_1 + \Delta_1, \ a_2 < X_2 \le a_2 + \Delta_2] \approx f_{X_1 X_2}(a_1, a_2) \Delta_1 \Delta_2} \tag{5.13}$$

This formula is illustrated in Fig. 5.5. The probability that random variables X_1 and X_2 are in the small rectangular region in the vicinity of a_1 and a_2 is equal to the density evaluated at (a_1, a_2) multiplied by the area of the small region. This interpretation is not a new result, but actually follows from the other equations presented so far. Given this result, it is easy to see that if $\mathcal{R}$ is any region of the x_1, x_2 plane, then the probability that the point (X_1, X_2) is in that region can be obtained by integration,

---

[2] By a line singularity we refer to a function that is singular in one dimension only; for example: $F(x_1, x_2) = \delta(x_1 - a)$. On the other hand, an *impulse* in the two-dimensional plane at coordinates (a_1, a_2) would have the form $F(x_1, x_2) = \delta(x_1 - a_1)\delta(x_2 - a_2)$.

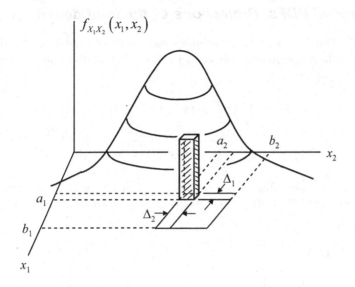

Figure 5.5 Interpretation of joint PDF as probability.

that is,

$$\Pr[(X_1, X_2) \in \mathcal{R}] = \int\int_{\mathcal{R}} f_{X_1 X_2}(x_1, x_2) dx_2 dx_1 \qquad (5.14)$$

For example, if (a_1, a_2) and (b_1, b_2) define the corners of a rectangular region (see Fig. 5.5), then

$$\Pr[a_1 < X_1 \le b_1, \ a_2 < X_2 \le b_2] = \int_{a_1}^{b_1} \int_{a_2}^{b_2} f_{X_1 X_2}(x_1, x_2) dx_2 dx_1$$

Two fundamental properties of the joint PDF are listed in Table 5.1. These are *defining* properties in the sense that any function satisfying these conditions can be interpreted as a PDF. First, the joint density can never be negative. Otherwise, you

non-negative	$f_{X_1 X_2}(x_1, x_2) \ge 0$
unit volume	$\int_{-\infty}^{\infty} \int_{-\infty}^{\infty} f_{X_1 X_2}(x_1, x_2) dx_2 dx_1 = 1$

Table 5.1 Fundamental properties of the joint PDF.

could integrate the density over the region where it is negative and obtain negative probability! Second, the probability obtained by integrating the density over the entire x_1, x_2 plane is equal to 1. From (5.12), integrating the density from $-\infty$ to ∞ in both variables is equivalent to computing $F_{X_1 X_2}(\infty, \infty)$, which is also 1.

5.2.2 Marginal PDFs: Projections of the joint density

Since both X_1 and X_2 are random variables, each of these can be described by its own PDF $f_{X_i}(x_i)$. These *marginal* densities for the random variables can be computed from the joint density as follows:

$$f_{X_1}(x_1) = \int_{-\infty}^{\infty} f_{X_1 X_2}(x_1, x_2)\,dx_2 \quad \text{(a)}$$

$$\text{(5.15)}$$

$$f_{X_2}(x_2) = \int_{-\infty}^{\infty} f_{X_1 X_2}(x_1, x_2)\,dx_1 \quad \text{(b)}$$

The marginals are computed by integrating over, or in a geometric sense, *projecting* along the other variable (see Fig. 5.6). To see why these last relations are valid, consider

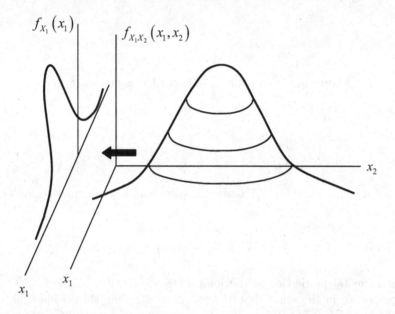

Figure 5.6 Interpretation of marginal PDF as a projection.

the event "$x_1 < X_1 \le x_1 + \Delta_1$" where Δ_1 is a small increment in x_1. The probability of this event is the same as the probability that (X_1, X_2) is in a very narrow strip of width Δ_1 in the x_1 direction and extending from $-\infty$ to ∞ in the x_2 direction. Thus using (5.14), we can write

$$\begin{aligned}
\Pr[x_1 < X_1 \le x_1 + \Delta_1] &= \Pr[x_1 < X_1 \le x_1 + \Delta_1,\ -\infty < X_2 < \infty] \\
&= \int_{x_1}^{x_1+\Delta_1} \int_{-\infty}^{\infty} f_{X_1 X_2}(u_1, u_2)\,du_2\,du_1 \\
&\approx \left[\int_{-\infty}^{\infty} f_{X_1 X_2}(x_1, u_2)\,du_2 \right] \Delta_1
\end{aligned}$$

But we also know that

$$\Pr[x_1 < X_1 \le x_1 + \Delta_1] \approx f_{X_1}(x_1)\,\Delta_1$$

and this probability is unique for sufficiently small values of Δ_1. Therefore the two expressions must be equal and the PDF for X_1 is as given in (5.15)(a). A similar argument applies to computing the other marginal density function.

The example below illustrates the use of the joint PDF for computing probabilities for two random variables, and computation of the marginal PDFs.

Example 5.3: Two random variables are uniformly distributed over the region of the x_1, x_2 plane shown below. Find the constant C and the probability of the event "$X_1 > \frac{1}{2}$." Also find the two marginal PDFs.

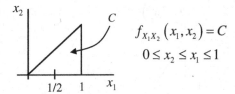

$$f_{X_1 X_2}(x_1, x_2) = C$$
$$0 \le x_2 \le x_1 \le 1$$

The constant C is found by integrating the density over the region where it is nonzero and setting that equal to 1.

$$\int_{-\infty}^{\infty} \int_{-\infty}^{\infty} f_{X_1 X_2}(x_1, x_2) dx_2 dx_1 = \int_0^1 \int_0^{x_1} C \, dx_2 dx_1$$
$$= C \int_0^1 x_1 \, dx_1 = C \left. \frac{x_1^2}{2} \right|_0^1 = \frac{C}{2} = 1$$

Hence $C = 2$.

The probability that $X_1 > 1/2$ is obtained by integrating the joint density over the appropriate region.

$$\Pr[X_1 > 1/2] = \int_{1/2}^1 \int_0^{x_1} 2 \, dx_2 dx_1 = \int_{1/2}^1 2x_1 \, dx_1 = \left. x_1^2 \right|_{1/2}^1 = 3/4$$

Finally, the marginals are obtained by integrating the joint PDF over each of the other variables. To obtain $f_{X_1}(x_1)$ the joint density function is integrated as follows:

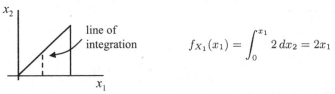

$$f_{X_1}(x_1) = \int_0^{x_1} 2 \, dx_2 = 2x_1$$

The density function is sketched below.

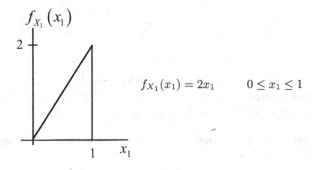

$$f_{X_1}(x_1) = 2x_1 \qquad 0 \le x_1 \le 1$$

To obtain $f_{X_2}(x_2)$ the integration is as shown below:

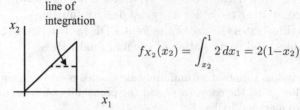

$$f_{X_2}(x_2) = \int_{x_2}^{1} 2\, dx_1 = 2(1-x_2)$$

The density function is sketched below.

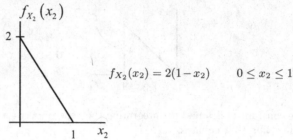

$$f_{X_2}(x_2) = 2(1-x_2) \qquad 0 \le x_2 \le 1$$

Notice that for both of these marginal densities the limits of definition $0 \le x_1 \le 1$ or $0 \le x_2 \le 1$ are very important.

□

5.2.3 Conditional PDFs: Slices of the joint density

Two random variables X_1 and X_2 are defined to be *independent* if their joint PDF is the product of the two marginals:

$$f_{X_1 X_2}(x_1, x_2) = f_{X_1}(x_1) \cdot f_{X_2}(x_2) \qquad \text{(for independent random variables)}$$

$$(5.16)$$

The definition follows from the analogous definition (2.12) for events. To show this, let us define the events:[3]

$$A_1 \quad : \quad x_1 < X_1 \le x_1 + \Delta_1$$
$$A_2 \quad : \quad x_2 < X_2 \le x_2 + \Delta_2$$

If A_1 and A_2 are *independent* then

$$\Pr[A_1 A_2] = \Pr[A_1]\Pr[A_2] \approx (f_{X_1}(x_1)\Delta_1)(f_{X_2}(x_2)\Delta_2) = f_{X_1}(x_1)f_{X_2}(x_2)\,\Delta_1\,\Delta_2$$

But from (5.13) the probability of $A_1 A_2$ is also given by

$$f_{X_1 X_2}(x_1, x_2)\,\Delta_1\,\Delta_2$$

These last two equations motivate the definition (5.16).

It is important to note that (5.16) is a definition and *test* for independence. It is *not* a condition that holds for all random variables. If two random variables are

---

[3] The 'events' A_1 and A_2 are most correctly thought of as events in the sample space such that the indicated conditions on the X_i obtain.

not independent, then there is no way to compute a unique joint density from the marginals. Papoulis and Pillai [1, p. 218] give an example of two different joint PDFs that have the same marginals.

Although the joint PDF is not always a product of the marginals, the joint PDF for any two random variables can always be expressed as the product of a marginal and a *conditional* density function:

$$f_{X_1 X_2}(x_1, x_2) = f_{X_1 | X_2}(x_1 | x_2) \, f_{X_2}(x_2) \tag{5.17}$$

The conditional density function $f_{X_1 | X_2}$ is thus *defined* by[4]

$$f_{X_1 | X_2}(x_1 | x_2) \stackrel{\text{def}}{=} \frac{f_{X_1 X_2}(x_1, x_2)}{f_{X_2}(x_2)} \tag{5.18}$$

Recall that the conditional PMF for discrete random variables can be interpreted as a (normalized) discrete *slice* through the joint PMF. Likewise, for any given value of x_2 the conditional PDF, $f_{X_1 | X_2}(x_1 | x_2)$, can be interpreted as a slice through the joint PDF normalized by the value $f_{X_2}(x_2)$ (see Fig. 5.7). The conditional density can thus

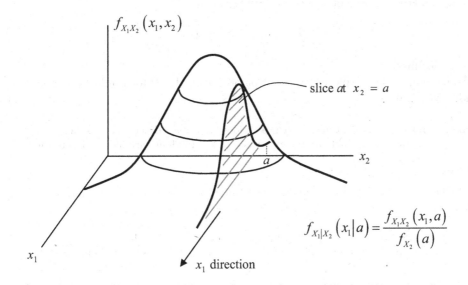

Figure 5.7 Interpretation of conditional PDF as a slice through the joint density.

be interpreted in the following way: the quantity

$$f_{X_1 | X_2}(x_1 | x_2) \, \Delta_1$$

is the probability that X_1 is in a small region $x_1 < X_1 \le x_1 + \Delta_1$ *given* that X_2 is in a correspondingly small region $x_2 < X_2 \le x_2 + \Delta_2$. Let us see why this interpretation is valid.

For convenience, define the events A_1 and A_2 as above, but let's *not* assume that these events are independent. Then according to the definition (2.13) for conditional

---

[4] Notice that when X_1 and X_2 are independent, (5.16) applies so that (5.18) reduces to $f_{X_1 | X_2}(x_1 | x_2) = f_{X_1}(x_1)$. In other words, when the random variables are independent, the conditional density is the same as the unconditional density.

probability,

$$\Pr[A_1|A_2] = \frac{\Pr[A_1 A_2]}{\Pr[A_2]} \approx \frac{f_{X_1 X_2}(x_1, x_2)\,\Delta_1\,\Delta_2}{f_{X_2}(x_2)\,\Delta_2}$$

The ratio of PDFs in this equation is the conditional density defined in (5.18). Thus, after cancelling the common term Δ_2 we have the result

$$\boxed{\Pr[x_1 < X_1 \le x_1 + \Delta_1 \,|\, x_2 < X_2 \le x_2 + \Delta_2] \approx f_{X_1|X_2}(x_1|x_2)\,\Delta_1} \qquad (5.19)$$

as claimed.

If it seems strange that Δ_2 does not appear in the above equation, just remember that the conditional density $f_{X_1|X_2}$ is a PDF for the random variable X_1 but not for the random variable X_2 (again, see Fig. 5.7). In this PDF, X_2 acts like a *parameter* that may control size and shape of the function, but has no other role. In particular, the conditional PDF for X_1 satisfies the condition (see Prob. 5.24)

$$\int_{-\infty}^{\infty} f_{X_1|X_2}(x_1|x_2)dx_1 = 1 \qquad (5.20)$$

but the integral over the other (conditioning) variable

$$\int_{-\infty}^{\infty} f_{X_1|X_2}(x_1|x_2)dx_2$$

has no meaning.

Example 5.4 below explores the concept of independence for random variables and conditional densities.

Example 5.4: Let us continue with the random variables described in Example 5.3. In particular, let us check for independence of the random variables and determine the two conditional PDFs.

The product of the marginal densities computed in the previous example is

$$f_{X_1}(x_1) \cdot f_{X_2}(x_2) = 2x_1 \cdot 2(1 - x_2)$$

This is clearly not equal to the joint density $f_{X_1 X_2}(x_1, x_2)$, which is a constant ($C = 2$) over the region of interest. Therefore the random variables are *not* independent.

The conditional density for X_1 can be computed from the definition (5.18).

$$f_{X_1|X_2}(x_1|x_2) = \frac{f_{X_1 X_2}(x_1, x_2)}{f_{X_2}(x_2)} = \frac{2}{2(1 - x_2)} = \frac{1}{1 - x_2} \qquad x_2 \le x_1 \le 1 \quad (1)$$

The conditional PDF is sketched below. It is found to be a *uniform* PDF with

limits that depend upon the conditioning variable x_2. Notice that the region of definition $x_2 \le x_1 \le 1$ in Equation (1) above is extremely important. These limits are derived from the original limits on the joint PDF: $0 \le x_2 \le x_1 \le 1$ (see the sketch of

$f_{X_1X_2}$ in Example 5.3). Without these limits, the density function would not appear as shown above and would not integrate to 1.

The conditional density for X_2 is computed in a similar manner:

$$f_{X_2|X_1}(x_2|x_1) = \frac{f_{X_1X_2}(x_1, x_2)}{f_{X_1}(x_1)} = \frac{2}{2x_1} = \frac{1}{x_1} \qquad 0 \le x_2 \le x_1$$

The result is sketched below.

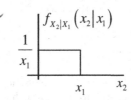

□

5.2.4 Bayes' rule for continuous random variables

A form of Bayes' rule can be derived for random variables that is very useful for signal detection and classification problems in electrical engineering. First, the joint PDF can be written using the conditional density for X_2 as

$$f_{X_1X_2}(x_1, x_2) = f_{X_2|X_1}(x_2|x_1)\, f_{X_1}(x_1) \tag{5.21}$$

Substituting this result in (5.18) yields

$$f_{X_1|X_2}(x_1|x_2) = \frac{f_{X_2|X_1}(x_2|x_1)\, f_{X_1}(x_1)}{f_{X_2}(x_2)} \tag{5.22}$$

The denominator of this equation can be computed from (5.21) and 5.15(b):

$$f_{X_2}(x_2) = \int_{-\infty}^{\infty} f_{X_1X_2}(x_1, x_2)dx_1 = \int_{-\infty}^{\infty} f_{X_2|X_1}(x_2|x_1)\, f_{X_1}(x_1)dx_1$$

Thus, substituting this result in (5.22) yields

$$f_{X_1|X_2}(x_1|x_2) = \frac{f_{X_2|X_1}(x_2|x_1)\, f_{X_1}(x_1)}{\displaystyle\int_{-\infty}^{\infty} f_{X_2|X_1}(x_2|x_1)\, f_{X_1}(x_1)dx_1} \tag{5.23}$$

This is Bayes' rule for continuous random variables.

The use of this formula for density functions can become a little involved. The following example illustrates the procedure.

Example 5.5: The power transmitted by station KNPS is rather erratic. On any particular day the power transmitted is a continuous random variable X_1 uniformly distributed between 2 and 10 kilowatts. A listener in the area typically experiences fading. Thus the signal strength at the receiver is a random variable X_2 which depends on the power transmitted. To keep the problem simple, let us assume that $0 < X_2 \le X_1$, although in reality there should be a scale factor, and assume that the density for X_2 conditioned on X_1 is given by

$$f_{X_2|X_1}(x_2|x_1) = \frac{2}{x_1^2}\, x_2, \qquad 0 < x_2 \le x_1$$

The two density functions are shown below.

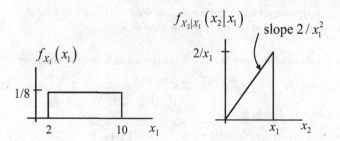

It is desired to find the conditional density $f_{X_1|X_2}(x_1|x_2)$ for the power transmitted, given the power received.

This is an application of Bayes' rule for densities, where the result can be computed directly from (5.23). The procedure is clearer, however, if we break the problem down into the individual steps leading to (5.23).

First we compute the joint density function $f_{X_1X_2}(x_1, x_2)$ as in (5.21)

$$f_{X_1X_2}(x_1, x_2) = f_{X_2|X_1}(x_2|x_1) f_{X_1}(x_1)$$

$$= \frac{2}{x_1^2} x_2 \cdot \frac{1}{8} = \frac{x_2}{4x_1^2}, \qquad 0 < x_2 \le x_1; \; 2 < x_1 \le 10$$

Notice that the inequalities defining where the joint density is nonzero are extremely important. This region is shown shaded in the figure below.

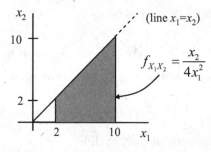

The marginal density function for x_2 can be computed by integrating the joint density over x_1. Notice from the above figure that there are two cases. For $0 < x_2 \le 2$ we have

$$f_{X_2}(x_2) = \int_2^{10} \frac{x_2}{4x_1^2} dx_1 = x_2 \left[-\frac{1}{4x_1} \right]_2^{10} = \frac{x_2}{10}, \qquad 0 < x_2 \le 2$$

while for $2 < x_2 \le 10$ we have

$$f_{X_2}(x_2) = \int_{x_2}^{10} \frac{x_2}{4x_1^2} dx_1 = x_2 \left[-\frac{1}{4x_1} \right]_{x_2}^{10} = \frac{10 - x_2}{40}, \qquad 2 < x_2 \le 10$$

The marginal density is plotted below.

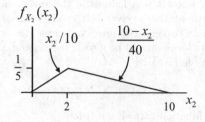

Finally, $f_{X_1|X_2}$ can be computed as in (5.22). (This is equivalent to (5.23).) Again, there are two cases. For $0 < x_2 \le 2$ we have

$$f_{X_1|X_2}(x_1|x_2) = \frac{x_2/4x_1^2}{x_2/10} = \frac{5}{2} \cdot \frac{1}{x_1^2} \qquad 0 < x_2 \le 2;\ 2 < x_1 \le 10$$

while for $2 < x_2 \le 10$ we have

$$f_{X_1|X_2}(x_1|x_2) = \frac{x_2/4x_1^2}{(10-x_2)/40} = \frac{10x_2}{10-x_2} \cdot \frac{1}{x_1^2} \qquad 2 < x_2 \le 10;\ 2 < x_1 \le 10$$

The conditional density $f_{X_1|X_2}(x_1|x_2)$ is plotted below. The conditional density can be used to estimate the actual power transmitted. This use is demonstated in Example 5.7, which appears in the next subsection.

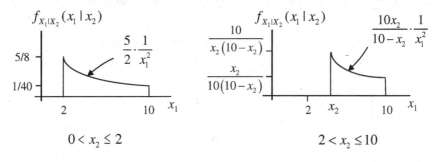

$0 < x_2 \le 2$ $\qquad\qquad\qquad\qquad\qquad\qquad$ $2 < x_2 \le 10$

□

5.3 Expectation and Correlation

Where two random variables are involved, expectation involves the *joint* PDF or PMF of the random variables. Without going back to general principles as in the previous chapter, let us simply state that for any function $g(X_1, X_2)$ the expectation is computed from the formula

$$\mathcal{E}\{g(X_1, X_2)\} = \int_{-\infty}^{\infty}\int_{-\infty}^{\infty} g(x_1, x_2) f_{X_1 X_2}(x_1, x_2)\, dx_2 dx_1 \qquad (5.24)$$

if the X_i are continuous, and

$$\mathcal{E}\{g(X_1, X_2)\} = \sum_{x_1}\sum_{x_2} g(x_1, x_2) f_{X_1 X_2}[x_1, x_2] \qquad (5.25)$$

if the X_i are discrete. Note that this definition of expectation is consistent with that defined for a single random variable if $g(X_1, X_2)$ is just a function of one of the variables but not the other. For example, if $g(X_1, X_2) = X_1$, then $\mathcal{E}\{X_1\}$ is the mean of X_1. Assuming X_1 is a continuous random variable, this is computed according to (5.24) as

$$\mathcal{E}\{X_1\} = \int_{-\infty}^{\infty}\int_{-\infty}^{\infty} x_1 f_{X_1 X_2}(x_1, x_2)\, dx_2 dx_1 = \int_{-\infty}^{\infty} x_1 \left(\int_{-\infty}^{\infty} f_{X_1 X_2}(x_1, x_2)\, dx_2\right) dx_1$$

Since the term in parentheses is just the marginal density $f_{X_1}(x_1)$, we are left with

$$\mathcal{E}\{X_1\} = \int_{-\infty}^{\infty} x_1 f_{X_1}(x_1)\, dx_1$$

which matches the expression (4.4) given for the mean in Chapter 4. From here on, we continue to treat expectation $\mathcal{E}\{\cdot\}$ as a linear operator (see discussion in Section 4.1.4), but keep in mind the one- or two-dimensional integrals or sums that the expectation operator represents.

5.3.1 Correlation and covariance

With two random variables, some additional types of moments are important. The quantity

$$r = \mathcal{E}\{X_1 X_2\} \tag{5.26}$$

is called the *correlation* of X_1 and X_2. The product $X_1 X_2$ is a second order term and thus the correlation is a type of second order moment that measures the similarity between X_1 and X_2, on the average. If the correlation is high, then in a large collection of independent experiments, X_1 and X_2 will have related values.

Correlation can be thought of as an (approximate) linear relationship between random variables. Figure 5.8 shows a scatter plot of the height data for fathers and sons

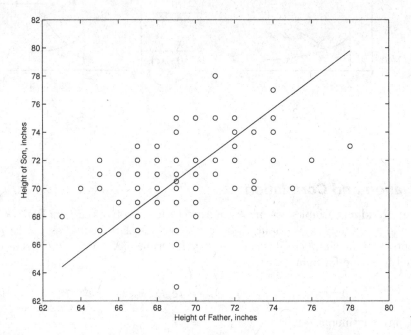

Figure 5.8 Scatter plot of father and son height data and regression line, exhibiting correlation.

collected by Sir Francis Galton in 1885. The straight line fitted to this data, which has a positive slope, indicates that a (positive) correlation exists between the two random variables (father height and son height). The straight line allows one to estimate the value of one random variable (son's height) from knowledge of the other (father's height). The subject of linear *regression analysis*, which deals with fitting such straight lines to data, was actually invented by Galton in his study of this data in the nineteenth century [2].

An important type of second moment related to the correlation is the *covariance c*, defined by

$$c = \mathrm{Cov}\,[X_1, X_2] \overset{\mathrm{def}}{=} \mathcal{E}\{(X_1 - m_1)(X_2 - m_2)\} \tag{5.27}$$

where m_1 and m_2 are the means of X_1 and X_2. This is a type of *central* moment

(moment about the mean) which is frequently more useful than the correlation. The correlation between two random variables may be large just due to the fact that the two random variables have large mean values. We may be more interested, however, in comparing the similarity of small *variations* about the mean. In this case the covariance is more relevant. For example, in an electronic circuit, amplifiers are often supplied with a large DC voltage, or bias, to put them into their desired region of operation. The information signal to be amplified rides on top of this DC bias and is rather small compared to the bias voltage, which just adds to the mean. In comparing two such waveforms within an amplifier, you would be more interested in the covariance, which focuses on just the information signal, rather than the correlation, which includes the bias voltage in the comparison.

The relation between correlation and covariance can be neatly expressed as

$$\text{Cov}\,[X_1, X_2] = \mathcal{E}\{X_1 X_2\} - m_1 m_2 \tag{5.28}$$

(The proof of this result is almost identical to the proof of (4.17) in Chapter 4.) This formula represents a convenient way to compute covariance. Moreover, by writing this equation in the form

$$\mathcal{E}\{X_1 X_2\} = \text{Cov}\,[X_1, X_2] + m_1 m_2$$

it is clear how the mean value of the random variables influences correlation.

Another very useful quantity in the analysis of two random variables is the *correlation coefficient* ρ. This quantity is the result of normalizing the covariance c by the standard deviations σ_1 and σ_2 of the two random variables:

$$\rho \stackrel{\text{def}}{=} \frac{\text{Cov}\,[X_1, X_2]}{\sigma_1 \sigma_2} = \frac{c}{\sigma_1 \sigma_2} \tag{5.29}$$

It has the unique property that

$$-1 \le \rho \le 1 \tag{5.30}$$

(see Prob. 5.28). The correlation coefficient is useful because it provides a *fixed scale* on which two random variables can be compared. Correlation and covariance on the other hand, can have any values from $-\infty$ to ∞. Thus the numerical value of r or c can only be judged in a relative sense as it relates to the particular random variables.

The three types of joint second moments that have been discussed are summarized in Table 5.2 for comparison. Each of these has uses in the appropriate type of analysis.

quantity	definition	symbol
Correlation	$\mathcal{E}\{X_1 X_2\}$	r
Covariance, $\text{Cov}\,[X_1, X_2]$	$\mathcal{E}\{(X_1 - m_1)(X_2 - m_2)\}$	c
Correlation Coefficient	$\dfrac{\text{Cov}\,[X_1, X_2]}{\sigma_1 \sigma_2}$	ρ

Table 5.2 Three related joint second moments.

Some terminology related to correlation is appropriate to discuss, because it can be somewhat confusing. Two terms are listed in Table 5.3 that pertain to random variables. Confusion sometimes arises because the random variables are said to be

term	meaning
uncorrelated	$\rho = c = 0$
orthogonal	$r = 0$

Table 5.3 Terms related to correlation.

uncorrelated when the *covariance* (c) is 0. When the *correlation* (r) is 0, the random variables are said to be *orthogonal*. Fortunately all of these distinctions disappear, however, when at least one of the random variables has 0 mean.

The term "orthogonal" is derived from a vector space representation of random variables [3] and is beyond the scope of this text. The use of the term "uncorrelated" when the covariance is 0 can be understood by setting (5.28) to 0 and writing that equation as

$$\mathcal{E}\{X_1 X_2\} = m_1 m_2 = \mathcal{E}\{X_1\}\mathcal{E}\{X_2\} \qquad \text{(for *uncorrelated* random variables)} \quad (5.31)$$

In this case the expectation of the product is seen to distribute over the two random variables. Equation (5.31) can in fact be taken as the *definition* of the term "uncorrelated."

It is important to note that *independent random variables are automatically uncorrelated*. To see this, just observe that the PDF for independent random variables factors to a product of the marginals; therefore we can write

$$\mathcal{E}\{X_1 X_2\} = \int_{-\infty}^{\infty}\int_{-\infty}^{\infty} x_1 x_2 f_{X_1}(x_1) f_{X_2}(x_2) dx_2 dx_1$$

$$= \left(\int_{-\infty}^{\infty} x_1 f_{X_1}(x_1) dx_1\right)\left(\int_{-\infty}^{\infty} x_2 f_{X_2}(x_2) dx_2\right) = \mathcal{E}\{X_1\}\mathcal{E}\{X_2\}$$

Uncorrelated random variables are *not* necessarily independent, however. A special situation occurs in the case of jointly Gaussian random variables (see Section 5.4). For the Gaussian case, uncorrelated random variables are also independent.

The following example demonstrates the calculation of the entire set of first- and second-moment parameters for two random variables.

Example 5.6: Two random variables are described by the joint density function shown below.

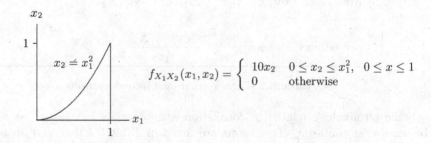

$$f_{X_1 X_2}(x_1, x_2) = \begin{cases} 10x_2 & 0 \le x_2 \le x_1^2, \ 0 \le x \le 1 \\ 0 & \text{otherwise} \end{cases}$$

The mean of random variable X_1 is given by

$$m_1 = \mathcal{E}\{X_1\} = \int_0^1 \int_0^{x_1^2} x_1 \cdot 10x_2 \, dx_2 \, dx_1 = 10 \int_0^1 x_1 \int_0^{x_1^2} x_2 \, dx_2 \, dx_1 = \frac{5}{6}$$

Likewise the mean of X_2 is given by

$$m_2 = \int_0^1 \int_0^{x_1^2} x_2 \, 10x_2 \, dx_2 \, dx_1 = 10 \int_0^1 \int_0^{x_1^2} x_2^2 \, dx_2 \, dx_1 = \frac{10}{21}$$

The variances of the two random variables are computed as

$$\mathcal{E}\{X_1^2\} = \int_0^1 \int_0^{x_1^2} x_1^2 \, 10x_2 \, dx_2 \, dx_1 = \frac{5}{7}$$

$$\sigma_1^2 = \mathcal{E}\{X_1^2\} - m_1^2 = \frac{5}{7} - \left(\frac{5}{6}\right)^2 = \frac{5}{252}$$

and

$$\mathcal{E}\{X_2^2\} = \int_0^1 \int_0^{x_1^2} x_2^2 \, 10x_2 \, dx_2 \, dx_1 = \frac{5}{18}$$

$$\sigma_2^2 = \mathcal{E}\{X_2^2\} - m_2^2 = \frac{5}{18} - \left(\frac{10}{21}\right)^2 = \frac{5}{98}$$

The correlation r is given by

$$r = \mathcal{E}\{X_1 X_2\} = \int_0^1 \int_0^{x_1^2} x_1 x_2 \, 10x_2 \, dx_2 \, dx_1 = \frac{5}{12}$$

The covariance is then computed using (5.28) as

$$c = \mathcal{E}\{X_1 X_2\} - m_x m_y = \frac{5}{12} - \left(\frac{5}{6}\right)\left(\frac{10}{21}\right) = \frac{5}{252}$$

Finally, the correlation coefficient ρ is computed as the normalized covariance.

$$\rho = \frac{c}{\sigma_1 \sigma_2} = \frac{5/252}{\sqrt{5/252}\sqrt{5/98}} = \sqrt{\frac{7}{18}}$$

□

5.3.2 Conditional expectation

Sometimes it is useful to perform expectation in stages. This leads to the concept of conditional expectation, where the expectation operation is performed with one variable remaining fixed. Let us be careful about notation here, however. Suppose the random variable X_2 is observed to have some particular value x_2. Then we could define the expectation of X_1 *given* x_2 to be the mean of the conditional density:

$$\mathcal{E}\{X_1|x_2\} = \int_{-\infty}^{\infty} x_1 f_{X_1|X_2}(x_1|x_2) dx_1$$

More frequently, the conditioning variable is also considered to be a random variable, however. In that case we would write

$$\mathcal{E}\{X_1|X_2\} = \int_{-\infty}^{\infty} x_1 f_{X_1|X_2}(x_1|X_2) dx_1 \tag{5.32}$$

The conditional expectation as defined by (5.32) is a random variable because of its dependence on the *random variable* X_2.

More generally, for any function $g(X_1, X_2)$, we can define

$$E\{g(X_1, X_2)|X_2\} = \int_{-\infty}^{\infty} g(x_1, x_2) f_{X_1|X_2}(x_1|X_2) dx_1 \qquad (5.33)$$

Since the quantity on the left is a random variable, we can further define its expectation as

$$E\{E\{g(X_1, X_2)|X_2\}\} = \int_{-\infty}^{\infty} E\{g(X_1, X_2)|X_2\} f_{X_2}(x_2) dx_2$$

Substituting (5.33) in the last equation and recognizing (5.17) then yields the result

$$\boxed{E\{E\{g(X_1, X_2)|X_2\}\} = E\{g(X_1, X_2)\}} \qquad (5.34)$$

where the right-hand side is the usual expectation (5.24) using the joint density function.[5]

As an illustration of conditional expectation, consider the following brief example.

Example 5.7: In many problems in engineering, it is desired to infer the value of one random variable (which cannot be measured directly) from another related random variable, which *can* be measured. This is the problem of statistical *estimation*. As an example of this, consider the situation of station KNPS and the listener described in Example 5.5. It is shown there that the probability density function for the power transmitted (X_1) given the power received (X_2) is

$$f_{X_1|X_2}(x_1|x_2) = \begin{cases} \dfrac{5}{2}\dfrac{1}{x_1^2} & 0 < x_2 \leq 2; \ 2 < x_1 \leq 10 \\[2mm] \dfrac{10x_2}{10 - x_2}\dfrac{1}{x_1^2} & 2 < x_2 \leq 10; \ 2 < x_1 \leq 10 \\[2mm] 0 & \text{otherwise} \end{cases}$$

Suppose we want to estimate the power transmitted from the power received. One well-known way to do this is to find the mean of the conditional density function. Denote this estimated value of X_1 by $\hat{X}_1$. Then

$$\hat{X}_1 = E\{X_1|X_2\} = \int_{-\infty}^{\infty} x_1 f_{X_1|X_2}(x_1|x_2) dx_1$$

Substituting the expressions for the conditional density and integrating produces

$$\hat{X}_1 = \int_2^{10} x_1 \frac{5}{2}\frac{1}{x_1^2} dx_1 = \frac{5}{2}\int_2^{10} \frac{1}{x_1} dx_1 = \frac{5}{2}\ln 5 \approx 4.0236 \qquad 0 < X_2 \leq 2$$

and

$$\hat{X}_1 = \int_{X_2}^{10} x_1 \frac{10X_2}{10 - X_2} \cdot \frac{1}{x_1^2} dx_1 = \frac{10X_2}{10 - X_2}\int_{X_2}^{10} \frac{1}{x_1} dx_1$$

$$= \frac{10}{(10/X_2) - 1}\ln(10/X_2) \qquad 2 < X_2 \leq 10$$

---

[5] If the double expectation seems confusing, just notice that (5.34) can be written as

$$\int_{-\infty}^{\infty} \left(\int_{-\infty}^{\infty} g(x_1, x_2) f_{X_1|X_2}(x_1|x_2) dx_1 \right) f_{X_2}(x_2) dx_2 = \int_{-\infty}^{\infty}\int_{-\infty}^{\infty} g(x_1, x_2) f_{X_1 X_2}(x_1, x_2) dx_1 dx_2$$

which is its true meaning.

A plot of this estimated value as a function of x_2 is given below.

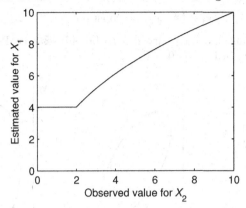

(Both random variables represent power.)

□

5.4 *Gaussian Random Variables*

One of the most common joint density functions is the bivariate density function for two jointly Gaussian random variables. The joint PDF for two jointly Gaussian random variables is given by the formidable expression[6]

$$f_{X_1 X_2}(x_1, x_2) \; = \; \frac{1}{2\pi\sigma_1\sigma_2\sqrt{1-\rho^2}} \times \qquad (5.35)$$

$$\exp -\frac{1}{2(1-\rho^2)} \left\{ \frac{(x_1-m_1)^2}{\sigma_1^2} - 2\rho\frac{(x_1-m_1)(x_2-m_2)}{\sigma_1\sigma_2} + \frac{(x_2-m_2)^2}{\sigma_2^2} \right\}$$

and depicted in Fig. 5.9. The bivariate Gaussian density is frequently represented in

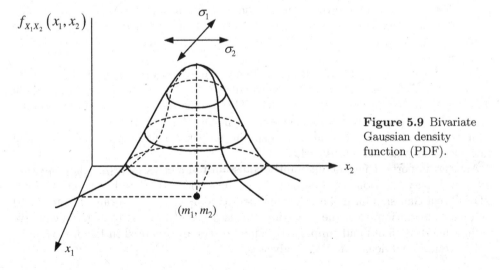

Figure 5.9 Bivariate Gaussian density function (PDF).

---

[6] A notationally more concise formula using vectors and matrices is given in Section 7.2.3 on random vectors.

terms of *contours* in the x_1, x_2 plane defined by

$$f_{X_1 X_2}(x_1, x_2) = \text{constant}$$

(for various chosen constants), which turn out to be ellipses (see Prob. 5.35). Some typical contours are shown in Fig. 5.10.

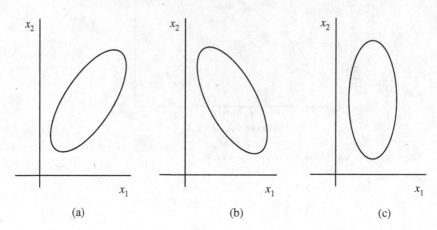

$$\qquad\qquad\quad \text{(a)} \qquad\qquad\qquad\qquad\qquad \text{(b)} \qquad\qquad\qquad\qquad\qquad \text{(c)}$$

Figure 5.10 Contours of the bivariate Gaussian density function. (a) $\rho > 0$ (b) $\rho < 0$ (a) $\rho = 0$

There are five parameters that characterize the bivariate Gaussian PDF. The parameters m_i and σ_i^2 are the means and variances of the respective Gaussian random variables. That is, the marginal densities have the form

$$f_{X_1}(x_1) = \frac{1}{\sqrt{2\pi\sigma_1^2}} e^{-\dfrac{(x_1-m_1)^2}{2\sigma_1^2}} \quad ; \quad f_{X_2}(x_2) = \frac{1}{\sqrt{2\pi\sigma_2^2}} e^{-\dfrac{(x_2-m_2)^2}{2\sigma_2^2}} \qquad (5.36)$$

which was introduced in Section 3.4. The remaining parameter ρ, is the *correlation coefficient*, defined in Section 5.3.1 of this chapter. Notice in Fig. 5.10 that when $\rho > 0$ the major axis of the ellipse representing the contours has a positive slope and when $\rho < 0$ the major axis of the ellipse has a negative slope. When $\rho = 0$, i.e., when the random variables are *uncorrelated*, the ellipse has its axes parallel to the x_1 and x_2 coordinate axes.

It is easy to show that for *Gaussian* random variables (only), "uncorrelated" implies "independent." To see this, notice that when $\rho = 0$ the product of the marginals in (5.36) is equal to the joint density (5.35). Thus, when $\rho = 0$, the random variables satisfy the condition (5.16) for independence. The fact that uncorrelated *Gaussian* random variables are also independent is a special property, unique to Gaussian random variables, that was mentioned earlier.

Another important fact about jointly Gaussian random variables, is that the *conditional* density functions are Gaussian. Thus both projections and slices of the joint PDF have a Gaussian form. The marginal densities (5.36) can be obtained from (5.35) by direct integration; then the conditional densities are obtained by algebraically dividing (5.35) by (5.36) and simplifying. These topics are explored in Prob. 5.34.

The conditional density for X_1 is given by

$$f_{X_1|X_2}(x_1|x_2) = \frac{1}{\sqrt{2\pi\sigma^2}} e^{-\dfrac{(x_1 - \mu(x_2))^2}{2\sigma^2}} \qquad (5.37)$$

where the parameters σ and $\mu(x_2)$ are given by

$$\sigma^2 = \sigma_1^2(1 - \rho^2) \tag{5.38}$$

$$\mu(x_2) = m_1 + \rho\frac{\sigma_1}{\sigma_2}(x_2 - m_2) \tag{5.39}$$

The general form of the conditional density is shown in Fig. 5.11 below.

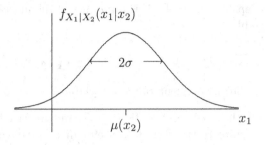

Figure 5.11 Conditional PDF for Gaussian random variables.

Now examine equation (5.38). Since the magnitude of ρ is less than 1 (as it must be if σ^2 is to remain positive) (5.38) shows that the variance σ^2 of the conditional density is less than the original variance σ_1^2. The variance σ^2 does not depend on the conditioning variable x_2 however, so the conditional density retains its same shape and size for all values of x_2. The mean μ of the conditional density is a function of x_2 however. Thus the position of the density along the x_1 axis depends (linearly) on the conditioning variable.

5.5 Multiple Random Variables

Sets of multiple random variables often occur in electrical and computer engineering problems. For example, a random signal may be sampled at N points in time and digitized. This section provides an introduction to methods dealing with multiple random variables. (A general more formal description is given in Chapter 7.) In addition, several topics associated with sums of random variables are discussed in detail. Sums of random variables occur frequently in applications, and the results developed here turn out to be very useful.

5.5.1 PDFs for multiple random variables

The probabilistic description for multiple random variables is similar (in principle) to that for just two random variables. The joint PDF, PMF, or CDF become a function of the n random variables $X_1, X_2, \ldots, X_n$. Marginal density functions are formed by integrating the joint density over the other variables, and conditional densities are defined as the ratio of the joint density to the marginals. Further, the concept of expectation is extended to multiple random variables as

$$\mathcal{E}\{g(x_1, x_2, \ldots, x_n)\} = \int_{-\infty}^{\infty} \cdots \int_{-\infty}^{\infty} f_{X_1 X_2 \cdots X_n}(x_1, x_2, \ldots, x_n)dx_1 \ldots dx_n \tag{5.40}$$

This is consistent with the earlier definitions of expectation and is seen to be a linear operation. Various moments can then be defined, but this is left as a topic for Chapter 7.

Multiple random variables can be cumbersome to deal with in practice because of the dimensionality. Nevertheless, three special cases commonly arise that simplify the procedure and make the analysis more tractable. These special cases are:

1. When the random variables are independent.

2. When the random variables form a Markov process.

3. When the random variables are jointly Gaussian.

Let us consider each of these cases briefly.

When the random variables are independent, the PDF is a product of the marginal PDFs for the individual random variables.

$$f_{X_1 X_2 \cdots X_n}(x_1, x_2, \ldots, x_n) = f_{X_1}(x_1) \cdot f_{X_2}(x_2) \cdots f_{X_n}(x_n) = \prod_{i=1}^{n} f_{X_i}(x_i)$$

(for independent random variables)   (5.41)

A similar statement can be made for the CDF and PMF (if the random variables are discrete). While this case seems quite restrictive, is is useful in a surprisingly large number of problems. With the assumption of independence, a whole variety of different PDFs (such as those described in earlier chapters) can be applied to problems involving multiple random variables.

The Markov model is useful when the random variables naturally form a sequence, as in samples of a signal or a time series. The model can be explained as follows. By using the definition of conditional probability, it is always possible to write (say, for $n = 4$)

$$f_{X_1 X_2 X_3 X_4}(x_1, x_2, x_3, x_4) = f_{X_4 | X_1 X_2 X_3}(x_4 | x_1, x_2, x_3) f_{X_1 X_2 X_3}(x_1, x_2, x_3)$$

Continuing this recursively produces

$$\begin{aligned} f_{X_1 X_2 X_3 X_4}(x_1, x_2, x_3, x_4) &= f_{X_4 | X_1 X_2 X_3}(x_4 | x_1, x_2, x_3) \\ &\times f_{X_3 | X_1 X_2}(x_3 | x_1, x_2) f_{X_2 | X_1}(x_2 | x_1) f_{X_1}(x_1) \end{aligned}$$

The Markov condition states that the conditional density for X_i depends only on the previous random variable in the sequence. That is $f_{X_i | X_1 X_2 \cdots X_{i-1}} = f_{X_i | X_{i-1}}$. Thus for Markov random variables, the joint density can be written as

$$f_{X_1 X_2 \cdots X_n}(x_1, x_2, \ldots, x_n) = f_{X_1}(x_1) \prod_{i=2}^{n} f_{X_i | X_{i-1}}(x_i | x_{i-1}) \qquad (5.42)$$

(for Markov random variables)

This can be a great simplification. The Markov model is most appropriate when the random variables are observed sequentially, but the conditions of the problem cannot justify independence. Markov chains are used extensively in the material on random processes in Chapter 11.

The last case listed is the Gaussian case. An expression for the joint Gaussian density for two random variables is given in Section 5.4. That joint density can be extended to the case of n random variables using vector and matrix notation. The fact that general expressions exist for the PDF of Gaussian random variables is fortunate, because signals and noise with Gaussian statistics are ubiquitous in electrical engineering applications. The jointly Gaussian PDF for multiple random variables is discussed in detail under the topic of *Random Vectors* in Chapter 7.

5.5.2 *Sums of random variables*

Certain properties of sums of random variables are important because sums of random variables occur frequently in applications.

Two random variables

Let us begin by considering the random variable

$$Y = X_1 + X_2$$

Since expectation is a linear operation, it follows that

$$\mathcal{E}\{Y\} = \mathcal{E}\{X_1\} + \mathcal{E}\{X_2\}$$

or

$$m_y = m_1 + m_2 \tag{5.43}$$

Thus the mean of the sum is the *sum of the means.*

Now consider the variance, which by definition is

$$\text{Var}\,[Y] = \mathcal{E}\left\{(Y - m_y)^2\right\} = \mathcal{E}\left\{X_1 + X_2 - (m_1 - m_2))^2\right\}$$

Regrouping terms and expanding yields

$$
\begin{aligned}
\text{Var}\,[Y] &= \mathcal{E}\left\{((X_1 - m_1) + (X_2 - m_2))^2\right\} \\
&= \mathcal{E}\left\{(X_1 - m_1)^2\right\} + 2\mathcal{E}\left\{(X_1 - m_1)(X_2 - m_2)\right\} + \mathcal{E}\left\{(X_1 - m_1)^2\right\}
\end{aligned}
$$

It follows that

$$\boxed{\text{Var}\,[Y] = \text{Var}\,[X_1] + \text{Var}\,[X_2] + 2\,\text{Cov}\,[X_1, X_2]} \tag{5.44}$$

An especially important case of this formula is when the random variables are independent (or just uncorrelated). In this case $\text{Cov}\,[X_1, X_2] = 0$ and *the variance of the sum is the sum of the variances.*

Now let us consider the PDF for the random variable Y. A simple way to approach this computation is via the CDF. In particular, $F_Y(y)$ can be computed as

$$
\begin{aligned}
F_Y(y) &= \Pr[Y \le y] = \Pr[X_1 + X_2 \le y] \\
&= \int_{-\infty}^{\infty} \int_{-\infty}^{y-x_1} f_{X_1 X_2}(x_1, x_2)dx_2 dx_1
\end{aligned}
$$

(The region of integration for this event is sketched in Fig. 5.12.) Since $f_Y(y)$ is the

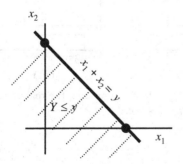

Figure 5.12 The event $X_1 + X_2 \le y$.

derivative of $F_Y(y)$, the PDF can be obtained by differentiating inside of the integral (assuming this is allowed):

$$f_Y(y) = \frac{dF_Y(y)}{dy} = \int_{-\infty}^{\infty} \left[\frac{d}{dy} \int_{-\infty}^{y-x_1} f_{X_1 X_2}(x_1, x_2) dx_2 \right] dx_1$$

or

$$f_Y(y) = \int_{-\infty}^{\infty} f_{X_1 X_2}(x_1, y - x_1) dx_1 \tag{5.45}$$

An analogous result holds for discrete random variables in terms of the joint PMF with the integral replaced by a summation.

A very important special case of (5.45) occurs when X_1 and X_2 are independent. In this case the joint density factors into a product of marginals and leaves

$$f_Y(y) = \int_{-\infty}^{\infty} f_{X_1}(x) f_{X_2}(y - x) dx \tag{5.46}$$

The subscript "1" on x has been dropped since x_1 is just a dummy variable of integration. The operation on the right-hand side can be recognized as *convolution* and written as

$$f_Y = f_{X_1} \circledast f_{X_2} \quad \text{(for independent random variables)} \tag{5.47}$$

where the special symbol in this equation is the convolution operator. The result also holds for independent discrete random variables using the PMF and the discrete form of convolution.

Let us illustrate the concepts discussed so far in this section with an example.

Example 5.8: Suppose that X_1 and X_2 are two independent uniform random variables with the density functions shown below.

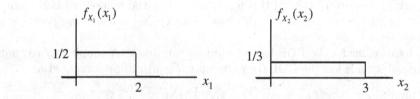

Since X_1 and X_2 are independent, (5.47) applies; the convolution is given by the integral (5.46) where the density f_{X_2} is reversed and slid over f_{X_1}. The arguments of the integral are shown below for a fixed value of y in three different cases.

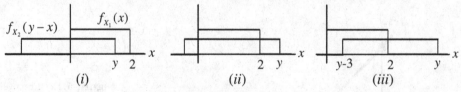

In each case we need to integrate only over the "overlap" region, i.e., the region where both functions are nonzero. The specific integrations are:

(i) for $0 \le y \le 2$ $\qquad$ $f_Y(y) = \int_0^y \frac{1}{2} \cdot \frac{1}{3} \, dx = \frac{1}{6} y$

(ii) for $2 < y \leq 3$

$$f_Y(y) = \int_0^2 \frac{1}{2} \cdot \frac{1}{3} \, dx = \frac{1}{3}$$

(iii) for $3 < y \leq 5$

$$f_Y(y) = \int_{y-3}^2 \frac{1}{2} \cdot \frac{1}{3} \, dx = \frac{1}{6}(5 - y)$$

For all other values of y the integrands have no overlap, so f_Y is 0. The complete density function for Y is sketched below.

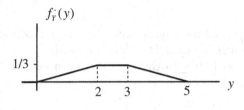

□

Two or more random variables

When Y is the sum of n random variables, i.e.,

$$Y = X_1 + X_2 + \cdots + X_n \tag{5.48}$$

the results of the previous discussion generalize. In particular, because expectation is a linear operator, it follows that

$$\mathcal{E}\{Y\} = \sum_{i=1}^n \mathcal{E}\{X_i\}$$

or

$$m_Y = \sum_{i=1}^n m_i \tag{5.49}$$

The expression for the variance follows from a procedure similar to that which was used to derive (5.44). The corresponding result for n random variables is

$$\text{Var}\,[Y] = \sum_{i=1}^n \text{Var}\,[X_i] + 2 \sum_{i=1}^n \sum_{j=1}^{i-1} \text{Cov}\,[X_i, X_j] \tag{5.50}$$

Again, probably the most important case is for the case of *independent* random variables, where (5.50) reduces to

$$\sigma_y^2 = \sum_{i=1}^n \sigma_i^2 \tag{5.51}$$

The joint density function for a sum of n *independent* random variables is the n-fold convolution of the densities. Probably the easiest way to see this is by using the moment generating function (MGF). The MGF for Y is defined as

$$M_Y(s) = \mathcal{E}\{e^{sY}\} = \mathcal{E}\left\{e^{s(X_1 + X_2 + \cdots + X_n)}\right\}$$

When the X_i are independent, this expectation of a product of random variables becomes a product of the expectations; that is,

$$M_Y(s) = \mathcal{E}\{e^{sY}\} = \mathcal{E}\{e^{sX_1}\} \cdot \mathcal{E}\{e^{sX_2}\} \cdots \mathcal{E}\{e^{sX_n}\} = \prod_{i=1}^n M_{X_i}(s) \tag{5.52}$$

The product of transforms is the convolution of the original functions, so that[7]

$$f_Y = f_{X_1} \circledast f_{X_2} \circledast \cdots \circledast f_{X_n} \qquad \text{(for independent random variables)} \qquad (5.53)$$

The n-fold convolution can be conveniently carried out in the transform domain, however, by using (5.52) and inverting.

IID random variables

A final case that deserves some special mention occurs when the random variables X_i are *independent* and *identically distributed* (IID). In other words, all of the random variables are described by identical PDFs. In this case the mean and variance of the sum of these random variables is given by

$$m_Y = nm_X \qquad (5.54)$$
$$\sigma_Y^2 = n\sigma_X^2 \qquad (5.55)$$

where m_X and σ_X^2 are the mean and variance of any one of the X_i. The standard deviation of Y then becomes

$$\sigma_Y = \sqrt{n}\,\sigma_X$$

The MGF of Y from (5.52) becomes

$$M_Y(s) = (M_X(s))^n$$

which can be inverted to obtain the PDF. A famous result applies here, however, which is known as the Central Limit Theorem (see Chapter 6). Loosely stated, this theorem claims in effect, that for any original densities $f_X(x_i)$ and any reasonably large values of n, the density function of Y approaches that of a Gaussian random variable!

5.6 Sums of some common types of random variables

As seen above, the density function for a sum of independent random variables is a convolution of densities. This leads to some interesting relations, discussed below.

5.6.1 Bernoulli random variables

Suppose Y in (5.48) is a sum of IID Bernoulli random variables. That is, each X_i is discrete and takes on only values of 1 or 0 with probability p and $q = 1-p$ respectively. The probability generating function (PGF) for the Bernoulli random variable reduces to

$$G_X(z) = \sum_{k=-\infty}^{\infty} f_X[k]z^k = q + pz$$

The PGF of Y is then given by

$$G_Y(z) = (G_X(z))^n = (q + pz)^n$$

This PGF is the nth power of a binomial, which can be expanded as

$$G_Y(z) = \sum_{k=0}^{n} \binom{n}{k} q^{(n-k)}(pz)^k = \sum_{k=0}^{n} \left[\binom{n}{k} p^k q^{(n-k)} \right] z^k$$

---

[7] A similar convolution for discrete random variables can be derived using the PGF.

The term in square brackets is thus $f_Y[k]$, which is the binomial PMF (see (3.5)).

It is really not surprising that Y turns out to be a binomial random variable. Since the X_i are either 1 or 0, Y simply counts the number of 1s in the sum. This was already seen when the binomial PMF was introduced to describe the probability of k 1s in a sequence of n binary digits.

5.6.2 Geometric random variables

Let K represent the sum of n IID geometric random variables of Type 1 (Equation (3.6)). Then the PMF for K can be found (using the PGF) to be

$$f_K[k] = \binom{k-1}{n-1} p^n (1-p)^{(k-n)} ; \qquad k \geq n \tag{5.56}$$

This is known as the *Pascal* PMF.

If the geometric PMF describes the number of bits to the *first* error in a stream of binary data, then the Pascal PMF describes the number of bits to the nth error in the stream. More generally, when observing discrete events, $f_K[k]$ is the probability that the nth event of interest occurs at the kth event. Some authors refer to this distribution or a variation of it as the *negative binomial*. However, while there is no dispute about the definition of the Pascal PMF, there does not seem to be precise agreement about the definition of the negative binomial [1, 4, 5, 6, 7].

Example 5.9:   Digital data is transmitted one byte at a time over a computer interface. Errors that may occur within each byte are detected with probability p. If six errors occur within a sequence of ten bytes, a retransmission is requested as soon as the sixth error is detected. What is the probability that a retransmission is requested?

In this example $n = 6$ and the probability of a retransmission is given by

$$P_r = \sum_{k=6}^{10} f_K[k] = \sum_{k=6}^{10} \binom{k-1}{5} p^6 (1-p)^{(k-6)}$$

Suppose that on the average, errors occur at the rate of 1 error in 100 bytes. Then by taking $p = 1/100$, the above expression evaluates to $P_r = 2.029 \times 10^{-10}$. This corresponds to one retransmission in about 5 megabytes of data.

$\square$

5.6.3 Exponential random variables

When Y is the sum of n IID exponential random variables, the PDF of Y is given by

$$f_Y(y) = \frac{\lambda^n y^{n-1}}{(n-1)!} e^{-\lambda y} ; \qquad y \geq 0 \tag{5.57}$$

This is known as the *Erlang* density function. To interpret this PDF, recall the use and interpretation of the exponential density function. In waiting for events that could occur at any time on a continuous time axis with average arrival rate λ, the exponential density characterizes the waiting time to the next such event. A sum of n exponential random variables X_i can therefore be interpreted as the total waiting time to the nth event. The Erlang density characterizes this total waiting time.

The Erlang PDF is most easily derived by starting with the MGF for the exponential,

$$M_X(s) = \frac{\lambda}{\lambda - s}$$

raising it to the nth power, and inverting. Further use of this density is made in Chapter 11 in the section on the Poisson process.

The integral of the density (5.57) is also a useful formula. In particular, the cumulative density function is given by

$$F_Y(y) = \int_0^y \frac{\lambda^n u^{n-1}}{(n-1)!} e^{-\lambda u} du = 1 - \sum_{k=0}^{n-1} \frac{(\lambda y)^k}{k!} e^{-\lambda y}; \qquad y \geq 0 \qquad (5.58)$$

The Erlang is a special case of the *Gamma* PDF which is given by

$$f_X(x) = \frac{\lambda^a x^{a-1}}{\Gamma(a)} e^{-\lambda x}; \qquad x \geq 0 \qquad (5.59)$$

where the parameter a is any positive real number and the Gamma function $\Gamma(a)$ is defined by

$$\Gamma(a) = \int_0^\infty u^{a-1} e^{-u} du \qquad (5.60)$$

When a is a positive integer n, this integral reduces to $\Gamma(n) = (n-1)!$, so the Gamma PDF reduces to the Erlang.

Example 5.10: Consider a supply of four light bulbs. The lifetime of each bulb is an exponential random variable with a mean of $1/\lambda = 2$ months. One light bulb is placed in service; when a bulb burns out it is immediately discarded and replaced with another. Let us find the probability that these four light bulbs last more than one year.

Let X_i represent the lifetime of the ith bulb. Then the total lifetime Y is given by

$$Y = X_1 + X_2 + X_3 + X_4$$

which is an Erlang random variable.

The probability that the total lifetime (in months) is more than 12 is given by

$$\begin{aligned} \Pr[Y > 12] &= 1 - \Pr[Y \leq 12] \\ &= 1 - F_Y(12) = \sum_{k=0}^{3} \frac{(12/2)^k}{k!} e^{-12/2} = 0.1512 \end{aligned}$$

where the last part of the equation comes from using (5.58) with $n = 4$, $\lambda = 1/2$, and $y = 12$.

You may notice the sum on the right side looks like a Poisson PMF, and indeed it is. You can think of "burnouts" as arrival of events. Then the probability of less than 4 burnouts in a period of 12 months is given by $\sum_{k=0}^{3} f_K[k]$ where f_K is the Poisson PMF given by (3.8) of Chapter 3.

$\square$

5.6.4 Gaussian random variables

In the case where Y is the sum of independent Gaussian random variables (not necessarily identically distributed), Y is also a Gaussian random variable. *This is an important fact to remember.* The proof of this result using the MGF is explored in Prob. 5.43 at the end of this chapter. If the random variables X_i have means m_i and variances σ_i^2 then Y has a Gaussian PDF with mean and variance

$$m_Y = \sum_{i=1}^{n} m_i \qquad \sigma_Y^2 = \sum_{i=1}^{n} \sigma_i^2$$

5.6.5 Squared Gaussian random variables

A final, slightly different sum of random variables is

$$Y = X_1^2 + X_2^2 + \cdots + X_n^2 \tag{5.61}$$

where the X_i are *zero-mean unit-variance Gaussian random variables* (i.e., $m_i = 0$, $\sigma_i^2 = 1$). In this case it can be shown that the PDF of Y is

$$f_Y(y) = \frac{y^{(n-2)/2}e^{-y/2}}{2^{n/2}\Gamma(n/2)} \tag{5.62}$$

where the Gamma function is defined by (5.60). This PDF is known as a *Chi-square* density function with "n degrees of freedom." This density is best known for its use in the "Chi-squared test," a statistical procedure used to test for Gaussian random variables. The Chi-squared density is also a special case of the Gamma density function given above.

5.7 Summary

Many problems in electrical and computer engineering involve sets of two or more random variables. A probabilistic description of these multiple random variables requires consideration of the joint PMF, the joint PDF, or joint CDF. If the random variables happen to be independent, then the joint distribution is the product of the marginals. Although the marginal PMFs or PDFs can always be derived from the joint PDF by summing or integrating, the joint PMF or PDF cannot always be derived from just the marginals. A joint PDF can be derived from marginals *and* conditional PDFs, however, e.g., $f_{X_1 X_2} = f_{X_1|X_2} f_{X_2}$. Conditional PDFs such as $f_{X_1|X_2}$ represent a density function for X_1 with X_2 acting as a parameter. Similar statements apply to the PMF of discrete random variables.

The extension of expectation to two or more random variables is straightforward and involves summation or integration with the joint density function. An important new concept however, is that of *correlation* between two random variables. Correlation, covariance, and the correlation coefficent ρ are all related quantities that describe probabilistic "similarity" between a pair of random variables. The PDF of Gaussian random variables is completely described by mean, variance, and covariance parameters.

Sums of multiple random variables occur frequently in practical applications. Important results exist for the mean, variance, and PDF of sums of random variables. Four new standard distributions that apply to the sums of random variables are the binomial PMF, the Pascal PMF, the Erlang PDF, and the Chi-squared PDF.

References

[1] Athanasios Papoulis and S. Unnikrishna Pillai. *Probability, Random Variables, and Stochastic Processes*. McGraw-Hill, New York, fourth edition, 2002.

[2] R. S. Cowan. *Sir Francis Galton and the Study of Heredity in the Nineteenth Century*. Garland, New York, 1985.

[3] Charles W. Therrien. *Discrete Random Signals and Statistical Signal Processing*. Prentice Hall, Inc., Upper Saddle River, New Jersey, 1992.

[4] William Feller. *An Introduction to Probability Theory and Its Applications - Volume I*. John Wiley & Sons, New York, second edition, 1957.

[5] Harold J. Larson and Bruno O. Shubert. *Probabilistic Models in Engineering Sciences*, volume I. John Wiley & Sons, New York, 1979.

[6] Alvin W. Drake. *Fundamentals of Applied Probability Theory*. McGraw-Hill, New York, 1967.

[7] Alberto Leon-Garcia. *Probability, Statistics, and Random Processes for Electrical Engineering*. Pearson, New York, third edition, 2008.

[8] Gilbert Strang. *Introduction to Applied Mathematics*. Wellesley Cambridge Press, Wellesley, MA, 1986.

Problems

Two discrete random variables

5.1 Given a joint PMF

$$f_{I_1 I_2}[i_1, i_2] = \begin{cases} Cp(1-p)^{i_1} & 0 \le i_1 \le 9, \ 0 \le i_2 \le 9 \\ 0 & \text{otherwise,} \end{cases}$$

Let $p = 1/8$.

(a) Find C to make this a valid PMF.

(b) Find the marginal density functions of I_1 and I_2.

(c) Determine the conditional density functions $f_{I_1|I_2}[i_1|i_2]$ and $f_{I_2|I_1}[i_2|i_1]$.

5.2 The joint probability mass function of two random variables I_1 and I_2 is given by

$$f_{I_1, I_2}(i_1, i_2) = \begin{cases} C(0.8)^{i_1}, & 0 \le i_1 \le i_2 \le 2 \\ 0, & \text{otherwise.} \end{cases}$$

(a) Find the value of C that makes $f_{I_1, I_2}(i_1, i_2)$ a valid joint PMF.

(b) Determine the marginal PMFs $f_{I_1}(i_1)$ and $f_{I_2}(i_1)$. Clearly specify the ranges of values for i_1 and i_2.

(c) Determine the conditional PMF $f_{I_1|I_2}(i_1|i_2)$.

5.3   Consider the joint PMF specified below:

$f_{I_1 I_2}[i_1, i_2]$	1	2	3	4
1	$\frac{6}{32}$	$\frac{2}{32}$	$\frac{4}{32}$	$\frac{3}{32}$
2	$\frac{2}{32}$	$\frac{3}{32}$	$\frac{2}{32}$	$\frac{1}{32}$
3	$\frac{1}{32}$	$\frac{1}{32}$	$\frac{2}{32}$	0
4	$\frac{1}{32}$	$\frac{1}{32}$	$\frac{1}{32}$	$\frac{2}{32}$

($i_1 \rightarrow$ across the top, $i_2 \downarrow$ down the side)

(a) Determine the marginal PMFs $f_{I_1}(i_1)$ and $f_{I_2}(i_2)$.

(b) Determine the conditional PMF $f_{I_1|I_2}(i_1|i_2)$.

(c) Find the CDF $F_{I_1, I_2}[i_1, i_2]$.

5.4   A joint PMF is defined by

$$f_{K_1 K_2}[k_1, k_2] = \begin{cases} C & 0 \le k_1 \le 5 \quad 0 \le k_2 \le k_1 \\ 0 & \text{otherwise} \end{cases}$$

(a) Find the constant C.

(b) Compute and *sketch* the marginal PMFs $f_{K_1}[k_1]$ and $f_{K_2}[k_2]$. Are the random variables K_1 and K_2 independent?

(c) Compute the conditional PMFs $f_{K_1|K_2}[k_1|k_2]$ and $f_{K_2|K_1}[k_2|k_1]$.

5.5   A thesis student measures some data K, which depends on setting a parameter α in the experiment. Unfortunately the student forgets to record the setting of the parameter α; but he knows that it was set to either 0.4 or 0.6 and that these values are equally likely. The data is described by the PMF

$$f_{K|\alpha}[k|\alpha] = \begin{cases} \alpha(1-\alpha)^{k-1} & k \ge 1 \\ 0 & \text{otherwise.} \end{cases}$$

From the advisor's point of view, what is the PDF for data, $f_K[k]$?

5.6   By using using the definitions (5.3) and (5.6) for marginal and conditional densities, show that (5.7) is true.

Two continuous random variables

5.7   The joint PDF $f_{X_1, X_2}(x_1, x_2)$ of two random variables X_1 and X_2 is shown below.

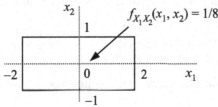

(a) Find the marginal PDFs.

(b) Determine the following probabilities:

(i) $\Pr[0 \leq X_1 \leq 1, -0.5 \leq X_2 \leq 2]$
(ii) $\Pr[X_1 < X_2]$
(iii) $\Pr[X_1^2 + X_2^2 \leq 1]$

5.8 A joint PDF is given to be

$$f_{X_1,X_2}(x_1,x_2) = \begin{cases} \frac{1}{24}, & -3 \leq x_1 \leq 3, -2 \leq x_2 \leq 2 \\ 0, & \text{otherwise.} \end{cases}$$

(a) Determine the probability: $\Pr[-2 \leq X_1 \leq 1, X_2 \geq 0.5]$.

(b) Find the probability: $\Pr[X_1 < 2X_2]$.

5.9 The joint PDF $f_{X_1,X_2}(x_1,x_2)$ of two random variables X_1 and X_2 is given by

$$f_{X_1,X_2}(x_1,x_2) = \begin{cases} C(4 - x_1 x_2), & 0 \leq x_1 \leq 4, 0 \leq x_2 \leq 1 \\ 0, & \text{otherwise.} \end{cases}$$

(a) Find C to make this a valid PDF.

(b) Find the marginal density functions of X_1 and X_2. Clearly define the ranges of values they take.

(c) Are the random variables independent?

5.10 The joint PDF for two random variables $f_{X_1 X_2}(x_1, x_2)$ is equal to a constant C for X_1 and X_2 in the region

$$0 \leq X_1 \leq 1 \quad \text{and} \quad 0 \leq X_2 \leq 2(1 - X_1)$$

and $f_{X_1 X_2}(x_1, x_2)$ is zero outside of this region.

(a) Sketch the region in the X_1, X_2 plane where the PDF is nonzero.

(b) What is the constant C?

(c) Compute the *marginal* density functions $f_{X_1}(x_1)$ and $f_{X_2}(x_2)$.

(d) Compute the *conditional* density functions $f_{X_1|X_2}(x_1|x_2)$ and $f_{X_2|X_1}(x_2|x_1)$.

(e) Are the two random variables statistically independent?

5.11 The joint probability density function of two random variables X_1 and X_2 is given by

$$f_{X_1,X_2}(x_1,x_2) = \begin{cases} K, & x_1 \geq 0, x_2 \geq 0, x_1 + x_2 \leq 2 \\ 0, & \text{otherwise.} \end{cases}$$

(a) Find the value of K that makes $f_{X_1,X_2}(x_1,x_2)$ a valid joint PDF.

(b) Find the joint cumulative distribution function for:

(i) $0 \leq x_1 \leq 2$ and $0 \leq x_2 \leq 2$;
(ii) $-1 \leq x_1 \leq 0$ and $-2 \leq x_2 \leq -1$;

(c) Using the joint CDF found in (b), find the probability $\Pr[X_1 < 0.5, X_2 > 0.5]$.

5.12 The joint probability density function of two random variables X_1 and X_2 is given by
$$f_{X_1,X_2}(x_1,x_2) = \begin{cases} C, & 0 \leq x_1 \leq x_2 \leq 1 \\ 0, & \text{otherwise.} \end{cases}$$

(a) Find C to make the joint PDF a valid one.

(b) Determine the conditional PDF $f_{X_1|X_2}(x_1|x_2)$. Be sure to specify the ranges of values for X_1 and X_2.

5.13 (a) Compute the joint CDF $F_{X_1X_2}(x_1,x_2)$ corresponding to the following joint PDF
$$f_{X_1X_2}(x_1,x_2) = \begin{cases} e^{-(x_1+x_2)} & 0 \leq x_1 < \infty \quad 0 \leq x_2 < \infty \\ 0 & \text{otherwise} \end{cases}$$

(b) The marginal cumulative distributions for the random variables are defined by $F_{X_i}(x_i) = \Pr[X_i \leq x_i]$. Find the marginal CDFs $F_{X_1}(x_1)$ and $F_{X_2}(x_2)$ for this joint density.

(c) Find the marginal PDFs $f_{X_1}(x_1)$ and $f_{X_2}(x_2)$ by differentiating the marginal CDFs.

(d) Are the random variables X_1 and X_2 independent?

5.14 Consider the joint PDF given in Prob. 5.7.

(a) Determine the joint CDF.

(b) From the CDF, compute $\Pr[X_1 > 1/2, X_2 < 1/2]$.

5.15 The joint PDF of two random variables is given by
$$f_{X_1X_2}(x_1,x_2) = \begin{cases} C(3 - x_1x_2) & 0 \leq x_1 \leq 1; \ 0 \leq x_2 \leq 3 \\ 0 & \text{otherwise} \end{cases}$$

(a) Find C to make this a valid PDF.

(b) Find the marginal density functions.

(c) Determine the CDF $F_{X_1X_2}(x_1,x_2)$.

5.16 Consider the joint PDF:
$$f_{X_1,X_2}(x_1,x_2) = \begin{cases} (3 - x_1x_2)/8, & 0 \leq x_1 \leq 2, 0 \leq x_2 \leq 3 \\ 0, & \text{otherwise.} \end{cases}$$

Find the conditional density function $f_{X_1|X_2}(x_1|x_2)$.

5.17 Refer to the joint PDF $f_{X_1X_2}(x_1,x_2)$ shown in Prob. 5.7.

(a) Compute the conditional PDF $f_{X_2|X_1}(x_2|x_1)$.

(b) Are X_1 and X_2 independent?

5.18 Two communications signals X_1 and X_2 are described by the joint density function
$$f_{X_1,X_2}(x_1,x_2) = \begin{cases} C & 0 \leq x_1 \leq 2 \text{ and } 0 \leq x_2 \leq 1 \\ C & 0 \leq x_1 \leq 2 \text{ and } 2 \leq x_2 \leq 3 \\ 0 & \text{otherwise} \end{cases}$$

(a) Sketch $f_{X_1,X_2}(x_1,x_2)$ and determine the constant C.

(b) Find and sketch the marginal pdf for X_2, i.e., find $f_{X_2}(x_2)$.

(c) Find and sketch $f_{X_1|X_2}(x_1|x_2)$, the PDF for X_1 *given* X_2.

5.19 Given the joint PDF of two random variables X_1 and X_2:

$$f_{X_1 X_2}(x_1, x_2) = \begin{cases} x_1 e^{-x_1(C+x_2)} & x_1 > 0, x_2 > 0 \\ 0 & \text{otherwise,} \end{cases}$$

determine the following:

(a) The constant C.

(b) The marginal PDFs $f_{X_1}(x_1)$ and $f_{X_2}(x_2)$.

(c) If X_1 and X_2 are independent.

(d) The conditional PDF $f_{X_1|X_2}(x_1|x_2)$.

(e) The conditional PDF $f_{X_2|X_1}(x_2|x_1)$.

5.20 Given a joint PDF

$$f_{X_1 X_2}(x_1, x_2) = \begin{cases} x_1 x_2 + C & x_1 + x_2 \leq 1, \ x_1 \geq 0, \ x_2 \geq 0 \\ 0 & \text{otherwise.} \end{cases}$$

(a) Find C.

(b) Compute the marginal PDFs: $f_{X_1}(x_1)$ and $f_{X_2}(x_2)$.

(c) Are the random variables independent?

(d) Determine the conditional PDF $f_{X_1|X_2}(x_1|x_2)$.

5.21 The probability that the first message arrives at a certain network node at an absolute time X given that the node comes online at an absolute time Y is given by the conditional density

$$f_{X|Y}(x|y) = \begin{cases} e^{-(x-y)} & 0 \leq y \leq x < \infty \\ 0 & x < y \end{cases}$$

(a) Suppose that it is known that the node comes online at time $Y = 0.5$, what is the (approximate) probability that the first message arrives between $X = 2$ and $X = 2.01$? <u>Hint</u>: 0.01 is a "small increment."

(b) If the probability that the node comes online at time Y is uniformly distributed between 0 and 1, what is the joint PDF $f_{XY}(x, y)$? Specify the region where the density is nonzero and sketch this region in the xy plane.

(c) What is the unconditional density $f_X(x)$? Be sure to specify the region where the density is nonzero.

5.22 The joint PDF for random variables X_1 and X_2 is given by

$$f_{X_1, X_2}(x_1, x_2) = \begin{cases} 1/4 & \text{if } |x_1| + 2|x_2| \leq 2 \\ 0 & \text{otherwise} \end{cases}$$

(a) Sketch the region of the x_1, x_2 plane where the joint density is nonzero.

(b) Find and sketch the marginal density $f_{X_1}(x_1)$.

(c) Find and sketch the conditional density $f_{X_1|X_2}(x_1|x_2)$.

5.23 Dagwood Bumstead rides in a car pool to work in the city. Dagwood is always late for the car pool; in fact his arrival time T_1 at the car pool is an exponential random variable with parameter λ_1. The other members of the car pool do not have infinite patience to wait for Dagwood and the length of time T_2 that they will wait for him is also an exponential random variable with parameter λ_2.

Assume Dagwood's mean arrival time is $1/\lambda_1 = 5$ minutes and the car pool mean waiting time is $1/\lambda_2 = 6$ minutes.

(a) What is the probability that Dagwood misses the car pool, i.e., what is $\Pr[T_1 > T_2]$?

(b) If Dagwood collides with the mailman in his rush out the door he will be delayed by an extra 30 seconds (0.5 minutes). In this case what is the probability that Dagwood misses the car pool?

(c) Assume that the probability of Dagwood colliding with the mailman is a random event with probability 1/3. What is the (unconditional) probability that Dagwood misses the car pool?

(d) If Dagwood is out the door and realizes that he forgot to kiss Blondie goodbye, he will run back to give Blondie a kiss. While this does not add significantly to his late time T_1, it doubles the chance that he will run into the mailman. In this case what is the (unconditional) probability that Dagwood misses the car pool?

(e) If Dagwood misses the car pool, he has one chance to make a flying leap for the bus. Dagwood can run at a maximum speed of 6 miles/hour. The speed of the bus as it's moving away from the bus stop in miles/hour is a random variable S with pdf given below.

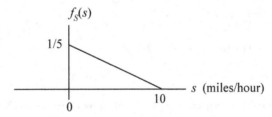

If Dagwood misses his car pool, what is the probability that he also misses the bus?

5.24 By using using the definitions (5.15) and (5.18) for marginal and conditional densities, show that (5.20) is true.

Expectation and correlation

5.25 The joint PDF for two random variables $f_{X_1 X_2}(x_1, x_2)$ is given by

$$f_{X_1 X_2}(x_1, x_2) = \begin{cases} 1 & 0 \leq x_1 \leq 1 \quad \text{and} \quad 0 \leq x_2 \leq 2(1 - x_1) \\ 0 & \text{otherwise} \end{cases}$$

Sketch the region where the density is nonzero and compute the following quantities:

(a) $m_{X_1} = \mathcal{E}\{X_1\}$ (f) $\sigma_{X_1}^2 = \text{Var}[X_1]$

(b) $m_{X_2} = \mathcal{E}\{X_2\}$ (g) $\sigma_{X_2}^2 = \text{Var}[X_2]$

(c) $\mathcal{E}\{X_1^2\}$ (h) $c_{X_1 X_2} = \text{Cov}[X_1, X_2]$

(d) $\mathcal{E}\{X_2^2\}$ (i) $\mathcal{E}\{X_1|X_2\}$

(e) $\mathcal{E}\{X_1 X_2\}$ (j) $\mathcal{E}\{X_2|X_1\}$

Hint: You may want to use some of the results from Prob. 5.10 to help in some of these computations.

5.26 Given the joint probability density function

$$f_{X_1,X_2}(x_1, x_2) = \begin{cases} 1/2, & 0 \le x_1 \le 2, 0 \le x_2 \le 2 - x_1 \\ 0, & \text{otherwise} \end{cases}$$

(a) Determine the conditional density function of X_1 given X_2, $f_{X_1|X_2}(x_1|x_2)$

(b) Find the conditional mean of X_1 given X_2, defined as $E[X_1|X_2]$.

5.27 Random variables X and Y are independent and described by the density functions shown below:

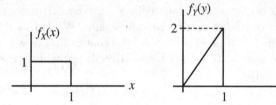

(a) Give an algebraic expression for the joint density $f_{XY}(x, y)$ (including limits) and sketch the region where it is nonzero below.

(b) What is $\Pr[X \le Y]$?

(c) What is the variance of Y?

(d) What is the correlation coefficient ρ_{XY}?

5.28 The Schwartz inequality [8] for functions of two variables can be written as

$$\left(\int_{-\infty}^{\infty} \int_{-\infty}^{\infty} h(x_1, x_2) g(x_1, x_2) dx_1 dx_2 \right)^2 \le$$

$$\left(\int_{-\infty}^{\infty} \int_{-\infty}^{\infty} h^2(x_1, x_2) dx_1 dx_2 \right) \left(\int_{-\infty}^{\infty} \int_{-\infty}^{\infty} g^2(x_1, x_2) dx_1 dx_2 \right)$$

where h and g are any two square integrable functions. The inequality holds with *equality* when h and g are linearly related, i.e., when $h(x, y) = a g(x, y)$ where a is a constant.

Apply the Schwartz inequality to the definition

$$c = \text{Cov}[X_1, X_2] = \int_{-\infty}^{\infty} \int_{-\infty}^{\infty} (x_1 - m_1)(x_2 - m_2) f_{X_1 X_2}(x_1, x_2) dx_1 dx_2$$

for the covariance of two random variables to show that

$$c^2 \le \sigma_1^2 \sigma_2^2$$

Use this to prove (5.30). Under what condition is ρ *equal* to ± 1?

5.29 Two random variables X_1 and X_2 are described by the joint density function shown below.

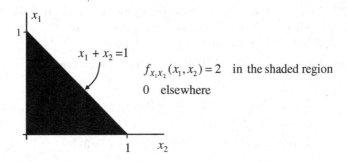

(a) Find and sketch $f_{X_2}(x_2)$.

(b) Find and sketch $f_{X_1|X_2}(x_1|x_2)$.

(d) Determine $\mathcal{E}\{X_1|X_2\}$.

Gaussian random variables

5.30 Two zero-mean unit-variance random variables X_1 and X_2 are described by the joint Gaussian PDF

$$f_{X_1 X_2}(x_1, x_2) = \frac{1}{\pi\sqrt{3}} e^{-(x_1^2 - x_1 x_2 + x_2^2)2/3}$$

(a) Show by integration that the marginal density for X_1 is Gaussian and is equal to

$$f_{X_1}(x_1) = \frac{1}{\sqrt{2\pi}} e^{-x_1^2/2}$$

Hint: Use the technique of "completing the square" to write

$$x_1^2 - x_1 x_2 + x_2^2 = \frac{3}{4}x_1^2 + \frac{1}{4}x_1^2 - x_1 x_2 + x_2^2 = \frac{3}{4}x_1^2 + \left(x_2 - \frac{1}{2}x_1\right)^2$$

Then form the definite integral over x_2 and make the substitution of variables $u = x_2 - \frac{1}{2}x_1$. Use the fact that the one-dimensional Gaussian form integrates to 1 in order to evaluate the integral.

(b) What is the conditional density $f_{X_2|X_1}(x_2|x_1)$? Observe that it has the form of a Gaussian density for X_2 but that the mean is a function of x_1. What *are* the mean and variance of this conditional density?

5.31 Two zero-mean unit-variance random variables X_1 and X_2 are described by the joint Gaussian PDF

$$f_{X_1 X_2}(x_1, x_2) = \frac{1}{2\pi\sqrt{1-\rho^2}} e^{-(x_1^2 - 2\rho x_1 x_2 + x_2^2)/2(1-\rho^2)}$$

Two new random variables Y_1 and Y_2 are defined through the transformation

$$\begin{bmatrix} Y_1 \\ Y_2 \end{bmatrix} = \begin{bmatrix} 1 & 0 \\ a & 1 \end{bmatrix} \begin{bmatrix} X_1 \\ X_2 \end{bmatrix}$$

(a) Find the constant, the Jacobian $J(\mathbf{y})$ of the transformation.

(b) Solve for X_1 and X_2 in terms of Y_1 and Y_2 and find $f_{Y_1 Y_2}(y_1, y_2)$.

(c) Let $\rho = 1/2$. For what value of a are the random variables Y_1 and Y_2 independent? What then are the variances of Y_1 and Y_2?

5.32 Consider the random variables described in Prob. 5.31. For a general value of ρ with $(-1 \leq \rho \leq +1)$ find a in terms of ρ so that Y_1 and Y_2 are independent. What are the variances of Y_1 and Y_2 in terms of ρ?

5.33 Determine the joint PDF of jointly Gaussian random variables for the following cases.

(a) $m_1 = m_2 = 0$, $\sigma_1 = \sigma_2 = 2$, and $\rho = 0.8$.

(b) $m_1 = m_2 = 0$, $\sigma_1 = \sigma_2 = s$, and $\rho = 0$.

(c) $m_1 = m_2 = 1$, $\sigma_1 = \sigma_2 = 3$, and $\rho = 0$.

5.34 By integrating (5.35) over one variable, show that the marginal densities for jointly Gaussian random variables are given by (5.36). Also show that the conditional densities are Gaussian and have the form (5.37).

5.35 By using the algebraic method of "completing the square" and setting the quadratic expression in the bivariate Gaussian density to a constant, show that the contours of the density can be written in the form

$$\frac{(y_1 - \alpha_1)^2}{\beta_1^2} + \frac{(y_2 - \alpha_2)^2}{\beta_2^2} = \gamma^2$$

which is the equation of an ellipse.

Multiple random variables

5.36 Let K be the discrete random variable representing the number rolled on a pair of dice. Compute the mean and variance of K using the fact that K is the sum of two random variables. Hint: The easy way to do this problem is to use the results of Prob. 4.8 (Chapter 4).

5.37 Stations A and B are connected by two parallel message channels which operate *independently*. A message is sent from A to B *over both channels at the same time*. Random variables X and Y, which are independent, represent the delays over these two channels. Each of these delays is an exponential random variable with mean of 1 hour.

A message is considered "received" as soon as it arrives on *any one* channel; the message is considered "verified" as soon as it has arrived over *both* channels.

(In order to answer the following questions, you may find it useful to draw the continuous sample space involving the two random variables X and Y and depict the events in the sample space.)

(a) What is the probability that a message is *received* within 15 minutes (1/4 hour) after it is sent?

(b) What is the probability that a message is *received and verified* within 15 minutes after it is sent?

(c) What is the probability that a message is verified within 15 minutes after it is received?

(d) Given that a message is received within 15 minutes after it is sent, what is the probability that it is verified within 15 minutes after it was sent?

[This problem is a modified version of a problem from Alvin W. Drake, *Fundamentals of Applied Probability Theory*, McGraw-Hill, New York, 1967. Reproduced by permission.]

5.38 Consider the situation in Prob. 5.37, but assume the delays X and Y are each *uniformly* distributed between 0 and 1 hour. Answer the following questions.

(a) What is the probability that a message is received within 15 minutes after it has been sent?

(b) What is the probability that a message is received but not verified within 15 minutes after it has been sent?

(c) Assume that a message is received on one channel and the chief in the communications room at point B goes on a 15 minute coffee break immediately following receipt of the message. What is the probability that the message will not be verified (i.e., that he will miss the second message)?

5.39 Random variables X, Y, and Z are described by the joint density function

$$f_{XYZ}(x, y, z) = \begin{cases} (x + 3y)z & \text{for } 0 \le x \le 1; \ \ 0 \le y \le 1; \ \ 0 \le z \le 1 \\ 0 & \text{otherwise} \end{cases}$$

Determine the following:

(a) The joint density for X and Y (without Z).

(b) The marginal density for Z.

(c) The conditional density for X and Y given Z.

(d) The probability that $X < 3$.

(e) The expected value of Z.

(f) Whether or not X and Y are independent.

5.40 K_1 and K_2 are two independent, identically-distributed (IID) Poisson random variables. Each PMF is given by

$$f_{K_i}[k] = \begin{cases} \dfrac{\alpha^k}{k!} e^{-\alpha} & k \ge 0 \\ 0 & \text{otherwise} \end{cases}$$

A new random variable J is equal to the sum $J = K_1 + K_2$.

(a) Show that the PGF for the random variable K_i is given by

$$G_{K_i}(z) = e^{\alpha(z-1)}$$

(b) What is the PGF $G_J(z)$ for the random variable J?

(c) What is the PMF $f_J[j]$? What form does it take?

(d) Find the mean and variance of J.

Sums of common random variables

5.41 Let X and Y be the number shown on each of two dice and let Z be the sum $(Z = X + Y)$. Assume that X and Y are independent and each is uniformly distributed over the integers 1 through 6. Using discrete convolution, show that the PMF for Z has a triangular shape. Compare it to Fig. 3.2 of Chapter 3.

5.42 Two random variables X_1 and X_2 are independent and each is uniformly distributed in the interval $0 \le X_i \le 1$.

(a) What is the PDF for the random variable $Y = X_1 + X_2$?

(b) Find the MGF $M_Y(s)$ for Y. Hint: What is the MGF for X_i? What happens to the MGFs when you convolve the densities?

(c) If a message must wait in two successive queues for service and X_i is the time in seconds for service in the ith queue, then Y is the total service time. What is the probability that the total service time is longer than 1.5 seconds?

5.43 In this problem you will show that the sum of two or more independent Gaussian random variables is Gaussian. Let

$$Y = X_1 + X_2 + \cdots + X_n$$

where the X_i are *independent* Gaussian random variables with means m_i and variances σ_i^2. The density for Y is obtained from the n-fold convolution of the individual Gaussian density functions (see (5.53)).

(a) What is the MGF for the ith random variable X_i? Hint: See Prob. 4.22.

(b) Since convolution of densities corresponds to multiplication of the MGFs, find the MGF for Y by multiplying the MGFs from part (a).

(c) Notice that the MGF for Y has the form of a Gaussian MGF. Thus Y must be Gaussian. What are the mean and variance of Y?

5.44 The sum of two independent random variables X_1 and X_2 is given by:

$$X = X_1 + X_2$$

where X_1 is a uniform random variable in the range $[0, 1]$, and X_2 is a uniform random variable $[-1, 0]$.

(a) Determine the mean and the variance of X.

(b) Determine and sketch the PDF of X.

(c) Using the PDF for X obtained in (b), compute its mean and variance. On your sketch of the PDF of X, indicate the mean and the standard deviation.

5.45 The sum of two independent random variables X_1 and X_2 is given by

$$X = X_1 + X_2$$

where X_1 is an exponential random variable with parameter $\lambda = 2$, and X_2 is another exponential random variable with parameter $\lambda = 3$.

(a) Find the mean and the variance of X.

(b) Determine the PDF of X.

5.46 In a certain computer network, you observe that your message is fifth in line for service, i.e., there are four others ahead of you. The processing time for each message is described by an exponential random variable with expected value $1/\lambda = 2.5$ sec.

(a) What is the mean waiting time until your meassage has been processed?

(b) What is the *variance* of the random variable that represents the waiting time until completion of processing?

5.47 A sum is formed of n independent and identically-distributed (IID) random variables:

$$T = T_1 + T_2 + \cdots T_n$$

where T_i is exponentially distributed with parameter λ. Under these conditions, the sum T is an Erlang random variable.

(a) Using the fact that T is the sum of n independent exponential random variables, determine the mean and the variance of the Erlang random variable T.

(b) The MGF of an individual exponential random variable T_i is given by

$$M(s) = \frac{\lambda}{\lambda - s}.$$

What is the MGF of T?

(c) Use the result of (b) to compute the mean and the variance of T.

5.48 The average life of a set of batteries used in an electronic flash unit is 6 months. What is the probability that four sets of batteries will last longer than 2 years?

Computer Projects

Project 5.1

In this project, you will use some data sets from the data package for this book to make scatter plots and plot their trendlines.

Consider the following three data sets, each containing two files. Treat the sequence of measurements in each file as a realization of a random variable.

1. The files x501.dat and x502.dat contain 500-points long realizations of filtered noise.

2. The data contained in files x503.dat and x504.dat are power plant heat input in million British thermal units (MMBTU) and the plant annual carbon dioxide emissions in tons, respectively, from 54 California power plants. (Source: Emissions and Generation Resource Integrated Database (EGRID), US Environmental Protection Agency, 1997.)

3. The data contained in files x505.dat and x506.dat are 132-points long measurements of modern automobile engine sizes in liters and their highway mileage in miles per gallon of fuel. (Source: Fuel Economy Guide, US Department of Energy, 1999.)

For each data set listed above, determine the following:

1. Estimate the means m_1, m_2, the variances σ_1^2, σ_2^2 and the correlation coefficient ρ of these random variables.

2. Make a scatter plot of the data in the two files.

3. Determine and plot the trendline of the data points.

4. Comment on the nature of correlation and trend between the mesurements in the two files.

Project 5.2

This project, based on Example 5.8, requires measuring the probability density function of a random variable obtained by adding two uniformly distributed random variables.

1. Generate sequences, at least 10000 points long, of the following two random variables.

   (a) A uniform random variable X_1 in the range of $[0\ 3]$.
   (b) A uniform random variable X_2 in the range of $[0\ 5]$.

2. Obtain a new random variable as a sum: $Y = X_1 + X_2$.

3. Measure the PDFs of X_1, $X2$ and Y.

4. Plot the measured PDFs along with the corresponding theoretical PDFs. Compare and comment on the results.

6 Inequalities, Limit Theorems, and Parameter Estimation

This chapter deals with a number of topics that on one hand seem abstract and theoretical but on the other hand are essential for the practical application of probability to real life problems. It begins with the topic of inequalities. The inequalities developed here are well known and are used to provide bounds on the probability for random variables when only partial information (e.g., mean and variance) is known for the distribution. As such, they are useful tools for calculating or bounding probability from limited real data and for proving general results that make as few assumptions as possible about the properties of the random variables.

The chapter then moves on to the topic of stochastic convergence and limit theorems. While an advanced treatment of convergence is well beyond the scope of this chapter, knowledge of some types of convergence for a sequence of random variables is of practical importance when dealing with a large number of measurements that may be taken in an engineering application. After discussing types of convergence, we develop two well known asymptotic results: the law of large numbers and the Central Limit Theorem. The former provides theoretical assurance that in practice, expectations can be replaced by averages of a sufficiently large number of random variables. The latter states the incredibly powerful result that the distribution of any sum of n independent random variables, for large enough n, is *Gaussian*.

The last half of this rather short chapter deals with the estimation of parameters. Parameter estimation is required when the conditions of a problem dictate a certain type of probabilistic distribution but the numerical values of the parameters are unknown. For example, it may be known that a Poisson model is the correct model for describing "hits" on a web page but the numerical value of the arrival rate (λ) for the model is typically unknown. Although parameter estimation usually falls in the realm in statistics, the basic ideas presented here are useful in practical applications of probability. We first discuss properties of estimates, and illustrate these via two fundamental statistics known as the sample mean and sample variance. The chapter then introduces the topic of an *optimal* estimate known as the maximum likelihood estimate and illustrates that this estimate often satisfies intuition about how the data should be processed to find the desired parameter.

The discussion of estimation is continued by introducing the concept of confidence intervals. Confidence intervals provide a way to report the quality of an estimate by providing an *interval* that contains the true parameter with certain probability. The chapter closes with an application of estimation to signal processing. The discussion illustrates that it is possible to detect the presence of very weak signals and measure their values if a sufficient number of observations is taken.

6.1 Inequalities

Suppose that the mean and variance for a given random variable X are known; but neither the CDF nor the PDF are known. Given sizeable measured data belonging to this random variable, estimated values can be computed for the mean and variance according to methods discussed later in this chapter even if the distribution is completely unknown. In general, knowing only the mean and variance of a random variable, we cannot determine the exact probabilities of particular events; but bounds on these probabilities can be obtained using the Markov and Chebyshev inequalities.[1]

6.1.1 Markov inequality

For a random variable X that takes only non-negative values, the Markov inequality is stated as

$$\Pr[X \geq a] \leq \frac{\mathcal{E}\{X\}}{a}, \tag{6.1}$$

where $a > 0$.

In order to prove this statement let us start with

$$\mathcal{E}\{X\} = \int_0^\infty x\, f_X(x)\, dx = \int_0^a x\, f_X(x)\, dx + \int_a^\infty x\, f_X(x)\, dx$$

(The lower limit on the first integral is zero because X is assumed to take on only non-negative values.) Since all integrals have finite *positive* values, discarding the first integral on the right, produces the inequality

$$\mathcal{E}\{X\} \geq \int_a^\infty x\, f_X(x)\, dx$$

Then replacing x in the integrand by the smaller constant value a yields

$$\mathcal{E}\{X\} \geq \int_a^\infty a\, f_X(x)\, dx = a \int_a^\infty f_X(x)\, dx = a\, \Pr[X \geq a].$$

which proves the result.

The Markov bound can be illustrated with a short example.

Example 6.1:   Average occupancy of packets in the buffer of a switch in the Internet is known to be 40%. Buffer occupancy is a random variable X.

Without knowing the PDF of X, what can be said about the probability that buffer occupancy equals or exceeds 60%?

The Markov inequality (6.1) provides the following result:

$$\Pr[X \geq 0.60] \leq \frac{\mathcal{E}\{X\}}{0.60} = \frac{0.40}{0.60} = 0.6667.$$

□

Observe that the result produced by the bound is reasonable, but not necessarily close to the correct probability since the bound covers a wide range of possible distributions. To illustrate that fact, consider the following example.

Example 6.2:   A voltage measured across a certain resistor is a uniform random variable in the range [0,5] volts. The probability that the voltage measured across this resistor

---

[1] In special cases, such as when the PDF is Gaussian or exponential, the probabilities can be computed exactly. In other cases, especially when the distribution is unknown, the inequalities may be the best mathematical tools available.

exceeds 4.0 volts is 0.2 (see figure following). Using the Markov inequality, the upper bound for this probability is

$$\Pr[X \geq 4.0] \leq \frac{E\{X\}}{4.0} = \frac{2.5}{4.0} = 0.625,$$

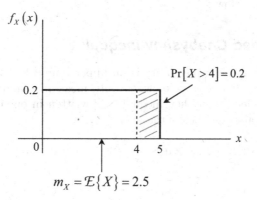

$$m_X = E\{X\} = 2.5$$

which is nowhere close to the true probability! Nevertheless, the bound is correct, in that the true value of 0.2 is below the upper bound. Note that a bound takes the worst case scenario into account.

□

6.1.2 Chebyshev inequality

The Chebyshev inequality for a random variable X is given by

$$\Pr[|X - m_X| \geq a] \leq \frac{\sigma_X^2}{a^2} \tag{6.2}$$

where $a > 0$.

To prove the Chebyshev inequality note that $(X - m_X)^2$ is a random variable that takes on only non-negative values. Then apply the Markov inequality to this quantity to obtain:

$$\Pr[(X - m_X)^2 \geq a^2] \leq \frac{E\{(X - m_X)^2\}}{a^2} = \frac{\sigma_X^2}{a^2}$$

By observing that the events $(X - m_X)^2 \geq a^2$ and $|X - m_X| \geq a$ are equivalent, the last equation is seen to be the same as (6.2).

The Chebyshev bound is illustrated using the random variable X of Example 6.2.

Example 6.3: The mean of the uniform variable described in Example 6.2 is $m_X = 2.5$. The variance is computed to be $\sigma_X^2 = 25/12$. Strictly speaking, we cannot compute a bound on $\Pr[X \geq 4.0]$ using the Chebyshev inequality. If someone tells us that the distribution is symmetric about the mean, however, we could notice that the probability of $X \geq 4.0$ must be one-half the probability of $|X - 2.50| \geq 1.5$. The upper bound for this latter probability is

$$\Pr[|X - 2.5| \geq 1.50] \leq \frac{\sigma_X^2}{(1.5)^2} = \frac{25}{12 \times 2.25} = 0.926$$

Thus the bound (assuming a symmetric distribution) would be $\Pr[X \geq 4.0] \leq 0.463$, which is closer to the true probability.

□

The Chebyshev inequality is generally more versatile than the Markov inequality because it does not make specific assumptions about the data. It presumes that the mean *and* variance are known, however, which helps tighten the bound. A version of the Chebyshev inequality pertaining to random variables on one side or the other of the mean is given below.

6.1.3 One-sided Chebyshev inequality

The one-sided Chebyshev inequality is an upper bound for probabilities of the form $\Pr[X - m_X \geq a]$, where $a > 0$, when only mean m_X and variance σ_X^2 of X are given. The one-sided Chebyshev inequality is written in one of two forms depending on whether X is above or below the mean:

$$\Pr[X - m_X \geq a] \leq \frac{\sigma_X^2}{\sigma_X^2 + a^2} \quad \text{for } X > m_X \quad (a)$$

$$\Pr[m_X - X \geq a] \leq \frac{\sigma_X^2}{\sigma_X^2 + a^2} \quad \text{for } X \leq m_X \quad (b) \tag{6.3}$$

Example 6.4:   Continuing with Example 6.2, let us determine the upper bound that the voltage is greater than 4.0 using the one-sided Chebyshev inequality. The probability of $X \geq 4.0$ is the same as the probability of $X - m_X \geq 1.5$ where $m_X = 2.5$ is the mean of the random variable. The upper bound using (6.3)(a) is thus

$$\Pr[X - 2.5 \geq 1.5] \leq \frac{\sigma_X^2}{\sigma_X^2 + (1.5)^2} = \frac{25}{25 + 27} = 0.481.$$

which is somewhat tighter than the Markov result (0.625), but not as tight as the earlier Chebyshev bound if symmetry is assumed.

□

The results for these three examples are summarized in Table 6.1.

inequality	bound
Markov	0.625
One-sided Chebyshev	0.481
Chebyshev / symmetry	0.463

Table 6.1 Probability bounds on a random variable. True probability is 0.200.

6.1.4 Other inequalities

There are two other important inequalities, Hölder's inequality and Minkowski's inequality, that are frequently encountered in the literature. Proofs can be found in several more advanced texts (e.g., [1]).

Hölder's inequality.[2] Consider two random variables X_1 and X_2 with means m_1 and m_2 and variances σ_1^2 and σ_2^2, respectively. Given that $k, l > 1$ and $k^{-1} + l^{-1} = 1$, Hölder's inequality states that

$$|\mathcal{E}\{X_1 X_2\}| \le \left(\mathcal{E}\{|X_1^k|\}\right)^{1/k} \left(\mathcal{E}\{|X_2^l|\}\right)^{1/l} \tag{6.4}$$

For $k = l = 2$, Hölder's inequality becomes the Cauchy-Schwartz inequality, which provides a bound on the correlation term $\mathcal{E}\{X_1 X_2\}$ in terms of the second moments of X_1 and X_2:

$$|\mathcal{E}\{X_1 X_2\}| \le \sqrt{\mathcal{E}\{X_1^2\}\mathcal{E}\{X_2^2\}},$$

By subtracting the mean from each random variable, the Cauchy-Schwartz inequality can also be applied to the covariance:

$$|\mathcal{E}\{(X_1 - m_1)(X_2 - m_2)\}| \le \sqrt{\mathcal{E}\{(X_1 - m_1)^2\}\mathcal{E}\{(X_2 - m_2)^2\}},$$

or

$$\mathrm{Cov}\,[X_1, X_2] \le \sigma_1 \sigma_2.$$

This provides the following bound on the correlation coefficient:

$$\rho_{12} = \frac{\mathrm{Cov}\,[X_1, X_2]}{\sigma_1 \sigma_2} \le 1$$

The correlation coefficient can be negative, however, in which case the last equation is a useless tautology. The complete bound, which is given by

$$\boxed{|\rho_{12}| \le 1} \tag{6.5}$$

can be obtained by redefining one of the random variables as its negative and repeating the last procedure.

Minkowski's inequality.[3] Given that $k \ge 1$, Minkowski's inequality states that

$$\left(\mathcal{E}\{|X_1 + X_2|^k\}\right)^{1/k} \le \left(\mathcal{E}\{|X_1^k|\}\right)^{1/k} + \left(\mathcal{E}\{|X_2^k|\}\right)^{1/k}. \tag{6.6}$$

For $k = 2$, Minkowski's inequality relates the second moment of $X_1 + X_2$ to those of X_1 and X_2:

$$\sqrt{\mathcal{E}\{(X_1 + X_2)^2\}} \le \sqrt{\mathcal{E}\{X_1^2\}} + \sqrt{\mathcal{E}\{X_2^2\}}.$$

Again, by subtraction of the mean we can extend the result to covariances

$$\sqrt{\mathcal{E}\{((X_1 - m_1) + (X_2 - m_2))^2\}} \le \sqrt{\mathcal{E}\{(X_1 - m_1)^2\}} + \sqrt{\mathcal{E}\{(X_2 - m_2)^2\}},$$

which can be written as

$$\boxed{(\mathrm{Var}\,[X_1 + X_2])^{1/2} \le \sigma_1 + \sigma_2} \tag{6.7}$$

By employing (5.44) of Chapter 5, this result can also be used to show the bound (6.5) on the correlation coefficient (see Prob. 6.8).

6.2 Convergence and Limit Theorems

A sequence of (non-random) numbers $\{x_n\}$ converges to a limit x if for any $\epsilon > 0$ there exists some sufficiently large integer n_c such that

$$|x_n - x| < \epsilon, \quad \text{for every } n > n_c$$

In this section, however, convergence is extended to sequences of *random variables.* and a number of different modes of convergence are cited.

Consider the experiment where a sequence of random variables is formed:

$$X_1, X_2, \ldots, X_n, \ldots$$

Since each outcome s of the random experiment results in a sequence of numbers $X_1(s)$, $X_2(s), \ldots, X_n(s), \ldots$, the sequence of random variables $X_1, X_2, \ldots, X_n, \ldots$ forms a family of such sequences of numbers. The convergence of a sequence of random variables, therefore, must consider criteria for convergence of not just one realization but the entire family of these sequences. Consequently, there are a number of modes of convergence of sequences of random variables, some strong and some weak. Our interest here is whether convergence of a random sequence obeys one of: convergence with probability 1, convergence in probability, convergence in the mean-square sense, and convergence in distribution. These types of convergence are defined below.

Convergence with probability 1. This type of convergence, also known as "convergence almost everywhere," states that the sequence of random variables X_n converges to the random variable X with probability 1 for all outcomes $s \in \mathcal{S}$:

$$\lim_{n \to \infty} \Pr[X_n(s) \to X(s)] = 1, \quad \text{for all } s \in \mathcal{S}. \tag{6.8}$$

Mean-square convergence. The sequence of random variables X_n converges to the random variable X in the mean-square sense if:

$$\lim_{n \to \infty} E[|X_n(s) - X(s)|^2] = 0, \quad \text{for all } s \in \mathcal{S}. \tag{6.9}$$

Convergence in probability. This type of convergence is also known as stochastic convergence or convergence in measure. The sequence of random variables X_n converges to the random variable X in probability if for any $\epsilon > 0$:

$$\lim_{n \to \infty} \Pr[|X_n(s) - X(s)| > \epsilon] = 0, \quad \text{for all } s \in \mathcal{S}. \tag{6.10}$$

From the Chebyshev inequality (6.2), we have

$$\Pr[|X_n - X| > \epsilon] \le \frac{E[|X_n - X|^2]}{\epsilon^2},$$

which implies that if a sequence of random variables converges in the mean square sense, then it also converges in probability.

Convergence in distribution. The sequence of random variables X_n with cumulative distribution functions $F_{X_n}(x)$ converges in distribution to the random variable X with cumulative distribution function $F_X(x)$ if:

$$\lim_{n \to \infty} F_{X_n}(x) = F_X(x) \tag{6.11}$$

for all x for which $F_X(x)$ is continuous. The Central Limit Theorem discussed later in this chapter is an example of convergence in distribution. Convergence in distribution makes no reference to the sample space, $\mathcal{S}$, but is only a condition on the cumulative distribution functions of the random variables in the sequence.

A Venn diagram representing the different convergence modes is shown in Figure 6.1. Each oval indicates all sequences of random variables that converge according to the indicated mode; no sequence converges outside of the largest oval. You may notice that convergence in distribution is the weakest mode of convergence. If a sequence converges in probability, then it also converges in distribution (but not vice versa).

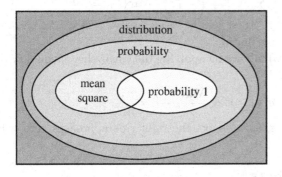

Figure 6.1 Comparison of different convergence modes.

6.2.1 Laws of large numbers

We now consider an important theoretical result known as the "law of large numbers." This result is important because it links the mean of a random variable to the average value of a number of realizations of the random variable, and thus justifies computing the mean experimentally as an average. Specifically, assume that we have outcomes or "observations" $X_1, X_2, \ldots, X_n$ resulting from n independent repetitions of the experiment that produces a certain random variable X. Thus the X_i are independent and identically-distributed (IID). The *sample mean* is defined as the average of these n random variables:

$$M_n = \frac{X_1 + X_2 + \ldots + X_n}{n} = \frac{1}{n}\sum_{i=1}^{n} X_i \tag{6.12}$$

Note that since each of the X_i is a random variable, *the sample mean is also a random variable*. The law of large numbers states that the sample mean converges to the mean m_X of the distribution as n approaches infinity. The law is usually stated in the two forms given below.

Weak law of large numbers

The weak law of large numbers provides a precise statement that the sample mean M_n is *close to* the distribution mean m_X in the following sense:

$$\lim_{n\to\infty} \Pr\left[|M_n - m_X| \geq \epsilon\right] = 0 \tag{6.13}$$

where ϵ is any small positive number. In other words, as n gets large, the sample mean converges *in probability* to the true mean m_X.

To prove this result, first consider the mean and variance of M_n. The mean is given by

$$\mathcal{E}\{M_n\} = \mathcal{E}\left\{\frac{1}{n}\sum_{i=1}^{n} X_i\right\} = \frac{1}{n}\sum_{i=1}^{n}\mathcal{E}\{X_i\} = m_X \tag{6.14}$$

(The last equality follows from the fact that all of the X_i have the same mean m_X.) It is seen from (6.14) that the mean of the sample mean is equal to the distribution mean m_X. The variance can also be easily computed by noting that the X_i are IID.

Thus (see Table 4.2 and equation 5.50) the variance is given by

$$\text{Var}\,[M_n] = \text{Var}\left[\frac{1}{n}\sum_{i=1}^{n}X_i\right] = \left(\frac{1}{n}\right)^2\sum_{i=1}^{n}\text{Var}\,[X_i] = \frac{\sigma_X^2}{n} \tag{6.15}$$

where σ_X^2 is the variance of the distribution.

The proof of the weak law (6.13) follows by applying the Chebyshev inequality (6.2) to M_n:

$$\text{Pr}\,[|M_n - m_X| \geq \epsilon] \leq \frac{\text{Var}(M_n)}{\epsilon^2} = \frac{\sigma_X^2}{n\epsilon^2}.$$

Clearly, for any chosen value of ϵ, as $n \to \infty$, the right side approachs zero, which proves the result.

Strong law of large numbers

The strong law of large numbers states that as $n \to \infty$, the sample mean M_n converges to the distribution mean m_X with probability one:

$$\text{Pr}\left[\lim_{n\to\infty} M_n = m_X\right] = 1. \tag{6.16}$$

A proof of the strong law will not be provided here. Note, however, that the weak law shows that the sample mean remains very close to the distribution mean but does not take into account the question of what actually happens to the sample mean as $n \to \infty$ which is considered by the strong law of large numbers. Figure 6.1 provides a comparison of the two modes of convergence.

6.2.2 Central limit theorem

The Central Limit Theorem is an extremely important result in probability theory. The essence of the Central Limit Theorem is that the distribution of the sum of a large number of independent random variables approaches a Gaussian. A statement of the Central Limit Theorem is presented here without proof.

Consider a sequence of independent and identically-distributed random variables X_1, X_2 ... X_n, each with mean m_X and variance σ_X^2. The sum $Y_n = X_1 + X_2 + ... + X_n$ has mean nm_X and variance $n\sigma_X^2$. Define a normalized random variable

$$Z_n = \frac{Y_n - nm_X}{\sigma_X\sqrt{n}},$$

which has mean 0 and variance 1. Its cumulative distribution function is

$$F_{Z_n}(z) = \text{Pr}[Z_n \leq z].$$

The CDF of the normalized random variable Z_n approaches that of a Gaussian random variable with zero mean and unit variance:

$$\lim_{n\to\infty} F_{Z_n}(z) = \Phi(z) = \frac{1}{\sqrt{2\pi}}\int_{-\infty}^{z} e^{-\frac{v^2}{2}}\,dv, \tag{6.17}$$

Since $\Phi(z)$ is the normalized Gaussian CDF discussed in Chapter 3, the sequence of normalized random variables Z_n *converges in distribution* to a Gaussian random variable.

The Central Limit Theorem holds for a sum of a large number of any sequence of IID random variables (having finite mean and finite variance) from any distribution. The theorem, in fact, is a result of convolving a large number of positive-valued functions and is used in a wide variety of applications. Extensions to the basic theorem are possible where various conditions stated above are relaxed.

Example 6.5:   Consider a sequence of 100 binary values (IID Bernoulli random variables) with distribution parameter $p = \Pr[1] = 0.6$. It is desired to determine (i) the probability of more than 65 1s in the sequence, and (ii) the probability that the number of 1s is between 50 and 70.

Let Y_{100} represent the number of 1s in the sequence and recall that Y_{100} is a binomial random variable; the PMF is given by equation 3.5 of Chapter 3. The probabilities can be found directly using the binomial PDF, for example

$$\Pr[Y_{100} > 65] = \sum_{k=66}^{100} \binom{100}{k} (0.6)^k (0.4)^{100-k}$$

This computation is lengthy however. It is simpler to observe that Y_{100} is the sum of 100 random variables X_i and apply the Central Limit Theorem.

The mean and the variance of each binary random variable are found to be $m_X = 0.6$ and $\sigma_X^2 = 0.24$ while the number of 1s in this sequence is given by the sum

$$Y_{100} = \sum_{i=1}^{100} X_i.$$

The mean of the sum is $100\, m_X = 60$ while the variance is $100\, \sigma_X^2 = 24$.

Since the normalized representation of Y_{100} is

$$Z_{100} = \frac{Y_{100} - 60}{\sqrt{24}},$$

we have

$$\Pr[Y_{100} > 65] = \Pr\left[Z_{100} > \frac{65 - 60}{\sqrt{24}}\right]$$

$$= Q\left(\frac{5}{\sqrt{24}}\right) \approx Q(1.02) \approx 0.154,$$

where the value for $Q(1.02)$ is obtained from the table provided on the inside cover of the text.

To determine the probability that Y_{100} is between 50 and 70, we write

$$\Pr[50 \leq Y_{100} \leq 70] = \Pr\left[\frac{50 - 60}{\sqrt{24}} \leq Z_{100} \leq \frac{70 - 60}{\sqrt{24}}\right]$$

$$\approx \Pr[-2.04 \leq Z_{100} \leq 2.04] = 1 - 2Q(2.04)$$

$$\approx 1 - 2 \times 0.021 = 0.958.$$

□

6.3 Estimation of Parameters

In order to apply probabilistic ideas in practice, it is necessary to have numerical values for the various parameters that describe the distribution. For example, for a Gaussian random variable, we need to know the mean and variance; for a binomial random variable we need to know the parameter p, and so on. In most cases, these parameters are not known *a priori* but must be *estimated* experimentally from observations of the random variable.

The *sample mean* has already been introduced as an estimate of the mean of a random variable. In the remainder of this chapter estimation procedures are developed for other parameters as well, and some basic theoretical results about the properties of these estimates are established.

The topic of estimation is treated only briefly here but the treatment is sufficient for our purposes. A more extensive treatment, also from an engineering point of view, can be found in a number of references (e.g., [4, 5, 6, 7]).

6.3.1 Estimates and properties

Let us first establish some notation. Consider a parameter θ pertaining to the distribution for some particular random variable X. As in Section 6.2.1 let $X_1, X_2, \ldots, X_n$ represent the sequence of outcomes resulting from n independent repetitions of the experiment that produces the random variable. We define a *statistic* or an *estimate* for the parameter θ by[2]

$$\Theta_n = \Theta(X_1, X_2, \ldots, X_n)$$

where Θ on the right represents a known function of random variables $X_1, X_2, \ldots, X_n$. Note that a capital letter is used for the estimate Θ_n since it depends on the random variables $X_1, X_2, \ldots, X_n$ and is thus a random variable itself. A subscript "n" is also attached to the estimate to indicate dependence on the number of observations n. Most authors also use a "hat" ($\hat{}$) over the symbol to indicate an estimate, but this additional notation is not necessary for most of the discussion in this chapter.

Three important properties of estimates are considered here:

1. An estimate Θ_n is *unbiased* if

$$\mathcal{E}\{\Theta_n\} = \theta$$

   where θ is the true value of the parameter. Otherwise the estimate is *biased* with bias $b(\theta) = \mathcal{E}\{\Theta_n\} - \theta$.

   An estimate is *asymptotically unbiased* if

$$\lim_{n \to \infty} \mathcal{E}\{\Theta_n\} = \theta$$

2. An estimate is said to be *efficient* with respect to another estimate if it has a lower variance. An estimate Θ_n is simply said to be efficient (without any qualification), if it is efficient with respect to Θ_{n-1} for all n.

3. An estimate Θ_n is *consistent* if

$$\lim_{n \to \infty} \Pr[|\Theta_n - \theta| < \epsilon] = 1$$

   for any arbitrarily small number ϵ. Thus the estimate is consistent if it converges in probability to the true value of the parameter θ.

---

[2] A *statistic* is any function of the random variables $X_1, X_2, \ldots, X_n$. An *estimate* is a statistic that is intended to approximate some particular parameter θ.

Consistency is obviously a desirable property.[3] It implies that the estimate becomes closer and closer to the true parameter as the number of observations, n, is increased. Observe that properies 1 and 2 together imply property 3, that is

> If Θ_n is unbiased and efficient, then Θ_n is a consistent estimate.

The proof of this statement follows from the Chebyshev inequality (6.2). Since Θ_n is unbiased, the mean is equal to θ. Equation 6.2 in this case becomes

$$\Pr[|\Theta_n - \theta| \geq \epsilon] \leq \frac{\text{Var}[\Theta_n]}{\epsilon^2}$$

Thus if the variance of Θ_n decreases with n, the probability that $|\Theta_n - \theta| \geq \epsilon$ approaches 0 as $n \to \infty$. In other words the probability that $|\Theta_n - \theta| < \epsilon$ approaches 1. This last property is illustrated in Fig. 6.2. As n becomes large, the PDF for Θ_n

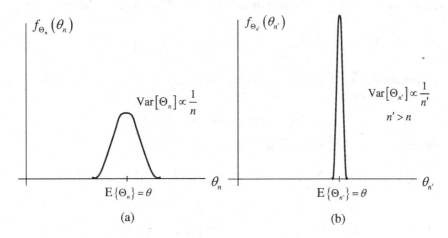

(a)   (b)

Figure 6.2 Illustration of consistency for an unbiased estimate whose variance decreases with n. (a) PDF of the estimate Θ_n. (b) PDF of the estimate $\Theta_{n'}$ with $n' > n$.

becomes more concentrated about the true parameter value θ. In the limit, the density takes on the character of an impulse located at θ.

The properties discussed in this section are illustrated below for some specific statistics.

6.3.2 Sample mean and variance

The sample mean M_n (already introduced) and the sample variance Σ_n^2 are statistics commonly used to estimate the distribution mean $m_X = E\{X\}$ and variance $\sigma_X^2 = E\{(X - m_X)^2\}$ of a random variable when those theoretical parameters are not known. Let us examine the properties of these particular statistics according to the definitions provided in the previous subsection.

---

[3] In general, an estimate is consistent if it converges in *some* sense to the true value of the parameter. Other possible (stronger) ways to define consistency are convergence with probability 1 or convergence in mean-square.

Sample mean

The sample mean M_n is defined by (6.12). It is shown in Section 6.2.1 that the mean of M_n is equal to the distribution mean m_X (see (6.14)). Therefore the sample mean is *unbiased*.

The variance of the sample mean is also derived in Section 6.2.1 (see (6.15)) and is given by σ_X^2/n. Since the variance decreases as a function of n, the estimate is *efficient*. Finally, since the estimate is unbiased and efficient, the sample mean is a *consistent estimate* for the mean.

Sample variance

. The sample variance is defined as

$$\Sigma_n^2 = \frac{(X_1 - M_n)^2 + \ldots + (X_n - M_n)^2}{n} = \frac{1}{n}\sum_{i=1}^{n}(X_i - M_n)^2 \qquad (6.18)$$

(Note that in estimating the variance, we use the sample mean M_n rather than the true mean m_X, since the latter is generally not known.) The mean of the sample variance is given by

$$\mathcal{E}\{\Sigma_n^2\} = \frac{1}{n}\sum_{i=1}^{n}\mathcal{E}\{(X_i - M_n)^2\}$$

This expectation is complicated by the fact that M_n, which appears on the right, is also a random variable. Let us first write

$$X_i - M_n = (X_i - m_X) - (M_n - m_X)$$

Substituting this in the previous equation yields

$$\mathcal{E}\{\Sigma_n^2\} = \frac{1}{n}\sum_{i=1}^{n}\mathcal{E}\{(X_i - m_X)^2\} + \frac{1}{n}\sum_{i=1}^{n}\mathcal{E}\{(M_n - m_X)^2\}$$

$$-\frac{2}{n}\mathcal{E}\left\{\sum_{i=1}^{n}(M_n - m_X)(X_i - m_X)\right\}$$

This equation can be simplified to

$$\mathcal{E}\{\Sigma_n^2\} = \sigma_X^2 + \mathcal{E}\{(M_n - m_X)^2\} - \frac{2}{n}\mathcal{E}\left\{(M_n - m_X)\sum_{i=1}^{n}(X_i - m_X)\right\}$$

where we have recognized that the first term is the distribution variance σ_X^2 and that the quantity $M_n - m_X$ does not depend on the summation variable i and can be removed from the sum.

Now observe that the last term can be written as:

$$\frac{2}{n}\mathcal{E}\left\{(M_n - m_X)\sum_{i=1}^{n}(X_i - m_X)\right\} = 2\mathcal{E}\left\{(M_n - m_X)\left(\frac{1}{n}\sum_{i=1}^{n}X_i - \frac{1}{n}\sum_{i=1}^{n}m_X\right)\right\}$$

$$= 2\mathcal{E}\{(M_n - m_X)(M_n - m_X)\} = 2\mathcal{E}\{(M_n - m_X)^2\}$$

Substituting this in the previous equation yields

$$\mathcal{E}\{\Sigma_n^2\} = \sigma_X^2 - \mathcal{E}\{(M_n - m_X)^2\}$$

Finally, by recognizing that $\mathcal{E}\{(M_n - m_X)^2\} = \text{Var}[M_n] = \sigma_X^2/n$ (see (6.15)), we arrive at the equation

$$\mathcal{E}\{\Sigma_n^2\} = \left(1 - \frac{1}{n}\right)\sigma_X^2,$$

This result indicates that the sample variance, although *not* unbiased, is nevertheless *asymptotically* unbiased since $\mathcal{E}\{\Sigma_n^2\} \to \sigma_X^2$ as $n \to \infty$.

An alternate form of the sample variance is sometimes used. This estimate is given by

$$(\Sigma_n^2)' = \frac{1}{n-1} \sum_{i=1}^{n} (X_i - M_n)^2 \qquad (6.19)$$

This differs from the previous definition (6.18) by a factor of $n/(n-1)$ and so is *unbiased*. Both forms of sample variance can usually be found in statistical analysis programs and on hand-held calculators.

The variance of the sample variance can also be computed and shown to approach 0 as n increases; thus the sample variance is a consistent estimate. The proof can be found in more advanced sources (e.g., [6, 7]).

6.4 Maximum Likelihood Estimation

Having now been introduced to at least a couple of different examples of estimates, you may wonder how the formulas for these estimates are derived. One answer to this question is through the procedure known as *maximum likelihood* (ML) estimation.

Consider the following hypothetical situation. We have a random variable X with a density function similar to one of those depicted in Fig. 6.3, but we do not know

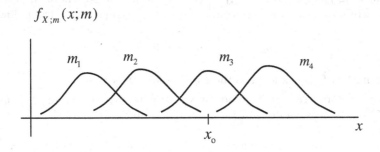

Figure 6.3 Choices in estimation of the mean.

the mean. Further, we have an observation of the random variable X; let's call that observation x_o. On the basis of that single observation, we want to estimate the mean of the density. Referring to Fig. 6.3, we could argue that the value selected for the mean should be m_3, because it makes the given observation (x_o) the *most likely* observation. After all, why would we choose m_1 or m_4? These choices would imply that the given observation x_o has low probability.

This reasoning can be extended to what we'll call the *principle of maximum likelihood*. This principle states that in choosing an estimate for a parameter, we should choose the value (of the parameter) that makes the observed data the *most likely* result. Let us formulate that principle mathematically.

First, in most practical estimation problems there is more than one observation of the data. Let us denote the various data samples or observations by $X_1, X_2, \ldots, X_n$, as in previous sections. For notational convenience, this set of observations is also represented by a random vector $\boldsymbol{X}$. If θ is the parameter that describes the density function for X, then the joint density function for all the observations is denoted by $f_{\boldsymbol{X};\theta}(\boldsymbol{x};\theta)$. The maximum likelihood (ML) estimate for the parameter θ is then defined by

$$\Theta_{ml}(\boldsymbol{X}) = \underset{\theta}{\mathrm{argmax}} \ \ f_{\boldsymbol{X};\theta}(\boldsymbol{X};\theta) \qquad (6.20)$$

In other words, it is the value of θ that maximizes $f_{\boldsymbol{X};\theta}$, given the observations. In line with the discussion in Section 6.3, we use capital letters for both Θ_{ml} and $\boldsymbol{X}$, since they are random variables. We will illustrate how this definition can be used to derive explicit formulas for estimates of particular parameters in a moment; however, let us first make some remarks about the procedure in general:

1. In maximum likelihood estimation the quantity $f_{\boldsymbol{x};\theta}(\boldsymbol{x};\theta)$ is considered to be a function of the parameter θ for fixed (known) values of the observations $\boldsymbol{X}$. When viewed *as a function of* θ, $f_{\boldsymbol{x};\theta}$ is called the "*likelihood function*" rather than the PDF.

2. The maximum likelihood estimate, resulting from the solution of (6.20), is the estimate with smallest variance (called a "minimum-variance" estimate). Thus the maximum likelihood estimate is an optimal estimate in that it is the *most efficient* estimate for the parameter θ.

3. Since the logarithm is a monotonic transformation, maximizing $\ln f_{\boldsymbol{x};\theta}$ is equivalent to maximizing $f_{\boldsymbol{x};\theta}$. Because many distributions involve powers of the variables, it is frequently most convenient to compute the ML estimate from the following:

$$\Theta_{ml}(\boldsymbol{x}) = \underset{\theta}{\mathrm{argmax}} \ \ \ln f_{\boldsymbol{x};\theta}(\boldsymbol{x};\theta) \qquad (6.21)$$

The quantity $\ln f_{\boldsymbol{x};\theta}$ is known as the *log likelihood functiion*.

With the foregoing results in mind, let us explore the procedure of computing a maximum likelihood estimate through a set of examples.

Example 6.6:   The time X between requests for service on a file server is a random variable with the exponential PDF

$$f_X(x) = \begin{cases} \lambda e^{-\lambda x} & x \geq 0 \\ 0 & \text{otherwise} \end{cases}$$

A total of $n+1$ requests are observed with independent interarrival times $X_1, X_2, \ldots X_n$. It is desired to find a maximum likelihood estimate for the arrival rate parameter λ.

Since the interarrival times are independent, the likelihood function for this problem is

$$f_{\boldsymbol{X};\lambda}(\boldsymbol{X};\lambda) = \prod_{i=1}^{n} \lambda e^{-\lambda X_i} = \lambda^n e^{-\lambda \sum_{i=1}^{n} X_i}$$

Since the form involves some exponential terms, it is easier to deal with the log likelihood function, which is given by

$$\ln f_{\boldsymbol{X};\lambda}(\boldsymbol{X};\lambda) = \ln \ \lambda^n e^{-\lambda \sum_{i=1}^{n} X_i} = n \ln \lambda - \lambda \sum_{i=1}^{n} x_i$$

To find the maximum likelihood estimate for λ, let us take the derivative of the log likelihood function and set it to 0. This yields

$$\frac{d}{d\lambda}\left[n \ln \lambda - \lambda \sum_{i=1}^{n} X_i \right] = \frac{n}{\lambda} - \sum_{i=1}^{n} X_i = 0$$

Solving this equation for λ produces the maximum likelihood estimate. Since the estimate is actually a random variable, we use a capital letter and write it as

$$\Lambda_{ml} = \frac{1}{\frac{1}{n}\sum_{i=1}^{n} X_i}$$

The estimate is seen to be the reciprocal of the average interarrival time.

□

When the distribution pertains to a discrete random variable, the procedure is similar. The likelihood function is derived from the PMF, however, rather than the PDF.

Example 6.7: The number of errors K occurring in a string of n binary bits is a binomial random variable with PMF

$$f_K[k] = \binom{n}{k} p^k (1-p)^{(n-k)} \qquad 0 \leq k \leq n$$

where the parameter p is the probability of a single bit error. Given an observation of k errors, what is the maximum likelihood estimate for the bit error parameter p?

The likelihood function for this problem is given by

$$f_{K;p}[K;p] = \binom{n}{K} p^K (1-p)^{(n-K)}$$

While the log likelihood function could be used here (see Prob. 6.23), it is easy enough in this case to deal with the likelihood function directly.

Taking the derivative with respect to the parameter and setting it to 0 yields

$$\frac{df_{K;p}}{dp} = \binom{n}{K}\left[Kp^{K-1}(1-p)^{n-K} - p^K(n-K)(1-p)^{n-K-1} \right]$$

$$= \binom{n}{K} p^{K-1}(1-p)^{n-K-1}\left[K(1-p) - p(n-K) \right] = 0$$

Solving for p, we find the maximum likelihood estimate to be[4]

$$P_{ml} = \frac{K}{n}$$

Now, recalling that the mean of the binomial distribution is $\mathcal{E}\{K\} = np$, we can write

$$\mathcal{E}\{P_{ml}\} = \frac{\mathcal{E}\{K\}}{n} = p$$

thus demonstrating that this estimate is *unbiased*.

□

---

[4] A capital letter is used for the estimate.

The following example is a little different from the previous two, and is intended to emphasize the distinction between the PDF and the likelihood function.

Example 6.8:   A random variable X has the uniform density function

$$f_X(x) = \begin{cases} 1/a & 0 \le x \le a \\ 0 & \text{otherwise} \end{cases}$$

Given n independent observations $X_1, X_2, \ldots X_n$, it is desired to find a maximum likelihood estimate for the parameter a.

This problem is tricky, because we have to think carefully about the meaning of the likelihood function. We cannot proceed blindly to take derivatives as in the previous two examples because the maximum of the likelihood function actually occurs at one of the boundaries of the region where the derivative is *not* zero.

To proceed with this example, consider the simplest case, $n = 1$. The likelihood function is obtained and plotted by writing the above equation for the density as a function of the variable a while the observation X is held fixed:

$$f_{X;a}(X;a) = \begin{cases} 1/a & a \ge X \\ 0 & \text{otherwise} \end{cases}$$

From the picture, it can be seen that the maximum of the likelihood function occurs at the point $a = X$; hence the ML estimate for $n = 1$ is given by $A_{ml} = X$.

Next consider $n = 2$. Since the observations are independent, the joint PDF is given by

$$f_{X_1 X_2}(x_1, x_2) = \begin{cases} 1/a^2 & 0 \le x_1 \le a, 0 \le x_2 \le a \\ 0 & \text{otherwise} \end{cases}$$

Writing and plotting this as a function of a, for fixed values of the observations, we obtain the likelihood function:

$$f_{X;a}(X;a) = \begin{cases} 1/a^2 & a \ge X_1, a \ge X_2 \\ 0 & \text{otherwise} \end{cases}$$

Notice that since *both* of the inequalities $a \ge X_1$ and $a \ge X_2$ must be satisfied for the likelihood function to be nonzero, the maximum of the function occurs at $A_{ml} = \max\{X_1, X_2\}$.

Continuing to reason in this way, we find that for an arbitrary number n of samples, the likelihood function is equal to $1/a^n$ for $a \geq \max\{X_1, X_2, \ldots, X_n\}$ and 0 otherwise. The peak of this function and thus the ML estimate occurs at

$$A_{ml} = \max\{X_1, X_2, \ldots, X_n\}$$

A little thought and reflection on this example should convince you that this is the intuitively correct answer.

□

A number of other cases arise where a maximum likelihood estimate is possible. For example, the sample mean, introduced in Section 6.2.1, is the ML estimate for the mean of the Gaussian distribution (and some other distributions as well). The proof is straightforward, so it is left to the problems at the end of this chapter.

6.5 Point Estimates and Confidence Intervals

The estimates that have been considered so far are known as *point estimates*. The parameter of interest has an unknown value and the data is used to estimate that value. This is illustrated in Fig. 6.4(a) where the estimate is denoted by Θ_n. Of course,

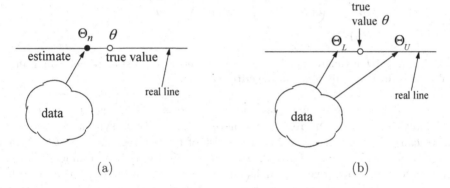

(a)	(b)

Figure 6.4 Estimation of a parameter. (a) Point estimation. (b) Interval estimation.

since the true value and the estimated value are real numbers, the estimate will never be *exactly* equal to the true value in practice. If the experiment were repeated some large number of times, however, we would hope that the resulting estimates occur on both sides of the true value and that the estimates would be tightly clustered. These results are expressed in terms of properties of the estimates discussed in Section 6.3.1.

Another way to approach the estimation problem is illustrated in Fig. 6.4(b). We consider an interval (Θ_L, Θ_U) on the real line such that the true parameter θ falls within that interval with certain probability. This statement is written as

$$\Pr[\Theta_L < \theta < \Theta_U] = 1 - \alpha$$

where α is a small number between 0 and 1.

The interval (Θ_L, Θ_U) is called a *confidence interval* [4, 8, 9] and Θ_L and Θ_U are called the lower and upper confidence limits. Like Θ_n, these are random variables and functions of the data. The variable α is called the *significance level* and the quantity $(1 - \alpha) \times 100\%$ is known as the *confidence level*. For example, if $\alpha = 0.05$ then (Θ_L, Θ_U) is called a 95% confidence interval.

The point estimate Θ_n and the confidence interval (Θ_L, Θ_U) are usually reported together. For example, we might say, "The signal is estimated to be 4.9 volts with a 95% confidence interval of $(4.7, 5.1)$." In graphical representation, the point estimate is plotted with a bar overlaid that represents the confidence interval (see Fig. 6.5).

Now consider the construction of a confidence interval more closely. Assume we have a point estimate Θ_n which is *unbiased* and draw the PDF for Θ_n as shown in Fig. 6.5. Note that since the estimate is unbiased, the mean of the PDF is the true parameter θ. Let us now choose a region that has probability α where α is the significance level

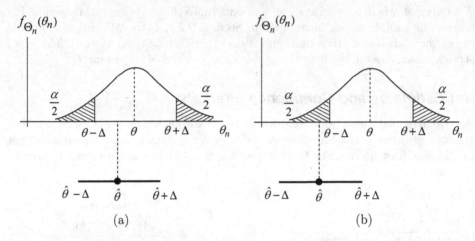

Figure 6.5 A point estimate and its confidence interval. (a) Estimate not falling in the critical region. (b) Estimate falling in the critical region.

of the test. This region is known as the *critical region*. Although there are many possible ways to choose this region, whenever the density is symmetric about the mean we can take the critical region to consist of two subregions $(-\infty, \theta - \Delta)$ and $(\theta + \Delta, \infty)$ each with probability $\alpha/2$. This procedure results in a confidence interval of smallest length. Now, given a particular numerical value for the estimate, call it $\hat{\theta}$, we construct the confidence interval as $(\hat{\theta} - \Delta, \hat{\theta} + \Delta)$. This confidence interval is depicted below the density function in Fig. 6.5 for two cases corresponding to whether or not the estimate falls in the critical region. Observe from Fig. 6.5(a) that when $\hat{\theta}$ is *not* in the critical region, the confidence interval includes the true parameter θ. This happens with probability $1 - \alpha$ (0.95 if $\alpha = 0.05$). On the other hand, if $\hat{\theta}$ does fall in the critical region (Fig. 6.5(b)), the confidence interval does *not* contain the true parameter. This happens with probability α. As an example, when a 95% confidence interval is constructed as shown, the true parameter θ will fall within that confidence interval 95% of the time.

Frequently the estimate Θ_n involves a sum of random variables so the density function f_{Θ_n} is Gaussian. In this case the parameter Δ can be expressed in standard deviations, $\Delta = z\sigma$ and the Q function can be used to find the value of the normalized variable z that yields the desired probability $\alpha/2$. The short table in Fig. 6.6 can be used to compute confidence intervals for Gaussian parameters. This is illustrated in the following example.

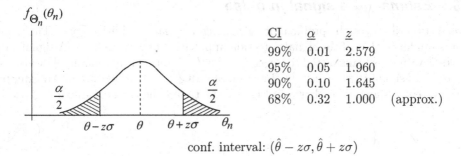

conf. interval: $(\hat{\theta} - z\sigma, \hat{\theta} + z\sigma)$

Figure 6.6 Confidence interval (CI) for estimate of a Gaussian random variable.

Example 6.9: One hundred observations of a certain random variable X with variance $\sigma_X^2 = 24$ are taken. The sample mean based on 100 samples is computed to be $M_{100} = 2.75$. Find a 95% confidence interval for this estimate.

The variance (6.15) and standard deviation of the sample mean are given by

$$\sigma_{100}^2 = 24/100 \quad \text{and} \quad \sigma_{100} = \sqrt{24/100} \approx 0.490$$

From the table in Fig. 6.6 the value of z that defines the critical region is given by $z = 1.960$. The figure here illustrates the computation of the upper confidence limit.

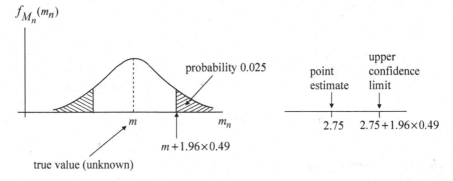

The diagram on the left shows how the critical region is chosen. The diagram on the right shows how the upper confidence limit is actually computed. (The lower confidence limit is found in a similar manner.) The complete 95% confidence interval is given by

$$(2.75 - 1.96 \cdot 0.490 \,,\, 2.75 + 1.96 \cdot 0.490) = (1.79, 3.71)$$

□

In practice, the variance σ_X^2 in this example may not be known. In this case the same procedure would be used with an *estimate* for the variance.

6.6 Application to Signal Estimation

We have seen in earlier chapters that many problems in communication and related areas involve probability in the form of hypothesis testing. This is the basis of most signal detection and classification schemes. A comparably large number of applications however, involve *estimation* of signals and other parameters. That is the topic of the present section.

6.6.1 *Estimating a signal in noise*

This section discusses the problem of estimating a signal in additive noise. The situation is similar to that of the signal detection application of Section 3.8.1. A signal with constant value s is transmitted to a receiver. There is noise in the channel, however, so the *received* signal is not constant, but random, due to the added noise. An example of the received signal X is shown in Fig. 6.7. At time t the received signal is given by

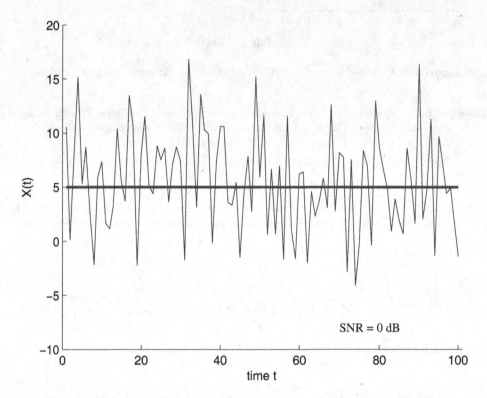

Figure 6.7 100 samples of a received signal with additive noise. Transmitted signal is shown by the dashed line ($s = 5$).

$$X_i = s + N_i \tag{6.22}$$

where N_i denotes the ith sample of the noise.

Assume that all of the random variables N_i are IID with mean 0 and variance σ^2. (In this context, σ^2 is the average noise *power*.) Then the observations X_i are IID random variables with mean s and variance σ^2. The signal-to-noise ratio in decibels is defined as

$$\text{SNR} = 10 \log_{10} \frac{s^2}{\sigma^2} = 20 \log_{10} \frac{s}{\sigma} \quad \text{(dB)} \tag{6.23}$$

It is desired to estimate the value of s using n samples of the received data, taking as many samples as necessary to get a good estimate. We will assume at first that the signal-to-noise ratio is equal to 0 dB. This implies that $\sigma = s$.

Let us form an estimate of the transmitted signal by averaging the samples of the received signal:

$$S_n = \frac{1}{n} \sum_{i=1}^{n} X_i \tag{6.24}$$

Notice that the estimate S_n is the *sample mean* and recall that the sample mean is unbiased

$$\mathcal{E}\{S_n\} = s \tag{6.25}$$

and the variance is given by

$$\sigma_S^2 = \frac{\text{Var}\,[X_i]}{n} = \frac{\sigma^2}{n} \tag{6.26}$$

We can interpret these last two equations with the following computer experiment. We compute the estimate (6.24) using $n = 100$ samples and repeat this computation 100 times using a different set of samples each time. The resulting 100 estimates are plotted in Fig. 6.8. Note that the estimates are clustered around the horizontal line

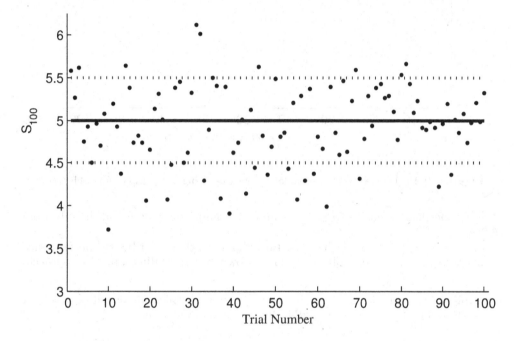

Figure 6.8 Plot of 100 trials in estimating a signal using 100 samples for each trial.

which represents the true value of the signal, $s = 5$. The standard deviation of the estimates as computed from (6.26) is $\sigma/\sqrt{n} = 5/10 = 0.5$. This is shown with broken lines in the figure.

To show the effect of increasing the number of samples in the estimate we repeat the experiment using $n = 400$. The results are shown in Fig. 6.9. This time the estimates are more tightly clustered about the true value $s = 5$ and the standard deviation is reduced to half of that in the previous figure ($\sigma/\sqrt{n} = 5/20 = 0.25$). The unbiased property of the estimate (6.24) is illustrated by the way samples cluster about the mean. The dependence of the variance on n (6.26) is illustrated by comparing Figures 6.8 and 6.9.

6.6.2 Choosing the number of samples

The previous section illustrates the effect of the number of samples on the variance of the estimate. Therefore if our criterion is to reduce the variance of the estimate to

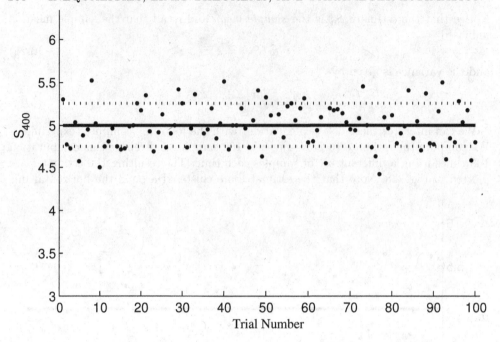

Figure 6.9 Plot of 100 trials in estimating a signal using 400 samples for each trial.

some prespecified value, the required number of samples can be found directly from (6.26).

A more quantitative approach can be taken however, that describes the performance of the estimate in terms of probability. The following example illustrates the approach.

Example 6.10: Consider the signal estimation problem described above and suppose that the estimate is desired to be within 10% of the true value, i.e.,

$$|S_n - s| > 0.1s$$

Since this is a random event we need to assign some probability. Let us require that $\Pr[|S_n - s| > 0.1s] = 0.05$. Since the sample mean involves a sum of IID random variables, it is assumed by the Central Limit Theorem that the distribution is Gaussian with mean and variance (6.25) and (6.26). The sketch below depicts the situation, where the two regions in the tails have equal probability of 0.025.

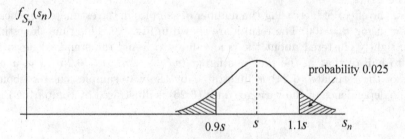

Using the Q function to evaluate probability we can write

$$Q\left(\frac{1.1s - s}{\sigma/\sqrt{n}}\right) = Q\left(\frac{0.1s\sqrt{n}}{\sigma}\right) = 0.025$$

Using MATLAB, or the table on the inside cover of this book, we find that the Q function is equal to 0.025 when its argument is approximately 1.96. Therefore, we have

$$\frac{0.1s\sqrt{n}}{\sigma} \approx 1.96$$

Since we have specified a 0 dB SNR, we have $s/\sigma = 1$ and therefore $n = (19.6)^2 \approx 384$ (samples).

□

The result of this example can be generalized somewhat. If the original SNR is something other than 0 dB then the ratio s/σ in the last equation of the example changes correspondingly. Then by taking logarithms and using the definition (6.23) for SNR, this last equation can be put in the form

$$\log_{10} n + (1/10)\, \text{SNR} = 2.443$$

This function appears as a straight line when plotted in semilog form, as shown in the graph below. Note that when the SNR is equal to approximately 24 dB, n is equal

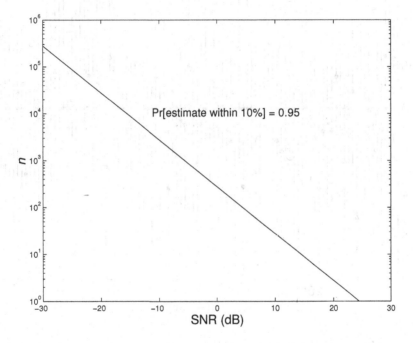

Figure 6.10 Plot of number of samples needed to estimate signal as a function of SNR.

to 1. This means that the signal with noise stays within ±10% of the true value with probability 0.95, so no averaging is required.[5] On the other hand, if the original SNR is very low, for example −20 dB, a large number of samples (approximately 28,000) is required to achieve the same performance. By using a sufficiently large value of n, however, the signal can be estimated to within ±10% for any SNR. This principle was used by naval facilities during the cold war to track submarines at very long distances and very low corresponding SNR values.

---

[5] Strictly speaking, our analysis is only approximate for $n = 1$ unless the noise is Gaussian, because the Central Limit Theorem cannot be invoked for $n = 1$. In many cases of interest, however, the noise *is* Gaussian.

6.6.3 *Applying confidence intervals*

Confidence limits can be applied to the signal estimates discussed in this section. This provides another quantitative way to report the estimate. Let us choose a confidence level of 95%; then following the procedure of Example 6.9 we compute the confidence interval and report it along with the point estimate. Let's consider the case where $n = 100$ samples and recall from the discussion of Fig. 6.8 that the standard deviation of the estimate is $\sigma_S = 0.5$. Then to find the 95% confidence limits, the quantity $1.96\,\sigma_S$ is added to and subtracted from the point estimate.

Figure 6.11 shows the 100 point estimates previously computed in Fig. 6.8 with

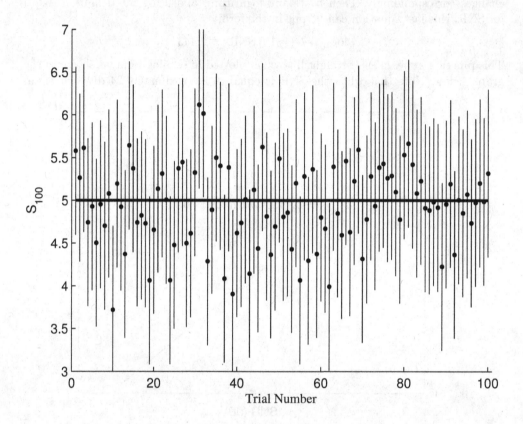

Figure 6.11 Plot of 100 estimates of a signal with confidence intervals overlaid.

confidence intervals overlaid on each estimate. The figure shows that 95% of the computed confidence intervals cover the horizontal line which represents the true signal value. Five intervals or 5% do not contain the true signal value.

As a final topic, let us compare the computation of confidence intervals to the procedure described in Section 6.6.2. Figure 6.12 shows the density function for S_n and the selection of a 95% confidence interval. (By now the procedure should be quite familiar.) Let us write the upper limit as some fraction β of the mean $(0 < \beta < 1)$, that is,

$$\beta s = 1.96\,\sigma_S \quad \text{or} \quad \beta = 1.96\,\frac{\sigma}{\sqrt{n}} \cdot \frac{1}{s}$$

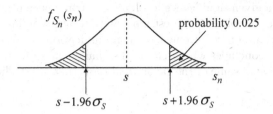

Figure 6.12 Choosing 95% confidence limits for the signal estimate.

For an SNR of 0 dB and $n = 100$ the ratio σ/s is equal to 1 and we find

$$\beta = \frac{1.96}{10} = 0.196$$

or about 20%. To reduce this number to about 10% we would need to take about 4 times as many samples. This agrees with the result of example 6.10 where it is concluded that 384 samples are needed to bring the estimate within 10% of the true signal value. In fact, if we take $n = 384$ in the above equation we find β is almost exactly equal to $1/10$.

6.7 *Summary*

The topics covered in this chapter are useful in the application of probability to practical engineering problems. The Markov and Chebyshev inequalities provide bounds for probabilities involving random variables when only the mean and variance of the distribution are known. While the bounds are typically not tight, they are sufficiently general to state or prove results about random variables that are independent of the actual distribution.

Several important results in the application of probability involve the convergence of a sequence of random variables as the number of terms becomes large. Some types of convergence that are commonly studied are convergence with probability 1, mean-square convergence, convergence in probability, and convergence in distribution. Two important results are discussed that depend on some of these modes of convergence. The law of large numbers states that an average of a large number of IID random variables (also known as the "sample mean") converges to the mean of these random variables. The Central Limit Theorem states that the distribution of a sum of a large number of random variables converges to the Gaussian distribution. This highlights the Gaussian distribution as a uniquely important distribution because many random variables of practical interest are defined as a sum of other elementary random variables.

In applying probability to real world problems, the numerical values of parameters in the distributions must usually be *estimated* from samples of observed or measured data. The properties of such estimates such as "bias" and "efficiency" are important characteristics. Moreover, convergence to the true parameter values as the number of observations becomes large (known as "consistency" of the estimate) is especially important. The concept of a *maximum likelihood* (ML) estimate is also introduced in this chapter and cited as the most efficient estimate when it exists. Finally the topic of confidence intervals is introduced as a way to report the trust we may have in a given point estimate. For a given significance level, a small confidence interval indicates that the estimate is more reliable than an estimate with a large confidence interval.

Estimation is a common requirement in communications, control and signal processing. The chapter concludes with a classical signal estimation problem that demonstrates the application of some of the theoretical tools that are studied in this chapter.

References

[1] Geoffrey Grimmitt and David Stirzaker. *Probability and Random Processes*. Oxford University Press, New York, third edition, 2001.

[2] O. Hölder. Über einen mittelwertsatz. *Göttingen Nachr.*, 38(7), 1889.

[3] H. Minkowski. *Geometrie der Zahlen*, volume 1. Leipzig, Germany, 1896.

[4] William Navidi. *Statistics for Engineers and Scientists*. McGraw-Hill, New York, 2006.

[5] Harry L. Van Trees. *Detection, Estimation, and Modulation Theory - Part I*. John Wiley & Sons, New York, 1968.

[6] Athanasios Papoulis and S. Unnikrishna Pillai. *Probability, Random Variables, and Stochastic Processes*. McGraw-Hill, New York, fourth edition, 2002.

[7] Charles W. Therrien. *Discrete Random Signals and Statistical Signal Processing*. Prentice Hall, Inc., Upper Saddle River, New Jersey, 1992.

[8] Athanasios Papoulis. *Probability and Statistics*. Prentice Hall, Inc., Upper Saddle River, New Jersey, 1990.

[9] Athanasios Papoulis. *Probability, Random Variables, and Stochastic Processes*. McGraw-Hill, New York, third edition, 1991.

[10] Alberto Leon-Garcia. *Probability, Statistics, and Random Processes for Electrical Engineering*. Pearson, New York, third edition, 2008.

Problems

Inequalities[6]

6.1 The magnitude of the voltage V across a component in an electronic circuit has a mean value of 0.45 volts. Given only this information, find a bound on the probability that $V \geq 1.35$.

6.2 The average time to print a job in the computer laboratory is known to be 30 seconds. The standard deviation of the printing time is 4 seconds. Compute a bound on the probability that the printing time is between 20 and 40 seconds.

6.3 The waiting time K until the first error in a binary sequence is described by the (type 1) geometric PDF. Let the probability of an error be $p = 1/8$.

   (a) Compute the mean and variance of K.

   (b) Apply the Markov inequality to bound the probability of the event $K \geq 16$.

   (c) Use the Chebyshev inequality to bound the probability of this event.

   (d) Use the one-sided Chebyshev inequality to bound the probability.

   (e) Compute the actual probability of the event $K \geq 16$.

---

[6] For the following problems, you may use the formulas in the front of the book for mean and variance of various distributions.

6.4 Given that X is an exponential random variable with parameter $\lambda = 0.8$, use the Markov inequality to compute the bound on $\Pr[X \geq 2.5]$.

6.5 Let X be an exponential random variable with parameter λ.

(a) Apply the Markov inequality to bound $\Pr[X \geq 2/\lambda]$.
(b) Use the Chebyshev inequality to compute the bound.
(c) Use the one-sided Chebyshev inequality to compute the bound.
(d) Compute the actual probability of this event.

6.6 Let X be a random variable with mean 0 and variance σ_X^2. It is known that if X is Gaussian, then
$$\Pr[|X| > 2\sigma_X] \approx 0.0455$$

(a) Compute the Chebyshev bound on this probability.
(b) Compute the probability of this event if X has the following distributions:

(i) Laplace.
(ii) Uniform.

6.7 The mean and standard deviation of a random variable X are given to be 1.6 and 1.2, respectively. Apply the one-sided Chebychev inequality to determine the upper bound for $\Pr[X \geq 3]$.

6.8 Derive the condition (6.5) ($|\rho_{12}| \leq 1$) by applying (5.44) of Chapter 5 to (6.7) and simplifying.

Convergence and limit theorems

6.9 A set of speech packets are queued up for transmission in a network. If there are twelve packets in the queue, the total delay D until transmission of the twelfth packet is given by the sum
$$D = T_1 + T_2 + \cdots + T_{12}$$
where the T_i are IID *exponential* random variables with mean $\mathcal{E}\{T_i\} = 8$ ms.

(a) What is the variance $\text{Var}[T_i]$? (You may refer to the formulas on the inside cover of the text.)
(b) Find the mean and variance of the total delay D.
(c) Speech packets will be lost if the total delay D exceeds 120 ms. Because D is the sum of IID exponential random variables, D is a 12-Erlang random variable. The CDF of this random variable is given by [10, p. 172]

$$F_D(d) = 1 - \sum_{k=0}^{11} \frac{(\lambda d)^k}{k!} e^{-\lambda d}$$

where λ is the parameter describing the exponential random variables T_i. Using this result, compute the probability that packets are lost, that is, $\Pr[D > 120]$.

(d) By applying the Central Limit Theorem and using the Gaussian distribution, compute the approximate value of $\Pr[D > 120]$. Compare this to the exact result computed in part (c).

(e) Finally evaluate the Markov bound for $\Pr[D > 120]$ and compare it to the two previous results.

Note: You may use MATLAB or some other program to evaluate the probabilities in parts (c) and (d).

6.10 A simple way to reduce noise is to perform what is called a moving average on a signal. Mathematically, the moving average operation can be written as

$$Y_n = \frac{1}{L} \sum_{k=n-L+1}^{n} X_k$$

where X_k are samples of the signal assumed to be IID random variables with mean 2 and variance 3.

(a) Find the mean and the variance of Y_n for $L = 12$.

(b) Based on the Central Limit Theorem, write an expression for the PDF of Y.

(c) How large does L have to be for the variance of Y to be less than 0.1?

(d) For the value of L obtained in (c), what is $\Pr[|Y - m_Y| > 0.1]$?

6.11 In a communication receiver, IID signal samples (random variables) X_i collected at a 25-element antenna array are linearly combined with equal weights, i.e.,

$$Y = \sum_{i=1}^{25} X_i.$$

The signal sample X_i collected at the ith antenna element is uniformly distributed over the interval $[-1, 1]$. Consider that the Central Limit Theorem applies. Calculate the probability $\Pr[Y \geq 0]$.

6.12 The lifetime of a power transistor operating under moderate radiation conditions is known to be an exponential random variable with an average lifetime of $\lambda^{-1} = 36$ months. Given 16 such transistors, use the Central Limit Theorem to determine the probability that the sum of their lifetimes is 50 years or less.

6.13 With all the handouts in this course, your binder fills up as an exponential random variable on average in a week, i.e., $\lambda^{-1} = 1$ week. What is the minimum number of binders you should buy at the beginning of an 11-week quarter, so the probability that you will not run out of binders during the quarter is 0.9? Use the Central Limit Theorem.

6.14 In a sequence of 800 binary values received at the output of a channel, 1s and 0s are equally likely. Assume that these are IID random variables. Calculate the probability that the number of 1s is between 300 and 500. What is the probability that the number of 1s is between 400 and 440?

6.15 The probability of a bit error in a binary channel is 0.1. Treat each transmission as a Bernoulli random variable; assume that successive transmissions are IID. If you transmit 1000 bits, what is the probability that 115 or fewer errors occur in the channel?

6.16 The lifetime of a light bulb is known to be an exponential random variable with average lifetime of $\lambda^{-1} = 12$ months. Given that eight such light bulbs are used sequentially, calculate the following probabilities.

(a) Treating the problem as an m-Erlang, what is the probability that their combined lifetime is less than 3 years?

(b) Using the Central Limit Theorem, determine the probability that their combined lifetime is less than 3 years.

Estimation of parameters

6.17 A random variable Θ is defined by

$$\Theta_n = \sum_{i=1}^{n} \Phi_i$$

where the Φ_i are independent random variables uniformly distributed from $-\pi$ to π.

(a) What are the mean and variance of Θ_n as a function of n?

(b) What is the PDF of Θ_n for large values of n?

6.18 Consider the two estimates Σ_n^2 and $\Sigma_n^{2'}$ described in Section 6.3.2.

(a) By expressing the sample variance Σ_n^2 in terms of $\Sigma_n^{2'}$, show that $\Sigma_n^{2'}$ is unbiased.

(b) Which of these two estimates is more *efficient*?

6.19 Let X be a Bernoulli random variable taking on values of 0 and 1 with probability p and $1 - p$. Given n independent samples $X_1 \ X_2 \ \cdots X_n$

(a) Suggest an estimate for the parameter p.

(b) Find the mean and variance of this estimate.

Maximum likelihood estimation

6.20 A random variable X is described by a Gaussian density function with unknown mean μ and variance σ_o^2. (You can assume that σ_o^2 is a *known* parameter.)

$$f_X(x) = \frac{1}{\sqrt{2\pi\sigma_o^2}} e^{-(x-\mu)^2/2\sigma_o^2}$$

Given n independent realizations of the random variable $X_1, X_2, \ldots X_n$, find the maximum likelihood estimate for the mean μ.

6.21 A random variable X is described by the Gaussian density function with mean 0 and unknown variance σ^2.

$$f_X(x) = \frac{1}{\sqrt{2\pi\sigma^2}} e^{-x^2/2\sigma^2}$$

Given n independent realizations of the random variable $X_1, X_2, \ldots X_n$, find the maximum likelihood estimate for the variance σ^2.

6.22 The mean of the exponential density function is given by $\mu = 1/\lambda$; therefore the PDF can be written as

$$f_X(x) = \begin{cases} \frac{1}{\mu} e^{-x/\mu} & x \geq 0 \\ 0 & \text{otherwise} \end{cases}$$

Using this parametric form of the density, and assuming n independent observations $X_1, X_2, \ldots X_n$, derive the ML estimate for μ.

6.23 Repeat Example 6.7, i.e., find the maximum likelihood estimate, using the *log likelihood function*.

Confidence Intervals

6.24 Five hundred observations of a random variable X with variance $\sigma_X^2 = 25$ are taken. The sample mean based on 500 samples is computed to be $M_{500} = 3.25$. Find 95% and 98% confidence intervals for this estimate.

Application to signal estimation

6.25 Plot the curve of n vs. SNR in Example 6.10 if it is required that:

  (a) the probability that the estimate is within 10% of the true value is 0.98.

  (b) the probability that the estimate is within 5% of the true value is 0.95.

  (c) the probability that the estimate is within 5% of the true value is 0.98.

6.26 The plot in Example 6.10 shows the value of n needed to ensure $\Pr[|S_n - s| \leq 0.1s]$ is equal to 0.95. Use the Chebyshev inequality to derive a lower bound for n as a function of SNR and compare it to the curve in Example 6.10. (The lower bound means that for *all* distributions of the noise, n must be greater than the value of the bound in order to satisfy the probability 0.95 requirement.)

Computer Projects

▸ **Project 6.1**

In this project you will empirically study the sums of random variables and demonstrate the Central Limit Theorem. Specifically, you will write computer code to generate Gaussian random variables based on the Central Limit Theorem using uniform and exponential random variables. You will then plot the density (distribution) function of these Gaussian random variables. Also, by successively convolving the Bernoulli and exponential density functions, you will observe that the resulting density function has the shape of a Gaussian density.

1. Generate six sequences of uniform random variables (length $\approx$ 10,000) in the range $[-0.5, 0.5]$. Each element in the ith sequence can be considered to be one realization of the random variable X_i for $i = 1, 2, \ldots, 6$. Note that these six uniform random variables are independent and identically distributed. Compute new random variables as follows:

$$Y_1 = \sum_{i=1}^{6} X_i$$

$$Y_2 = X_1 + 0.5X_2 + 0.8X_3 + 1.8X_4 + 0.3X_5 + 0.5X_6$$

This results in two new sequences, the elements of which can be thought of as $\approx$ 10,000 separate realizations of the random variables Y_1 and Y_2.

  (a) Experimentally determine the PDF and CDF of Y_1 and Y_2 by developing a normalized histogram and integrating. (It is recommended that you use about 20 to 30 intervals or "bins" for this work.)

  (b) Estimate the mean and variance of Y_1 and Y_2 and compare the estimated values to those from theoretical calculations.

  (c) Plot the estimated PDF and CDF of Y_1 and Y_2 and compare their shapes to those of the Gaussian. Do this by plotting a Gaussian curve of the correct mean and variance on top of your PDF result.

2. Generate 20 sequences of IID exponential random variables X_i using parameter $\lambda = 0.5i$ for the ith sequence. Obtain a new random variable Y as follows:

$$Y = \sum_{i=1}^{20} X_i$$

(Note that if you have 10,000 realizations of X_i in the ith sequence, you will have 10,000 realizations of the random variable Y.)

(a) Experimentally determine the PDF and CDF for Y.
(b) Compute the theoretical mean and variance for Y.
(c) Plot the PDF and CDF and compare their shapes to those of the Gaussian.

3. Successively convolve 50 Bernoulli PMFs (not the random variables). Choose $p = 0.8$ for all of the PMFs. Plot and compare the shape (only the shape) of the result to that of the Gaussian PDF. What is the theoretical mean and standard deviation of a sum of 50 Bernoulli random variables? How does it compare to what you see in the plot?

MATLAB programming notes

The MATLAB functions "mean" and "std" can be used in steps 1(b) and 2(b) to estimate mean and variance. The function "pdfcdf" from the software package can be used for Step 1(a); the function "expon" can be used for Step 2.

Project 6.2

The objective in this project is to experimentally investigate the mean and variance of the sample mean statistic using random variables of different kinds. This project builds upon Computer Project 3.1 of Chapter 3 and Computer Project 4.1 of Chapter 4.

1. Generate sequences of random variables as specified in Project 3.1 (a)–(f).
2. For each random variable in Project 3.1 (a)–(f), calculate the sample mean

$$M_n = \frac{1}{n} \sum_{i=1}^{n} X_i$$

using $n = 10$ variables in the sequence.

3. Repeat the last step a total of $N = 10,000$ times each time using a different set of the X_i so that you have N estimates of the sample mean M_n.

4. Estimate the mean and the variance of the *sample mean* using the following formulas:

$$\mathcal{E}\{M_n\} \approx \frac{1}{N} \sum_{k=1}^{N} M_n^{(k)} \qquad \text{Var}[M_n] \approx \frac{1}{N} \sum_{i=1}^{N} X_i^2 - (\mathcal{E}\{M_n\})^2$$

where $M_n^{(k)}$ represents the kth estimate of the sample mean.   (Over)

5. Repeat Steps 2, 3, and 4 by increasing the length n to 100, 1,000, and 10,000 random variables. Plot the experimental mean and variance of M_n versus the length n on a logarithmic scale. How well do the experimental results correspond to the theory?

7 Random Vectors

When two or more random variables sharing similar properties appear in a random experiment it often makes sense to represent these random variables by a vector whose components are the random variables. A good example of this are the random variables resulting from samples of a digital signal or time series. When this is the case we can carry out the analysis of the problem using concise vector and matrix notation instead of the otherwise cumbersome scalar notation that tends to obscure the fundamental nature of the problem.

This chapter develops the vector/matrix notation and the essential concepts pertaining to random vectors. It is seen that the probabilistic ideas developed in the earlier chapters of this book extend quite naturally to random vectors. While the treatment provided here is quite basic from the point of view of linear algebra, it is nevertheless sufficient to discuss a few practical applications that occur in electrical and computer engineering.

7.1 Random Vectors

A vector whose components are random variables is called a *random vector*. In this section a random vector is denoted by a bold upper case symbol such as[1]

$$\boldsymbol{X} = [X_1, X_2, \ldots, X_K]^T$$

where the X_i are random variables, and the *realization* of the random vector (i.e., the value that it takes on) is denoted by a bold italic lower case symbol

$$\boldsymbol{x} = [x_1, x_2, \ldots, x_K]^T$$

The notation is thus consistent with that used in earlier parts of the text. To keep the section brief, random vectors are discussed the context of continuous random variables, although a similar development could be done for purely discrete random variables if needed.

7.1.1 Cumulative distribution and density functions

Let us define the notation $\boldsymbol{X} \leq \boldsymbol{x}$ to mean

$$\boldsymbol{X} \leq \boldsymbol{x}: \quad X_1 \leq x_1, X_2 \leq x_2, \ldots, X_K \leq x_K \tag{7.1}$$

The cumulative distribution function (CDF) for the random vector is then defined as

$$F_{\boldsymbol{X}}(\boldsymbol{x}) = F_{X_1 X_2 \ldots X_K}(x_1, x_2, \ldots, x_K) \stackrel{\text{def}}{=} \Pr[\boldsymbol{X} \leq \boldsymbol{x}] \tag{7.2}$$

As in the one- and two-dimensional cases, this function is monotonic and continuous from the right in each of the K arguments. $F_{\boldsymbol{X}}$ equals or approaches 0 as *any* of the

---

[1] Throughout this text, vectors are represented by column vectors and the superscript T denotes vector or matrix transpose.

arguments X_i approach minus infinity and equals or approaches 1 when *all* of the X_i approach infinity.

The PDF for the random vector (which is in reality a *joint* PDF among all of its components) is given by

$$f_X(x) = \frac{\partial}{\partial x_1} \frac{\partial}{\partial x_2} \cdots \frac{\partial}{\partial x_K} F_X(x) \tag{7.3}$$

The PDF for a random vector has the probabilistic interpretation

$$\Pr[x < X \le x + \Delta] \approx f_X(x) \Delta_1 \Delta_2 \ldots \Delta_K \tag{7.4}$$

where $x < X \le x + \Delta$ denotes the event

$$x < X \le x + \Delta: \quad x_1 < X_1 \le x_1 + \Delta_1, \ldots, x_K < X_K \le x_K + \Delta_K \tag{7.5}$$

and where the Δ_i are small increments in the components of the random vector.

The probability that X is any region $\mathcal{R}$ of the K-dimensional space is given by the multidimensional integral

$$\Pr[X \in \mathcal{R}] = \int_{\mathcal{R}} f_X(x) dx \stackrel{\text{def}}{=} \int \int \cdots \int_{\mathcal{R}} f_X(x) dx_K \ldots dx_2 \, dx_1 \tag{7.6}$$

In particular, if $\mathcal{R}$ is the entire K-dimensional space then

$$\int_{-\infty}^{\infty} f_X(x) \, dx = \int_{-\infty}^{\infty} \cdots \int_{-\infty}^{\infty} \int_{-\infty}^{\infty} f_X(x) \, dx_K \ldots dx_2 \, dx_1 = 1 \tag{7.7}$$

Example 7.1:   To solidify concepts we repeat the illustration of a multidimensional CDF and PDF for a pair of random variables X_1 and X_2 that appears in as Fig. 5.4 of Chapter 5. The CDF F_X defined in (7.2) is illustrated below for the case $K = 2$.

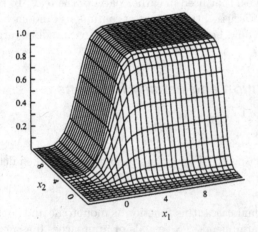

Notice that it remains positive and increases monotonically in both arguments from an initial value of 0 to a final value of 1. The multidimensional PDF, f_X, shown facing for $K = 2$ is defined by (7.3).

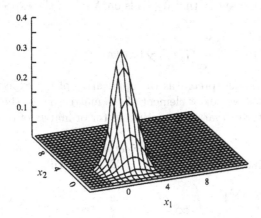

It satisfies the properties given in equations (7.4) through (7.7). Scalar versions of all of these equations appear in Chapter 5, Section 5.2.1.

□

7.1.2 Random vectors with independent components

When the components of a random vector are independent, the distribution and density functions appear as products of the marginals. In particular, the CDF can be written as

$$F_{\boldsymbol{X}}(\boldsymbol{x}) = \prod_{i=1}^{n} F_{X_i}(x_i) \tag{7.8}$$

while the PDF has the similar form

$$f_{\boldsymbol{X}}(\boldsymbol{x}) = \prod_{i=1}^{n} f_{X_i}(x_i) \tag{7.9}$$

Most often the component distributions or densities are all of the same form, so we can drop the subscript i on these terms and write $F_X(x_i)$ or $f_X(x_i)$.

Vectors with independent components may occur when sampling a random process such as noise (see Part 2 of this text) when the observations are known to be independently generated and derived from the same phenomenon. Frequently it is necessary only to assume that the components of the random vector are uncorrelated and not necessarily independent (see Section 7.2.4).

7.2 Analysis of Random Vectors

Frequently the density function describing random variables is not known. This is especially true for random vectors. In most cases the analysis is therefore based on first and second order moments and frequently these moment parameters are estimated from data. In the Gaussian case these moments are all that are required to determine the density function.

7.2.1 Expectation and moments

If $\mathbf{G}$ is any (scalar, vector, matrix) quantity that depends on $\mathbf{X}$, then the expectation of $\mathbf{G}(\mathbf{X})$ can be defined as

$$\mathcal{E}\{\mathbf{G}(\mathbf{X})\} = \int_{-\infty}^{\infty} \mathbf{G}(\mathbf{x}) f_{\mathbf{X}}(\mathbf{x}) \, d\mathbf{x} \tag{7.10}$$

If $\mathbf{G}$ is a vector or matrix, (7.10) is interpreted as the application of the expectation operation to every component of the vector or element of the matrix. Thus, the result of taking the expectation of a vector or matrix is another vector or matrix of the same size.

The quantities usually of most interest for a random vector are the first and second moments. The *mean vector* is defined by taking $\mathbf{G}(\mathbf{X}) = \mathbf{X}$ and writing

$$\mathbf{m}_{\mathbf{X}} = \mathcal{E}\{\mathbf{X}\} = \int_{-\infty}^{\infty} \mathbf{x} f_{\mathbf{X}}(\mathbf{x}) d\mathbf{x} \tag{7.11}$$

The mean vector is a vector of constants

$$\mathbf{m}_{\mathbf{X}} = \begin{bmatrix} m_1 \\ m_2 \\ \vdots \\ m_K \end{bmatrix} \quad \text{with} \quad m_i = \mathcal{E}\{X_i\} \tag{7.12}$$

The *correlation matrix* is defined by

$$\mathbf{R}_{\mathbf{X}} = \mathcal{E}\{\mathbf{X}\mathbf{X}^T\} \tag{7.13}$$

Note that the term $\mathbf{X}\mathbf{X}^T$ is an *outer* product rather than an inner product of vectors; thus the result is a matrix, not a scalar. The correlation matrix has the form

$$\mathbf{R}_{\mathbf{X}} = \begin{bmatrix} \mathcal{E}\{X_1^2\} & \mathcal{E}\{X_1 X_2\} & \cdots & \mathcal{E}\{X_1 X_K\} \\ \mathcal{E}\{X_2 X_1\} & \mathcal{E}\{X_2^2\} & \cdots & \mathcal{E}\{X_2 X_K\} \\ \vdots & \vdots & & \vdots \\ \mathcal{E}\{X_K X_1\} & \mathcal{E}\{X_K X_2\} & \cdots & \mathcal{E}\{X_K^2\} \end{bmatrix}$$

The off-diagonal terms represent the correlations $r_{ij} = \mathcal{E}\{X_i X_j\}$ between all pairs of vector components while the diagonal terms are the second moments of the components. The mean vector and correlation matrix thus provide a complete description of the random vector using moments up to second order. The correlation matrix from its definition has the symmetry property

$$\mathbf{R}_{\mathbf{X}} = \mathbf{R}_{\mathbf{X}}^T \tag{7.14}$$

It also has the important property

$$\mathbf{a}^T \mathbf{R}_{\mathbf{X}} \mathbf{a} \geq 0 \quad \text{for any vector } \mathbf{a} \tag{7.15}$$

This *positive semidefinite* property is easily proven by substituting (7.13) in (7.15):

$$\mathbf{a}^T \mathcal{E}\{\mathbf{X}\mathbf{X}^T\} \mathbf{a} = \mathcal{E}\{(\mathbf{a}^T \mathbf{x})^2\} \geq 0$$

The *covariance matrix* is the matrix of second *central* moments of the random vector and is defined by

$$\mathbf{C}_{\mathbf{X}} = \mathcal{E}\{(\mathbf{X} - \mathbf{m}_{\mathbf{X}})(\mathbf{X} - \mathbf{m}_{\mathbf{X}})^T\} \tag{7.16}$$

This matrix has the form

$$
\mathbf{C_X} = \begin{bmatrix} \sigma_1^2 & c_{12} & \cdots & c_{1K} \\ c_{21} & \sigma_2^2 & \cdots & c_{2K} \\ \vdots & \vdots & & \vdots \\ c_{K1} & c_{K2} & \cdots & \sigma_K^2 \end{bmatrix} \qquad \text{where} \qquad \begin{array}{ccl} \sigma_i^2 & = & \text{Var}\,[X_i] \\ c_{ij} & = & \text{Cov}\,[X_i, X_j] \end{array}
$$

and also has the symmetry property

$$
\mathbf{C_X} = \mathbf{C_X^T} \tag{7.17}
$$

The covariance and correlation matrices are related as

$$
\mathbf{C_X} = \mathbf{R_X} - \mathbf{m_X m_X^T} \tag{7.18}
$$

The proof of this is similar to the proof of (4.17) of Chapter 4. For the case of a two-dimensional random vector this equation has the form

$$
\begin{bmatrix} \sigma_1^2 & c_{12} \\ c_{21} & \sigma_2^2 \end{bmatrix} = \begin{bmatrix} \mathcal{E}\{X_1^2\} & \mathcal{E}\{X_1 X_2\} \\ \mathcal{E}\{X_2 X_1\} & \mathcal{E}\{X_2^2\} \end{bmatrix} - \begin{bmatrix} m_1 \\ m_2 \end{bmatrix} \begin{bmatrix} m_1 & m_2 \end{bmatrix}
$$

This is just the matrix embodiment of the relations (4.17) and (5.28), so it is not anything new.

As in the case of a single random variable, it is often easier to compute the covariance matrix using (7.18) than to compute it from the definition (7.16). The following example illustrates this computation.

Example 7.2: The two jointly-distributed random variables described in Chapter 5, Example 5.6 are taken to be components of a random vector

$$
\mathbf{X} = [X_1, X_2]^T
$$

The mean vector and correlation matrix for this random vector are given by

$$
\mathbf{m_X} = \begin{bmatrix} \mathcal{E}\{X_1\} \\ \mathcal{E}\{X_2\} \end{bmatrix} = \begin{bmatrix} 5/6 \\ 10/21 \end{bmatrix}
$$

and

$$
\mathbf{R_X} = \begin{bmatrix} \mathcal{E}\{X_1^2\} & \mathcal{E}\{X_1 X_2\} \\ \mathcal{E}\{X_2 X_1\} & \mathcal{E}\{X_2^2\} \end{bmatrix} = \begin{bmatrix} 5/7 & 5/12 \\ 5/12 & 5/18 \end{bmatrix}
$$

where the numerical values of the moments are taken from Example 5.6. The covariance matrix can then be computed from (7.18) as

$$
\mathbf{C_X} = \begin{bmatrix} 5/7 & 5/12 \\ 5/12 & 5/18 \end{bmatrix} - \begin{bmatrix} 5/6 \\ 10/21 \end{bmatrix} \begin{bmatrix} 5/6 & 10/21 \end{bmatrix} = \begin{bmatrix} 5/252 & 5/252 \\ 5/252 & 5/98 \end{bmatrix}
$$

The elements σ_1^2, c, and σ_2^2 correspond to the numerical values computed in the earlier Example 5.6.

□

7.2.2 Estimating moments from data

When the mean vector, correlation and covariance matrices are not known, these parameters can be estimated from data. Given a set of sample vectors $\mathbf{X}_1, \mathbf{X}_2, \ldots, \mathbf{X}_n$

we can define a vector version of the sample mean as

$$M_n = \frac{1}{n}\sum_{i=1}^{n} X_i \qquad (7.19)$$

Here, n is the number of sample vectors where each vector has K components. This estimate, like the scalar version, is unbiased and consistent.

An estimate for the correlation matrix, known as the sample correlation matrix, can be obtained by replacing the expectation in (7.13) by an averaging operation as follows:

$$R_n = \frac{1}{n}\sum_{i=1}^{n} X_i X_i^T \qquad (7.20)$$

It can be shown that this estimate is also unbiased and consistent.

The estimate for the covariance matrix is a little trickier because it involves the sample mean, which is also an estimate. The unbiased form of the covariance estimate is

$$C_n = \frac{1}{n-1}\sum_{i=1}^{n}(X_i - M_n)(X_i - M_n)^T \qquad (7.21)$$

(See the discussion of this point in Chapter 6 for the scalar case.) This sample covariance matrix can be computed more easily by using (7.20) and the relation

$$C_n = \frac{n}{n-1}(R_n - M_n M_n^T) \qquad (7.22)$$

(see Prob. 7.4).

An example of the computation is given below.

Example 7.3: Given a set of four two-dimensional sample vectors

$$X_1 = \begin{bmatrix} 2 \\ -2 \end{bmatrix}, \quad X_2 = \begin{bmatrix} -2 \\ 1 \end{bmatrix}, \quad X_3 = \begin{bmatrix} 1 \\ 0 \end{bmatrix}, \quad X_4 = \begin{bmatrix} 0 \\ 2 \end{bmatrix}$$

the sample mean is computed as

$$M_n = \frac{1}{n}\sum_{i=1}^{n} X_i = \frac{1}{4}\left\{ \begin{bmatrix} 2 \\ -2 \end{bmatrix} + \begin{bmatrix} -2 \\ 1 \end{bmatrix} + \begin{bmatrix} 1 \\ 0 \end{bmatrix} + \begin{bmatrix} 0 \\ 2 \end{bmatrix} \right\} = \begin{bmatrix} \frac{1}{4} \\ \frac{1}{4} \end{bmatrix}$$

The sample correlation matrix is

$$\begin{aligned}
R_n &= \frac{1}{n}\sum_{i=1}^{n} X_i X_i^T \\
&= \frac{1}{4}\left\{ \begin{bmatrix} 2 \\ -2 \end{bmatrix}\begin{bmatrix} 2 & -2 \end{bmatrix} + \begin{bmatrix} -2 \\ 1 \end{bmatrix}\begin{bmatrix} -2 & 1 \end{bmatrix} + \begin{bmatrix} 1 \\ 0 \end{bmatrix}\begin{bmatrix} 1 & 0 \end{bmatrix} + \begin{bmatrix} 0 \\ 2 \end{bmatrix}\begin{bmatrix} 0 & 2 \end{bmatrix} \right\} \\
&= \begin{bmatrix} \frac{9}{4} & -\frac{3}{2} \\ -\frac{3}{2} & \frac{9}{4} \end{bmatrix}
\end{aligned}$$

Finally, the (unbiased) sample covariance matrix is

$$\begin{aligned}
C_n &= \frac{1}{n-1}\sum_{i=1}^{n}(X_i - M_n)(X_i - M_n)^T \\
&= \frac{1}{3}\left\{ \begin{bmatrix} \frac{7}{4} \\ -\frac{9}{4} \end{bmatrix}\begin{bmatrix} \frac{7}{4} & -\frac{9}{4} \end{bmatrix} + \begin{bmatrix} -\frac{9}{4} \\ \frac{3}{4} \end{bmatrix}\begin{bmatrix} -\frac{9}{4} & \frac{3}{4} \end{bmatrix} + \begin{bmatrix} \frac{3}{4} \\ -\frac{1}{4} \end{bmatrix}\begin{bmatrix} \frac{3}{4} & -\frac{1}{4} \end{bmatrix} + \begin{bmatrix} -\frac{1}{4} \\ \frac{7}{4} \end{bmatrix}\begin{bmatrix} -\frac{1}{4} & \frac{7}{4} \end{bmatrix} \right\} \\
&= \begin{bmatrix} \frac{35}{12} & -\frac{25}{12} \\ -\frac{25}{12} & \frac{35}{12} \end{bmatrix}
\end{aligned}$$

or, using (7.21)

$$C_n = \frac{n}{n-1}(R_n - M_n M_n^T)$$

$$= \frac{4}{3}\left(\begin{bmatrix} \frac{9}{4} & -\frac{3}{2} \\ -\frac{3}{2} & \frac{9}{4} \end{bmatrix} - \begin{bmatrix} \frac{1}{4} \\ \frac{1}{4} \end{bmatrix}\begin{bmatrix} \frac{1}{4} & \frac{1}{4} \end{bmatrix}\right) = \begin{bmatrix} \frac{35}{12} & -\frac{25}{12} \\ -\frac{25}{12} & \frac{35}{12} \end{bmatrix}$$

□

In this example, the diagonal terms of the correlation and covariance matrices turn out to be equal. This is a property of this particular data set and is not always the case. The terms on opposite sides of the diagonal are guaranteed to be equal, however, due to the symmetry of these matrices and their estimates.

7.2.3 The multivariate Gaussian density

A well known and very useful example of a multidimensional probability density function is the multivariate Gaussian density. The multivariate Gaussian density for an K-dimensional random vector X is defined by

$$f_X(x) = \frac{1}{(2\pi)^{\frac{K}{2}}|C_X|^{\frac{1}{2}}} \exp -\tfrac{1}{2}(x - m_X)^T C_X^{-1}(x - m_X) \qquad (7.23)$$

where $|C_X|$ represents the determinant of the covariance matrix. In the two-dimensional case m_X and C_X are given by

$$m_X = \begin{bmatrix} m_1 \\ m_2 \end{bmatrix} \qquad C_X = \begin{bmatrix} \sigma_1^2 & c_{12} \\ c_{21} & \sigma_2^2 \end{bmatrix} = \begin{bmatrix} \sigma_1^2 & \rho\sigma_1\sigma_2 \\ \rho\sigma_1\sigma_2 & \sigma_2^2 \end{bmatrix}$$

where ρ is the correlation coefficient defined by equation (5.29) of Chapter 5. In this case the explicit inverse of the matrix can be found as

$$C_X^{-1} = \frac{1}{1-\rho^2}\begin{bmatrix} \dfrac{1}{\sigma_1^2} & -\dfrac{\rho}{\sigma_1\sigma_2} \\ -\dfrac{\rho}{\sigma_1\sigma_2} & \dfrac{1}{\sigma_2^2} \end{bmatrix}$$

while the determinant is given by

$$|C_X| = \sigma_1^2\sigma_2^2(1-\rho^2)$$

Substituting these last two equations in (7.23) yields the bivariate density function for two jointly Gaussian random variables given as equation (5.35) of Chapter 5.

In Section 5.4 of Chapter 5 it is shown that the marginal and conditional densities derived from the bivariate density are all Gaussian. This property also extends to the multivariate Gaussian density with $K > 2$. In particular, any set of components of the random vector has a jointly Gaussian PDF, and any set of components conditioned on any other set is also jointly Gaussian. The multivariate Gaussian density is especially important for problems in signal processing and communications. Further analysis of the Gaussian density and its covariance matrix involves the use of eigenvalues and eigenvectors, however, which is beyond the scope of this text. A more extensive treatment including the case of complex random variables can be found in [1].

7.2.4 *Random vectors with uncorrelated components*

If the components of a random vector are uncorrelated, this means that the off-diagonal terms of the covariance matrix c_{ij} for $i \neq j$ are zero. In this case the covariance matrix has the diagonal form

$$\mathbf{C_X} = \begin{bmatrix} \sigma_1^2 & 0 & \cdots & 0 \\ 0 & \sigma_2^2 & \cdots & 0 \\ \vdots & \vdots & & \vdots \\ 0 & 0 & \cdots & \sigma_K^2 \end{bmatrix}$$

If, further, the covariance terms are all equal, the covariance matrix then has the special form

$$\mathbf{C_X} = \sigma_o^2 \mathbf{I} \tag{7.24}$$

where $\mathbf{I}$ is the $K \times K$ identity matrix.

A random vector with independent components will also have uncorrelated components and thus a diagonal covariance matrix. A random vector with uncorrelated components will not have *independent* components except in the Gaussian case, however. Random vectors with uncorrelated components may arise from sampling a noise sequence. The noise in this case is referred to as "white noise" and the adjective "white" is often carried over to describe the random vector. More about this can be found in the discussion of white noise in Part 2.

7.3 *Transformations*

7.3.1 *Transformation of moments*

Translations and linear transformations have easily formulated effects on the first and second order moments of random vectors because the order of the moments is preserved. The results of nonlinear transformations are not easy to formulate because first and second order moments of the new random variable generally require higher order moments of the original random variable. Our discussion here is thus restricted to translations and *linear* transformations.

When a random vector is translated by adding a constant, this has the effect of adding the constant to the mean. Consider

$$\boldsymbol{Y} = \boldsymbol{X} + \mathbf{b} \tag{7.25}$$

where $\boldsymbol{X}$ and $\boldsymbol{Y}$ are random vectors and $\mathbf{b}$ is a constant vector. Then taking the expectation yields

$$\mathcal{E}\{\boldsymbol{Y}\} = \mathcal{E}\{\boldsymbol{X}\} + \mathcal{E}\{\mathbf{b}\}$$

or

$$\boxed{\mathbf{m_Y} = \mathbf{m_X} + \mathbf{b}} \tag{7.26}$$

It will be seen shortly, that while a translation such as (7.25) has an effect on the correlation matrix, it has *no* effect on the covariance matrix because the covariance matrix is defined by subtracting the mean.

Consider now a linear transformation of the form

$$\boldsymbol{Y} = \mathbf{A}\boldsymbol{X} \tag{7.27}$$

where the dimension of $\boldsymbol{Y}$ is not necessarily the same as the dimension K of $\boldsymbol{X}$. The mean of Y is easily found as

$$\mathcal{E}\{Y\} = \mathcal{E}\{AX\} = A\mathcal{E}\{X\}$$

or

$$\boxed{m_Y = Am_X} \tag{7.28}$$

The correlation matrix of Y can be found from the definition

$$R_Y = \mathcal{E}\{YY^T\} = \mathcal{E}\{(AX)(AX)^T\} = A\mathcal{E}\{XX^T\}A^T$$

or

$$\boxed{R_Y = AR_X A^T} \tag{7.29}$$

It can easily be shown that the covariance matrix satisfies an identical relation (see Prob. 7.9). Note that the matrix A in this transformation does not need to be invertible nor even need to be a square matrix. In other words, the linear transformation may result in a compression or expansion of dimensionality. The results are valid in any case.

The complete set of first and second order moments under a combination of linear transformation and translation is listed for convenience in Table 7.1. Notice that the

transformation	$Y = AX + b$
mean vector	$m_Y = Am_X + b$
correlation matrix	$R_Y = AR_X A^T + Am_X b^T + bm_X^T A^T + bb^T$
covariance matrix	$C_Y = AC_X A^T$

Table 7.1 Transformation of moments

formula for the correlation matrix is quite complicated when $b \neq 0$ and $m_X \neq 0$ while the expression for the covariance matrix is much simpler. This is because for the covariance we eliminate the mean value before taking the expectation (see (7.16)).

7.3.2 Transformation of density functions

The general case

Let us consider a general invertible nonlinear transformation of a random vector of the form[2]

$$Y = g(X) \tag{7.30}$$

---

[2] The notation in this equation for a vector-valued function is shorthand for the following:

$$\begin{bmatrix} Y_1 \\ Y_2 \\ \vdots \\ Y_K \end{bmatrix} = \begin{bmatrix} g_1(X_1, X_2, \ldots, X_K) \\ g_2(X_1, X_2, \ldots, X_K) \\ \vdots \\ g_K(X_1, X_2, \ldots, X_K) \end{bmatrix}$$

The g_i are the component functions of the vector-valued function g; since the transformation here is required to be invertible, the dimension of Y is assumed to be the same as that of X.

Since the transformation is invertible, X can be written as

$$X = g^{-1}(Y) \tag{7.31}$$

The relation between the densities f_X and f_Y is derived following essentially the same procedure used when dealing with single random variables (see Chapter 3). The probability that X is in a small rectangular region of the X space is given by (7.4). Let us denote the volume of this small region by

$$\Delta V_X = \Delta_1 \Delta_2 \ldots \Delta_K$$

and let ΔV_Y represent the volume of the corresponding region in the Y space (see Fig. 7.1). Note that this region need not be a rectangular region. Since the probability that

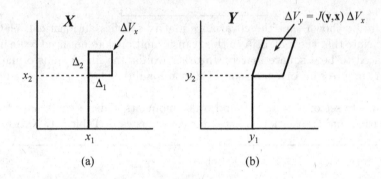

Figure 7.1 Mapping of small region of X space to Y space. (a) Small region of X space. (b) Corresponding region in Y space.

X is in the small region of volume ΔV_X is the same as the probability that Y is in the small corresponding region of volume ΔV_Y, the two probabilities can be equated:

$$f_X(x)\Delta V_X = f_Y(y)\Delta V_Y \tag{7.32}$$

The volume of the small region in the Y space is given by (see, e.g., [2])

$$\Delta V_Y = J(y, x)\Delta V_X \tag{7.33}$$

where $J(y, x)$ is the Jacobian of the transformation:

$$J(y, x) \stackrel{\text{def}}{=} \text{abs} \begin{vmatrix} \frac{\partial g_1(x)}{\partial x_1} & \frac{\partial g_2(x)}{\partial x_1} & \cdots & \frac{\partial g_K(x)}{\partial x_1} \\ \frac{\partial g_1(x)}{\partial x_2} & \frac{\partial g_2(x)}{\partial x_2} & \cdots & \frac{\partial g_K(x)}{\partial x_2} \\ \vdots & \vdots & & \vdots \\ \frac{\partial g_1(x)}{\partial x_K} & \frac{\partial g_2(x)}{\partial x_K} & \cdots & \frac{\partial g_K(x)}{\partial x_K} \end{vmatrix} \tag{7.34}$$

and "abs" denotes the absolute value of the determinant. Substituting (7.33) in (7.32) yields

$$f_X(x)\Delta V_X = f_Y(y)J(y, x)\Delta V_X$$

Then cancelling the term ΔV_X and applying (7.31) produces the desired relation

$$\boxed{f_Y(y) = \frac{1}{J(y, x)} f_X(x)\Big|_{x = g^{-1}(y)}} \tag{7.35}$$

The example below illustrates a *nonlinear* transformation of a Gaussian random vector using the formulas presented in this section. The resulting random variables have well-known distributions which are *not* Gaussian.

Example 7.4: In a communications problem the in-band noise causes an amplitude and phase modulation of the carrier. The noise signal is of the form

$$N(t) = A(t)\cos(2\pi f_c t + \Theta(t))$$

where f_c is the carrier frequency and $N(t)$ and $\Theta(t)$ represent the amplitude and phase (respectively) of the noise, which for any time "t" are random variables. If the trigonometric formula for the cosine of a sum of terms is used, the noise signal can be written in rectangular or "quadrature" form:

$$N(t) = X_1(t)\cos(2\pi f_c t) + X_2(t)\sin(2\pi f_c t)$$

where

$$X_1(t) = A(t)\cos\Theta(t) \quad \text{and} \quad X_2(t) = A(t)\sin\Theta(t)$$

Let us consider sampling this noise signal at a given time 't' and thus drop the dependence of all of the variables on time. It is known that the random variables X_1 and X_2 representing the quadrature components of the noise are typically independent zero-mean Gaussian random variables each with variance σ^2. Thus the joint density for X_1 and X_2 is given by

$$f_{X_1 X_2}(x_1, x_2) = \frac{1}{\sqrt{2\pi\sigma^2}}e^{-x_1^2/2\sigma^2} \cdot \frac{1}{\sqrt{2\pi\sigma^2}}e^{-x_2^2/2\sigma^2} = \frac{1}{2\pi\sigma^2}e^{-(x_1^2+x_2^2)/2\sigma^2} \quad (1)$$

It is desired to find the density functions for the random variables A and Θ and to determine if these random variables are independent.

We can proceed by defining the random vectors $\boldsymbol{X} = \begin{bmatrix} X_1 \\ X_2 \end{bmatrix}$ and $\boldsymbol{Y} = \begin{bmatrix} A \\ \Theta \end{bmatrix}$ and using the formula (7.35) for the transformation of the joint density. The transformation $\boldsymbol{Y} = \boldsymbol{g}(\boldsymbol{X})$ is given by

$$\begin{bmatrix} A \\ \Theta \end{bmatrix} = \begin{bmatrix} \sqrt{X_1^2 + X_2^2} \\ \tan^{-1}(X_2/X_1) \end{bmatrix} = \begin{bmatrix} g_1(\boldsymbol{X}) \\ g_2(\boldsymbol{X}) \end{bmatrix}$$

The Jacobian of the transformation is given by

$$J(\boldsymbol{y}, \boldsymbol{x}) = \text{abs} \begin{vmatrix} \frac{\partial g_1(\boldsymbol{x})}{\partial x_1} & \frac{\partial g_2(\boldsymbol{x})}{\partial x_1} \\ \frac{\partial g_1(\boldsymbol{x})}{\partial x_2} & \frac{\partial g_2(\boldsymbol{x})}{\partial x_2} \end{vmatrix} = \text{abs} \begin{vmatrix} \frac{1}{2}\frac{2x_1}{\sqrt{x_1^2+x_2^2}} & \frac{-x_2/x_1^2}{1+x_2^2/x_1^2} \\ \frac{1}{2}\frac{2x_2}{\sqrt{x_1^2+x_2^2}} & \frac{1/x_1}{1+x_2^2/x_1^2} \end{vmatrix}$$

$$= \text{abs} \begin{vmatrix} \cos\theta & -(\sin\theta)/a \\ \sin\theta & (\cos\theta)/a \end{vmatrix} = \frac{1}{a}(\cos^2\theta + \sin^2\theta) = \frac{1}{a}$$

From (7.35) (and using (1) above) we thus have

$$f_{A\Theta}(a, \theta) = a\frac{1}{2\pi\sigma^2}e^{-a^2/2\sigma^2}$$

This is the joint density function.

The marginal density of A is then given by

$$f_A(a) = \int_{-\pi}^{\pi} f_{A\Theta}(a, \theta)d\theta = \frac{a}{2\pi\sigma^2}e^{-a^2/2\sigma^2}\int_{-\pi}^{\pi} d\theta = \frac{a}{\sigma^2}e^{-a^2/2\sigma^2}; \qquad a \geq 0$$

This PDF is known as the Rayleigh probability density with parameter $\alpha = 1/\sigma$.

The marginal density for Θ is then

$$f_\Theta(\theta) = \int_{-\pi}^{\pi} f_{A\Theta}(a, \theta) da = \frac{1}{2\pi} \int_0^\infty \frac{a}{\sigma^2} e^{-a^2/2\sigma^2} da = \frac{1}{2\pi} \qquad -\pi < \theta \leq \pi$$

Therefore the phase is *uniformly* distributed. Observe that $f_{A\Theta}(a, \theta) = f_A(a) \cdot f_\Theta(\theta)$, so the random variables A and Θ are *independent*.

□

This example derives the well-known result that a signal which at any time "t" has identically-distributed zero-mean Gaussian components in a rectangular coordinate system, has independent magnitude and phase. The magnitude has a Rayleigh distribution and the phase has a uniform distribution.

The Gaussian case

Since the Gaussian density function is defined entirely by its mean vector and co-variance matrix, any linear transformation or translation considered in Section 7.3.1 applied to a Gaussian random vector results in another Gaussian random vector with the transformed parameters.

To see this explicitly, let's consider the case when the transformation $\mathbf{g}$ is nonsingular and of the form

$$\mathbf{Y} = \mathbf{A}\mathbf{X} + \mathbf{b} \tag{7.36}$$

In this case the transformation equation (7.35) for the density function becomes

$$f_\mathbf{Y}(\mathbf{y}) = \frac{1}{\text{abs}|\mathbf{A}|} f_\mathbf{X}(\mathbf{A}^{-1}(\mathbf{y} - \mathbf{b})) \tag{7.37}$$

where "abs$|\mathbf{A}|$" denotes the absolute value of the determinant and $\mathbf{A}^{-1}$ is the matrix inverse. The example below illustrates the steps in showing that the resulting density $f_\mathbf{Y}(\mathbf{y})$ is Gaussian.

Example 7.5: Consider an invertible transformation of the form (7.36) where $\mathbf{X}$ has the Gaussian PDF given by (7.23). Applying the formula (7.37) yields

$$\begin{aligned} f_\mathbf{Y}(\mathbf{y}) &= \frac{1}{\text{abs}|\mathbf{A}|} \frac{1}{(2\pi)^{\frac{K}{2}} |\mathbf{C}_\mathbf{X}|^{\frac{1}{2}}} \exp -\tfrac{1}{2} (\mathbf{A}^{-1}(\mathbf{y} - \mathbf{b}) - \mathbf{m}_\mathbf{X})^T \mathbf{C}_\mathbf{X}^{-1} (\mathbf{A}^{-1}(\mathbf{y} - \mathbf{b}) - \mathbf{m}_\mathbf{X}) \\ &= \frac{1}{(2\pi)^{\frac{K}{2}} [|\mathbf{A}|^2 |\mathbf{C}_\mathbf{X}|]^{\frac{1}{2}}} \exp -\tfrac{1}{2} (\mathbf{y} - \mathbf{b} - \mathbf{A}\mathbf{m}_\mathbf{X})^T (\mathbf{A}^{-1})^T \mathbf{C}_\mathbf{X}^{-1} \mathbf{A}^{-1} (\mathbf{y} - \mathbf{b} - \mathbf{A}\mathbf{m}_\mathbf{X}) \end{aligned}$$

This appears to have the form of a multivariate Gaussian PDF. Let us investigate further.

From Table 7.1, the mean and covariance of $\mathbf{Y}$ are given by

$$\mathbf{m}_\mathbf{Y} = \mathbf{A}\mathbf{m}_\mathbf{X} + \mathbf{b} \qquad \text{and} \qquad \mathbf{C}_\mathbf{Y} = \mathbf{A}\mathbf{C}_\mathbf{X}\mathbf{A}^T \tag{2}$$

The inverse covariance matrix is thus

$$\mathbf{C}_\mathbf{Y}^{-1} = (\mathbf{A}\mathbf{C}_\mathbf{X}\mathbf{A}^T)^{-1} = (\mathbf{A}^T)^{-1} \mathbf{C}_\mathbf{X}^{-1} \mathbf{A}^{-1} = (\mathbf{A}^{-1})^T \mathbf{C}_\mathbf{X}^{-1} \mathbf{A}^{-1}$$

while the determinant is

$$|\mathbf{C}_\mathbf{Y}| = |\mathbf{A}\mathbf{C}_\mathbf{X}\mathbf{A}^T| = |\mathbf{A}||\mathbf{C}_\mathbf{X}||\mathbf{A}^T| = |\mathbf{A}|^2 |\mathbf{C}_\mathbf{X}|$$

By comparing these results with the terms in (1) we can easily verify that $f_Y(y)$ in fact has the Gaussian form

$$f_Y(y) = \frac{1}{(2\pi)^{\frac{K}{2}}|C_Y|^{\frac{1}{2}}} \exp -\tfrac{1}{2}(y - m_Y)^T C_Y^{-1}(y - m_Y)$$

with parameters m_Y and C_Y are given in (2).

□

This example proves that (at least when the linear transformation is invertible) a linear transformation and/or translation of a Gaussian random vector is another Gaussian random vector. The result is true even when the linear transformation is not invertible, but the proof is a bit more complicated.

7.4 Cross Correlation and Covariance

In many applications it is necessary to deal with two random vectors, X and Y. Most often the relation between these vectors is expressed using second moments.

If X and Y are two random vectors, not necessarily of the same dimensions, the *cross-correlation* matrix is defined by

$$R_{XY} = \mathcal{E}\{XY^T\} \tag{7.38}$$

As before, observe that this is an outer product of vectors, not an inner product. If X has K components and Y has L components then R_{XY} is a $K \times L$ matrix with terms $\mathcal{E}\{X_i Y_j\}$.

The *cross-covariance* matrix is defined in a similar way by

$$C_{XY} = \mathcal{E}\{(X - m_X)(Y - m_Y)^T\} \tag{7.39}$$

where the mean is subtracted before taking the expectation. These two matrices are related as

$$C_{XY} = R_{XY} - m_X m_Y^T \tag{7.40}$$

The two random vectors are defined to be *uncorrelated* if the covariance matrix is identically zero. Thus it follows from (7.40) that for uncorrelated random vectors

$$R_{XY} = m_X m_Y^T$$

If the correlation matrix is zero, the random vectors are said to be *orthogonal*.

Two random vectors are *independent* if the joint distribution is the product of the marginals, i.e., $F_{XY} = F_X F_Y$. Random vectors that are independent are also uncorrelated, however the converse is not true except for jointly Gaussian random vectors.

As an example of the use of some of these properties, consider the sum of two zero-mean random vectors that are uncorrelated. Let us show that the correlation matrix of the sum is equal to the sum of the correlation matrices. Since the random vectors have zero mean, the correlation and covariance matrices are identical. Let $Z = X + Y$, then

$$\begin{aligned} R_Z &= \mathcal{E}\{ZZ^T\} = \mathcal{E}\{(X + Y)(X + Y)^T\} \\ &= \mathcal{E}\{XX^T\} + \mathcal{E}\{XY^T\} + \mathcal{E}\{YX^T\} + \mathcal{E}\{YY^T\} \\ &= R_X + R_{XY} + R_{YX} + R_Y \end{aligned}$$

Since the random vectors are uncorrelated (and orthogonal) $R_{XY} = R_{YX} = 0$.

Therefore

$$\mathbf{R}_Z = \mathbf{R}_X + \mathbf{R}_Y$$

7.5 Applications to Signal Processing

Random vectors appear in a host of practical signal processing applications having to do with communication, decision and control. This section expands the problem of optimal decision for signals treated in Section 3.8 of Chapter 3 to develop a more true-to-life model for the detection process in communications systems and radar or sonar. The chapter also describes the general problem of classification that occurs in pattern recognition and briefly discusses the data compression method known as vector quantization.

7.5.1 Digital communication

Consider the case of binary digital communication where one of a pair of waveforms (representing logical 1 or 0) is transmitted to a receiver, whose job it is to detect the signal and decide which of the two waveforms was sent. The signals to be detected are assumed to have a known form, such as a sine wave, and have been sampled and represented in discrete time. A typical pair of such signals representing 1 and 0 is shown in Fig. 7.2. These represent one cycle of the waveforms shown in Fig. 3.33 of Chapter 3, which are used in the method known as BPSK (see Section 3.8.1). It

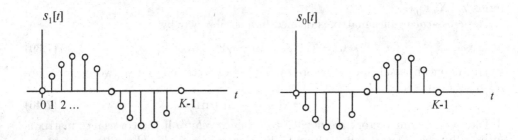

Figure 7.2 Depiction of signals to be detected in noise.

is assumed that time is normalized, so t takes on the values $0, 1, 2, \ldots$ and that K samples of each signal are taken and formed into a vector

$$\mathbf{s}_i = \begin{bmatrix} s_i[0] & s_i[1] & \cdots & s_i[K-1] \end{bmatrix}^T \qquad i = 1, 0$$

Random Gaussian noise is present, so the receiver observes a random vector $\mathbf{X}$ and must choose one of two hypotheses:

$$H_1: \quad \mathbf{X} = \mathbf{s}_1 + \mathbf{N} \quad \text{(the signal is } \mathbf{s}_1\text{)}$$

$$H_0: \quad \mathbf{X} = \mathbf{s}_0 + \mathbf{N} \quad \text{(the signal is } \mathbf{s}_0\text{)}$$

where $\mathbf{N}$ is a vector of the added noise samples. The noise vector has the Gaussian density function

$$f_{\mathbf{N}}(\boldsymbol{\eta}) = \frac{1}{(2\pi)^{\frac{K}{2}} |\mathbf{C}_{\mathbf{N}}|^{\frac{1}{2}}} \exp -\tfrac{1}{2} \boldsymbol{\eta}^T \mathbf{C}_{\mathbf{N}}^{-1} \boldsymbol{\eta}$$

with mean vector zero and known covariance matrix $\mathbf{C_N}$ that accounts for any correlation between samples.

Under each hypothesis, the PDF of X is the same as that of the noise except for a shift in the mean; that is,

$$f_{X|\mathrm{H}_i}(x|\mathrm{H}_i) = \frac{1}{(2\pi)^{\frac{K}{2}}|\mathbf{C_N}|^{\frac{1}{2}}} \exp -\tfrac{1}{2}(x - \mathbf{s}_i)^T \mathbf{C_N}^{-1}(x - \mathbf{s}_i) \qquad i = 1,0 \qquad (7.41)$$

Following the discussion in Section 3.8.1, the posterior probabilities can be written as

$$\Pr[\mathrm{H}_i|x] = \frac{f_{X|\mathrm{H}_i}(x|\mathrm{H}_i)\mathrm{P}_i}{f_X(x)} ; \qquad i = 0,1 \qquad (7.42)$$

where P_i are the prior probabilities and $f_X(x)$ is the unconditional density of the observations. Recall then, the procedure that minimizes the probability of error is to choose H_0 or H_1 according to

$$\Pr[\mathrm{H}_1|x] \underset{H_0}{\overset{H_1}{\gtrless}} \Pr[\mathrm{H}_0|x]$$

Upon substituting (7.42) into this equation, cancelling the common term $f_X(x)$, and rearranging, we arrive at the *likelihood ratio test*

$$\boxed{\frac{f_{X|\mathrm{H}_1}(x|\mathrm{H}_1)}{f_{X|\mathrm{H}_0}(x|\mathrm{H}_0)} \underset{H_0}{\overset{H_1}{\gtrless}} \frac{\mathrm{P}_0}{\mathrm{P}_1}} \qquad (7.43)$$

which represents the optimal decision rule.

This decision rule can be considerably simplified algebraically. First, by substituting (7.41) in (7.43), taking logarithms, and cancelling common terms, we can write

$$\ln \frac{f_{X|\mathrm{H}_1}(x|\mathrm{H}_1)}{f_{X|\mathrm{H}_0}(x|\mathrm{H}_0)} = -\tfrac{1}{2}(x - \mathbf{s}_1)^T \mathbf{C_N}^{-1}(x - \mathbf{s}_1) + \tfrac{1}{2}(x - \mathbf{s}_0)^T \mathbf{C_N}^{-1}(x - \mathbf{s}_0) \underset{H_0}{\overset{H_1}{\gtrless}} \ln \frac{\mathrm{P}_0}{\mathrm{P}_1}$$

Then by expanding terms, simplifying, and using the symmetry property of the covariance matrix, the decision rule can ultimately be written in the form

$$\boxed{\mathbf{w}^T x \underset{H_0}{\overset{H_1}{\gtrless}} \tau} \qquad (7.44)$$

where $\mathbf{w}$ is the weight vector

$$\mathbf{w} = \mathbf{C_N}^{-1}(\mathbf{s}_1 - \mathbf{s}_0) \qquad (7.45)$$

and τ is the threshold $\tau = \tfrac{1}{2}(\mathbf{s}_1^T \mathbf{C_N}^{-1}\mathbf{s}_1 - \mathbf{s}_0^T \mathbf{C_N}^{-1}\mathbf{s}_0) + \ln \frac{\mathrm{P}_0}{\mathrm{P}_1}$. The optimal decision rule thus forms an inner product of the observation vector with a weight vector $\mathbf{w}$ and compares the result to a scalar threshold τ. If the threshold is exceeded, then the receiver declares that a 1 was sent; otherwise the receiver declares that a 0 was sent.

In binary communication, the prior probabilities are usually equal ($\mathrm{P}_1 = \mathrm{P}_0 = 1/2$). Further, if the two signals satisfy a symmetry property where the quadratic forms $\mathbf{s}_i^T \mathbf{C_N}^{-1}\mathbf{s}_i$ are equal, then the threshold τ becomes zero and it is only necessary to check the sign of the inner product $\mathbf{w}^T x$ in (7.44). (This is the case for the signals depicted in Fig. 7.2.)

An important special case occurs when the samples of the noise are uncorrelated.

This is the so-called "white noise" case. The noise covariance matrix then has the form

$$\mathbf{C}_N = \begin{bmatrix} \sigma_o^2 & 0 & \cdots & 0 \\ 0 & \sigma_o^2 & \cdots & 0 \\ \vdots & \vdots & & \vdots \\ 0 & 0 & \cdots & \sigma_o^2 \end{bmatrix} = \sigma_o^2 \mathbf{I}$$

where σ_o^2 is the variance of one sample of the noise. In this case the weight vector (7.45) in the optimal receiver is proportional to the difference of the signal vectors.

$$\mathbf{w} = \frac{1}{\sigma_o^2}(\mathbf{s}_1 - \mathbf{s}_0)$$

Therefore, for $\tau = 0$, we can interpret the decision rule (7.44) as comparing inner products of $\boldsymbol{x}$ with two signal vectors:

$$\mathbf{s}_1^T \boldsymbol{x} \underset{H_0}{\overset{H_1}{\gtrless}} \mathbf{s}_0^T \boldsymbol{x} \tag{7.46}$$

One more thing is worth mentioning before closing this section. Since $\boldsymbol{x}$ is a Gaussian random vector, the quantity

$$Y = \mathbf{w}^T \boldsymbol{x}$$

computed in the optimal decision rule is a Gaussian random variable. Therefore the scalar quantity Y satisfies all the assumptions for the detection problem introduced in Section 3.8.1 and results derived there concerning the probability of error apply directly to the present case.

7.5.2 Radar/sonar target detection

The radar or sonar target detection problem introduced in Section 3.8.2 shares many features with the binary digital communication problem. Recall that the signal to be detected arises as a scattering (reflection) of a known waveform from a target when the target is present. Once again, the optimal decision rule involves a likelihood ratio test involving two hypotheses:

$$H_1 : \quad \boldsymbol{X} = \mathbf{s} + \boldsymbol{N} \quad \text{target is present (in additive noise)}$$

$$H_0 : \quad \boldsymbol{X} = \boldsymbol{N} \quad \text{no target is present (just noise)}$$

Typically, prior probabilities for the two hypotheses are not available (or in fact meaningful) in this application; rather we choose a threshold in the likelihood ratio test to fix the false alarm probability while maximizing the detection probability (see Section 3.8.2 of Chapter 3). The conditional density functions of the two hypotheses for this problem are

$$f_{\boldsymbol{X}|H_1}(\boldsymbol{x}|H_1) = \frac{1}{(2\pi)^{\frac{K}{2}}|\mathbf{C}_N|^{\frac{1}{2}}} \exp -\tfrac{1}{2}(\boldsymbol{x} - \mathbf{s})^T \mathbf{C}_N^{-1}(\boldsymbol{x} - \mathbf{s})$$

$$f_{\boldsymbol{X}|H_0}(\boldsymbol{x}|H_0) = \frac{1}{(2\pi)^{\frac{K}{2}}|\mathbf{C}_N|^{\frac{1}{2}}} \exp -\tfrac{1}{2}\boldsymbol{x}^T \mathbf{C}_N^{-1}\boldsymbol{x} \tag{7.47}$$

In this case, substituting (7.47) into (7.43) and simplifying leads to the optimal decision rule

$$\boxed{\mathbf{w}^T \boldsymbol{x} \underset{H_0}{\overset{H_1}{\gtrless}} \tau'} \tag{7.48}$$

where $\mathbf{w}$ is the weight vector

$$\mathbf{w} = \mathbf{C}_N^{-1}\mathbf{s} \tag{7.49}$$

and τ' is a threshold chosen to obtain the desired balance of detection and false alarm probability. (The ratio P_0/P_1 is ignored.)

Once again, the case of white noise leads to an important and practical simplification. In this case the weight vector is simply a scaled version of the signal:

$$\mathbf{w} = \frac{1}{\sigma_0}\mathbf{s}$$

In this form the receiver is sometimes referred to as a "correlation receiver" because the inner product on the left of (7.48) compares or "correlates" the known signal with the received observation.

An alternate and especially convenient realization for the receiver is to build a filter whose impulse response is the signal reversed or turned around in time. This realization is known as a "matched filter" and is shown in Fig. 7.3. The convolution of the reversed

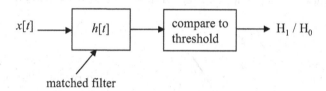

Figure 7.3 Matched filter realization of the optimal detector.

signal and the input performed by the matched filter implements the inner product in (7.44) which is then compared to the threshold.

The matched filter implementation has one further advantage. The expected time of arrival of the signal does not need to be known in order for the inner product to be computed. The filter processes the observation sequence continuously. When a signal arrives and causes the output to exceed the threshold, a detection is announced. Further, the estimated signal *time of arrival* corresponds to the time at which the matched filter has its *largest* output. This principle is used in all radar and sonar systems.

7.5.3 Pattern recognition

The most general cases involving for optimal decision involving Gaussian random vectors is the case of observations with different mean vectors *and* covariance matrices. This happens, for example, when the observations are samples of random signals with different first and second moment statistics.[3] Statistical pattern recognition deals with the general topic of classification: we form an observation vector $\boldsymbol{x}$ and want to decide (in the two-class case) if it came from class 1 (H_1) or class 0 (H_0).

If the two classes of observations have prior probabilities P_i and conditional densities $f_{\boldsymbol{X}|H_i}(\boldsymbol{x}|H_i)$ ($i = 1, 0$), then the decision rule that minimizes the probability of

---

[3] Random signals appear in many branches of electrical engineering and are discussed further in Part 2 of this text.

error is once again the likelihood ratio test

$$\frac{f_{\boldsymbol{X}|H_1}(\boldsymbol{x}|H_1)}{f_{\boldsymbol{X}|H_0}(\boldsymbol{x}|H_0)} \underset{H_0}{\overset{H_1}{\gtrless}} \frac{P_0}{P_1} \qquad (7.43)$$

(repeated here for convenience). In the Gaussian case the class conditional densities have the form

$$f_{\boldsymbol{X}|H_i}(\boldsymbol{x}|H_i) = \frac{1}{(2\pi)^{\frac{K}{2}}|\mathbf{C}_i|^{\frac{1}{2}}} \exp -\tfrac{1}{2}(\boldsymbol{x}-\mathbf{m}_i)^T \mathbf{C}_i^{-1}(\boldsymbol{x}-\mathbf{m}_i) \qquad i = 1, 0 \qquad (7.50)$$

where $\mathbf{m}_i$ and $\mathbf{C}_i$ are the mean vectors and covariance matrices for the classes.

Substituting (7.50) in (7.43) and taking the logarithm leads to the decision rule

$$(\boldsymbol{x}-\mathbf{m}_1)^T \mathbf{C}_1^{-1}(\boldsymbol{x}-\mathbf{m}_1) - (\boldsymbol{x}-\mathbf{m}_0)^T \mathbf{C}_0^{-1}(\boldsymbol{x}-\mathbf{m}_0) + \ln \frac{|\mathbf{C}_1|}{|\mathbf{C}_0|} \underset{H_0}{\overset{H_1}{\gtrless}} 2\ln \frac{P_1}{P_0} \qquad (7.51)$$

which is known as the *quadratic classifier*.[4] This classifier divides the space of observation vectors into two regions with quadratic boundaries as depicted in Fig. 7.4. Observation vectors falling in one region are declared to be in class 1 while observation vectors falling in the complementary region are declared to be class 0.

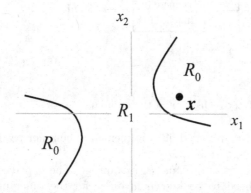

Figure 7.4 Illustration of decision boundaries for quadratic classifier with observation vector $\boldsymbol{x}$ falling in region R_0. (Hypothesis H_0 is chosen when the $\boldsymbol{x}$ falls in region R_0.) Note that a region may consist of two or more subregions as is the case for R_0 in this example.

A few comments are in order here.

1. The terms on the left $(\boldsymbol{x}-\mathbf{s}_i)^T \mathbf{C}_i^{-1}(\boldsymbol{x}-\mathbf{s}_i)$ are known as (squared) Mahalanobis distances. Each term measures the "distance" of the observation vector $\boldsymbol{x}$ from the appropriate mean vector, normalized by the covariance. (In the scalar case these represent the (squared) distance to each mean measured in standard deviations.)

2. When the components of the observation vector are uncorrelated, the covariance matrices are diagonal and the decision rule ends up comparing Euclidean distances with an appropriate threshold.

3. When the covariance matrices are identical, the decision rule simplifies to the form (7.44) derived in Section 7.5.1 for binary communication. In pattern recognition this decision rule is known as a *linear classifier*.

4. When the observations are uncorrelated and the covariances are identical, the decision rule involves a comparison of inner products with the two mean vectors $\mathbf{m}_i$ as in (7.46).

---

[4] Note that the inequality and the ratio of prior probabilities have been reversed.

The quadratic classifier is the optimal decision rule when the distributions of the two classes is Gaussian but performs reasonably well even when the Gaussian assumption is not satisfied. An interesting optimality property can be found by taking the expected value of (7.51) and applying a well-known inequality from information theory. It can then be shown that the inequality in (7.51) is satisfied in favor of H_1 whenever the obsevations are from class 1 and in favor of H_0 when the observation vector is from class 0. Therefore, one can say that regardless of the distribution, the decision rule (7.51) makes the correct decision "on the average."

7.5.4 Vector quantization

Quantization of a random input

In the area of digital signal processing a continuous-time signal is converted to a digital sequence by an analog to digital (A/D) converter or ADC (see Fig. 7.5 (a)). The input to the ADC is a continuous waveform while the output is a sequence of binary numbers.

The conversion occurs in three steps shown in Fig. 7.5(b). The signal is first sampled in time. Then the sample X is quantized into one of a finite number of levels and the

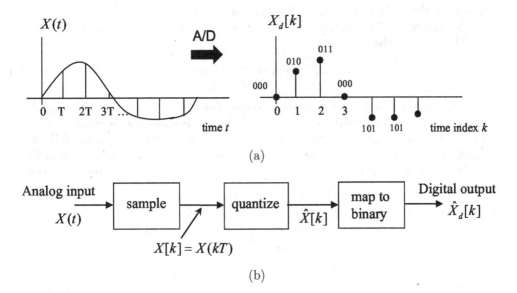

Figure 7.5 Analog to digital conversion. (a) Input and output of the ADC. (b) Steps performed by the converter. Square brackets denote a discrete-time signal.

quantized value $\hat{X}$ is mapped to a binary number. In all of this, quantization is an important step.

The input/output characteristic of the quantizer is depicted in Fig. 7.6. The x axis is divided into a set of quantization intervals with upper limits x_i. If the input signal falls in the ith interval $x_{i-1} < x \leq x_i$ then the quantizer ourput is r_i. The values r_i are called *reconstruction levels* because they can be used to (approximately) reconstruct the signal.

The question arises of how to choose the reconstruction levels. Let us define the *error* when the signal is in the ith interval as the difference $x - r_i$. Then we could choose r_i to minimize the average squared error (mean-square error). The sum over

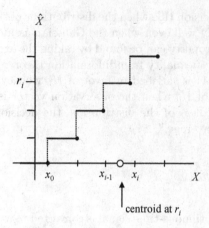

Figure 7.6 Input/output characteristic of the quantizer. The open
dot represents the centroid of the
interval.

all intervals is called the *distortion*

$$D = \sum_i \int_{x_{i-1}}^{x_i} (x - r_i)^2 f_X(x) dx$$

where f_X is the probability density function for X.

To find the optimal reconstruction levels r_i we can take the derivative and set it to zero. This leads to the condition

$$r_i = \int_{x_{i-1}}^{x_i} x f_X(x) dx \qquad \text{for all } i$$

Thus r_i should be chosen as the mean or *centroid* of the ith interval. If it is assumed that f_X is constant in each interval then r_i is the mid point of the interval.

Having chosen the reconstruction levels, we can view quantization in the following manner. The X axis is divided into cells with boundary points x_i as shown in Fig. 7.7. The reconstruction levels are at the centers of the cells and are shown with open

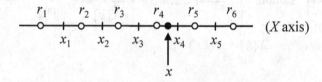

Figure 7.7 Quantization of a scalar input.

dots. When a particular value x falls within the ith cell it is mapped to r_i. (In the case shown $i = 4$.) The reconstruction levels are chosen such that x is mapped to the *closest* one.

Quantization of random vectors

The signals in the previous section are scalar-valued functions of time. With modern technology it is easy to think of signals that are vector-valued. For example, in digital imaging each pixel is represented by a set of three color components, (red, green, blue or some other representation [3]. Thus each pixel can be represented by a vector with three components.

As another example, a digital audio signal is often more than a single sequence of binary numbers. In stereo sound we have left and right channels so samples of the sequence can be thought of as two-dimensional vectors. In surround sound there are even more channels and more dimensionality.

Even when the signal *is* scalar valued it is common to process a block of data all at once. If there are K samples in the block, then we can think of this as a K-dimensional vector. Data compression leads to substantial savings in storage and transmission in all of these applications and vector "quantization" is an important part of the technique.

The idea behind vector quantization is straightforward although the details may be complex. Figure 7.8 shows the case for two dimensions (2-D). The vector space has

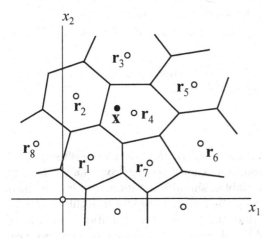

Figure 7.8 A set of cells and codewords for vector quantization in two dimensions. Vector **x** is mapped to codeword $\mathbf{r}_4$.

been divided into a set of non-overlapping cells with centroids $\mathbf{r}_i$ (open circles). An arbitrary vector such as **x** is "quantized" by replacing it with the centroid of the cell in which it resides. The $\mathbf{r}_i$ are the (2-D) reconstruction levels, now called "codewords," and the complete set of these codewords is called the "codebook."

Since the boundaries of the cells in Fig. 7.8 are irregular it would appear to be difficult to determine which cell contains the observation **x**. However, the boundary between any two cells is the locus of points where the distances to the two centroids are equal. Therefore, as long as there is a measure of "distance" in the vector space, **x** is in the cell whose centroid is *closest* to the observation. In the literature on vector quantization the distance between points in the vector space is referred to as the "distortion" and it is possible to use several different measures of distortion including the Mahalanobis distance mentioned earlier.

Training and developing the codebook

The development of the codebook for vector quantization is in theory a difficult endeavor. Recall that in one dimension the boundaries of the cells were predetermined and the representation levels were chosen to be centroids of each region. One could do a similar thing in 2-D but the reconstruction levels could become quite arbitrary and in higher dimensions one really tends to lose insight. Furthermore, we want to choose the codewords in a way that will tend to minimize the overall average distortion.

The practical solution is to use training data; i.e., samples of real data that are representative of the data that one eventually wants to process. If the training data is properly chosen, then it is hoped that the data points will form some natural group-ings or "clusters" in the vector space so that new data can be represented with low distortion. The centroids of these clusters will be used as codewords. A well-known iterative procedure for training via clustering is known variously as the "K-means" algorithm, the Linde-Buzo-Gray algorithm [4], or the Generalized Lloyd algorithm [5]. The essence of the procedure is given below [6, p. 218].

1. Begin with an arbitrary assignment of samples to clusters or begin with an arbitrary set of cluster centers and assign samples to nearest cluster centers.

2. Compute the centroid (sample mean) of each cluster.

3. Reassign the samples to the cluster with the nearest mean.

4. If the classification of the samples has not changed, stop; else go to step 2.

Various additional steps and heuristics can be added to this algorithm including criteria for splitting and merging clusters to better fit the data.

7.6 Summary

Sets of related random variables, such as samples of a random signal, are often treated as random vectors. With this formulation, well-known results from matrix algebra can be applied to describe the random variables, simplify notation, and help in their analysis. An introduction to methods of random vectors is provided in this chapter.

In principle, a random vector is described by the joint distribution and density function of its component random variables. However, in many applications we deal only with first and second moments; the mean of a random vector is itself a vector while the correlation and covariance are matrices. Together these quantities represent all moments of the vector components up to second order. It is common to estimate the mean vector and correlation or covariance matrices from sample data when these parameters are not known *a priori*.

In applications, random vectors are usually subject to transformations of various sorts. Translations and *linear* transformations are particularly easy to apply to the moments of a random vector using matrix notation. Applying these and other nonlinear transformations to the probability density function is difficult but still feasible in some cases. The Gaussian case is of particular interest and importance since the multivariate Gaussian density is specified completely by its mean vector and covariance matrix. When subject to a translation or a linear transformation, the moments are easily computed and the random vector remains Gaussian.

A number of engineering applications are best formulated using random vectors; some of these are discussed in this chapter. In the application of signal detection, introduced earlier in Chapter 3, we can view the problem in a more realistic context where multiple samples of the received signal are used.

References

[1] Charles W. Therrien. *Discrete Random Signals and Statistical Signal Processing*. Prentice Hall, Inc., Upper Saddle River, New Jersey, 1992.

[2] Wilfred Kaplan. *Advanced Calculus*. Addison-Wesley, Reading, Massachusetts, fifth edition, 2003.

[3] William K. Pratt. *Digital Signal Processing: A System Design Approach*. John Wiley & Sons, New York, second edition, 1991.

[4] Y. Linde, A. Buzo, and R. M. Gray. An algorithm for vector quantizer design. *IEEE Transactions on Communication*, COM-26.

[5] Allen Gersho and Robert M. Gray. *Vector Quantization and Signal Compression*. Kluwer, Boston, 1992.

[6] Charles W. Therrien. *Decision, Estimation, and Classification: An Introduction to Pattern Recognition and Related Topics*. John Wiley & Sons, New York, 1989.

Problems

Random vectors and analysis

7.1   The joint PDF of two random variables X_1 and X_2 is given by

$$f_{X_1 X_2}(x_1, x_2) = \begin{cases} 1/12 & -1 \le x_1 \le 1, \ -3 \le x_2 \le 3 \\ 0 & \text{otherwise.} \end{cases}$$

  (a) Find the first moment of the random variables in vector notation.

  (b) Determine the correlation matrix.

  (c) Determine the covariance matrix.

7.2   Show that the estimate (7.20) for the correlation matrix is unbiased.

7.3   Show that the estimate (7.21) for the covariance matrix is unbiased.

7.4   Using (7.19) through (7.21), show that the estimate for the covariance matrix can be put in the form (7.22).

7.5   Find the sample mean, sample correlation matrix, and sample covariance matrix (7.21) for the following data:

$$X_1 = \begin{bmatrix} 1 \\ 1 \end{bmatrix} \quad X_2 = \begin{bmatrix} 1 \\ 2 \end{bmatrix} \quad X_3 = \begin{bmatrix} 2 \\ 0 \end{bmatrix} \quad X_4 = \begin{bmatrix} 0 \\ 1 \end{bmatrix}$$

Check your results using (7.22).

7.6   (a) What are the mean vector and the correlation matrix for the random vector

$$X = \begin{bmatrix} X_1 \\ X_2 \end{bmatrix}$$

where X_1 and X_2 are described by

$$f_{X_1 X_2}(x_1, x_2) = \begin{cases} 1 & 0 \le x_1 \le 1, \ 0 \le x_2 \le 2(1 - x_1) \\ 0 & \text{otherwise.} \end{cases}$$

  (b) What is the covariance matrix?

7.7   The mean vector and covariance matrix for a Gaussian random vector X are given by

$$\mathbf{m}_X = \begin{bmatrix} 1 \\ 1 \end{bmatrix} \qquad \mathbf{R}_X = \begin{bmatrix} 4 & -1 \\ -1 & 3 \end{bmatrix}$$

  (a) Compute the covariance matrix $\mathbf{C}_X$.

  (b) What is the correlation coefficient $\rho_{X_1 X_2}$?

    (c) Invert the covariance matrix and write an explicit expression for the Gaussian density function for X.

7.8 Show that when a Gaussian random vector has uncorrelated components, the multivariate Gaussian density function factors into a product of one-dimensional Gaussian densities.

Transformations

7.9 Using the definition (7.16) for the covariance matrix and assuming the linear transformation $Y = AX$, show that

$$C_Y = AC_X A^T$$

Repeat the proof, this time using the relation $C_Y = R_Y - m_Y m_Y^T$ together with (7.28) and (7.29). Also show for a square matrix A that $|R_Y| = |A|^2 |R_X|$. (The symbol $|\cdot|$ denotes the determinant.)

7.10 A pair of random variables X_1 and X_2 is defined by

$$X_1 = 3U - 4V$$
$$X_2 = 2U + V$$

where U and V are independent random variables with mean 0 and variance 1.

    (a) Find a matrix A such that

$$\begin{bmatrix} X_1 \\ X_2 \end{bmatrix} = A \begin{bmatrix} U \\ V \end{bmatrix}$$

    (b) What are R_X and C_X of the random vector X?

    (c) What are the means and variances of X_1 and X_2?

    (d) What is the correlation $E\{X_1 X_2\}$?

7.11 Repeat Prob. 7.10 if $E\{U\} = 1$ and $E\{V\} = -1$. Everything else remains the same.

7.12 A random vector X has independent components. Each component has an exponential PDF $\lambda e^{-\lambda x_i} u(x_i)$ where u is the unit step function. A transformation is applied as follows

$$Y = AX \quad \text{where} \quad A = \frac{1}{\sqrt{2}} \begin{bmatrix} 1 & -1 \\ 1 & 1 \end{bmatrix}$$

    (a) Find and sketch $f_X(x)$ in the x-plane.

    (b) Find and sketch $f_Y(y)$ in the y-plane. Don't forget to show where the density is non-zero.

    (c) Are the components of Y uncorrelated? Show why or why not.

Cross correlation and covariance

7.13 Show that the cross correlation matrix and the cross covariance matrix for two random vectors satisfy the relation (7.40). Specialize your result for $Y = X$ and thus prove (7.18).

7.14 Let X and Y be two random vectors of the same dimension and form the sum

$$Z = X + Y$$

(a) Under what conditions is the correlation matrix for Z the sum of the correlation matrices for X and Y?

(b) Under what conditions is the covariance matrix for Z the sum of the covariance matrices for X and Y?

7.15 Let random vector Y be defined by

$$\begin{bmatrix} Y_1 \\ Y_2 \end{bmatrix} = \begin{bmatrix} 3 & 2 & 1 \\ 1 & -2 & 1 \end{bmatrix} \begin{bmatrix} X_1 \\ X_2 \\ X_3 \end{bmatrix}$$

The correlation matrix for X has the form

$$\mathbf{R}_X = \begin{bmatrix} 9 & 3 & 1 \\ 3 & 9 & 3 \\ 1 & 3 & 9 \end{bmatrix}$$

(a) Find the correlation matrix $\mathbf{R}_Y$ for Y.

(b) Find the cross correlation matrix $\mathbf{R}_{XY}$.
Hint: $\mathbf{R}_{XY} = E\{XY^T\}$. Substitute $Y = AX$ and simplify.

7.16 Random vectors are independent if $f_{XY}(x, y) = f_X(x)f_Y(y)$. Show that when X and Y are independent, they are also uncorrelated.

Application to signal processing

7.17 Suppose that in a signal detection problem, the signal vector consists of five samples

$$\mathbf{s} = \tfrac{1}{3}\begin{bmatrix} 1 & 2 & 3 & 2 & 1 \end{bmatrix}^T$$

The noise samples are uncorrelated with variance $\sigma_o^2 = 1$ and mean 0.

(a) Find the explicit form of the decision rule if the prior probability for the signal is $P_1 = 2/3$.

(b) How does the result change if $P_1 = 1/2$?

7.18 Through algebraic simplification, show that the optimal decision rule of (7.43) can be reduced to (7.44) and derive the expression for the threshold τ.

7.19 Show that when the covariance matrices $\mathbf{C}_1$ and $\mathbf{C}_0$ are equal, the quadratic classifier (7.51) reduces to a linear form

$$\mathbf{w}^T \mathbf{x} \underset{H_0}{\overset{H_1}{\gtrless}} \tau$$

where $\mathbf{w}$ is an appropriate weight vector and τ is a threshold. Sketch the form of the decision boundary in the 2-dimensional case.

7.20 The signal vector in a detection problem is given by

$$\mathbf{s} = \begin{bmatrix} 2 & 2 \end{bmatrix}^T$$

The noise has zero mean and a covariance matrix of the form

$$\mathbf{C}_N = \sigma^2 \begin{bmatrix} 1 & \rho \\ \rho & 1 \end{bmatrix}$$

where $\sigma^2 = 2$ and $\rho = 1/\sqrt{2}$.

(a) Find the explicit form of the decision rule as a function of the parameter $\eta = \ln(P_0/P_1)$.

(b) The contour for the density of the received vector when no signal is present is defined by

$$f_{\mathbf{X}|H_0}(\mathbf{x}|H_0) = f_{X_1 X_2|H_0}(x_1, x_2|H_0) = \text{constant}$$

From (7.47) this is equivalent to requiring

$$\mathbf{x}^T \mathbf{C}_N^{-1} \mathbf{x} = \text{constant}$$

Carefully sketch or plot this contour in the x_1, x_2 plane. You can set the constant equal to 1.

Hint: To put the equation for the contour in the simplest form, consider rotating the coordinate system by the transformation $y_1 = (x_1 + x_2)/\sqrt{2}$ and $y_2 = (x_2 - x_1)/\sqrt{2}$ (i.e., y_1 and y_2 become the new coordinate axes).

(c) Sketch or plot the contour for the density of $\mathbf{X}$ when the signal *is* present.

Hint: The form of the contour is the same except for a shift in the mean vector.

(d) The decision boundary between regions in the plane where we decide H_1 and where we decide H_0 is defined by the equation

$$\mathbf{w}^T \mathbf{x} = \tau$$

where $\mathbf{w}$ and τ are given in Section 7.5.1. Plot the decision boundary in the x_1, x_2 plane for $\eta = 0$, $\eta = 1/2$, and $\eta = -1/2$. Observe that since the decision boundary interacts with the two Gaussian density functions, changing η provides a trade-off between the probability of false alarm and the probability of detection.

7.21 A set of five codewords for a vector quantization algorithm is shown below. Draw the cells corresponding to the five codewords. Which codeword will be used to represent the observation $\mathbf{x} = [\,2 \quad 3\,]^T$?

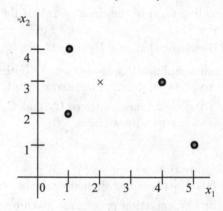

Computer Projects

Project 7.1

In this project you are asked to estimate the parameters (mean and covariance) of two-dimensional Gaussian data and plot the contours of the density functions.

The files x01.dat, x02.dat, and x03.dat from the data package for this book each contain 256 realizations of two-dimensional Gaussian random vectors. The kth line of the file contains the kth realization of the random vector:

$$\mathbf{x}^{(k)} = \begin{bmatrix} x_1^{(k)} & x_2^{(k)} \end{bmatrix}^T$$

An *estimate* for the mean vector and covariance matrix can be made using the formulas (7.19) and (7.22) of Section 7.2.2.

For each of the three data sets, do the following:

1. Estimate the mean vector and covariance matrix using the above formulas. Identify the parameters m_1, m_2, σ_1, σ_2, c_{12}, and ρ and list their values in a table.

2. Make a scatter plot of the data. That is, plot the vectors as points in the x_1, x_2 plane without any connecting lines.

3. The contours of a density function are defined by the condition

$$f_X(x) = \text{constant}$$

For the Gaussian case the contours are ellipses (see Section 5.4 of Chapter 5). On the same plot where you plotted the data, plot the ellipse representing the contour of the Gaussian density function. Choose the constant appropriately, and observe that most of the data fall within the ellipse.

4. Tell what the shape and orientation of the ellipse indicates for each data set.

MATLAB programming notes

You may use the function "plotcov" from the software package for this book to do the plotting. Try setting the parameter to 2 standard deviations.

Project 7.2

Computer Project 3.2 considered a detection problem where the "signal" transmitted consisted of a single value (A) observed in noise. In this project, you will consider a more realistic detection problem where the signal is an entire *sequence* of values. The signal is described by the discrete-time sequence

$$s[k] = A\cos(2\pi k/12) \quad k = 0, 1, \ldots, 11$$

where the amplitude A is to be specified.

Following the procedure in Section 7.5, this sequence is formed into a 12-dimensional vector $\mathbf{s}$ which is transmitted in Gaussian noise. The noise is assumed to have mean 0 and autocorrelation function

$$R_N[l] = 2 \cdot (0.85)^{|l|} \quad -\infty < l < \infty$$

The covariance matrix $\mathbf{C_N}$ thus has elements $c_{ij} = R_N[|i - j|] = 2 \cdot (0.85)^{|i-j|}$. Using these facts,

1. Develop and print the explicit form of the covariance matrix and its *inverse*.

2. Find the optimal weight vector $\mathbf{w}$ and threshold τ in the optimal decision rule (7.44) assuming $P_0 = P_1 = \frac{1}{2}$. Note that the weight vector depends on the signal amplitude A.

3. Observe that if we define $t = \mathbf{w}^T \mathbf{X}$, then t is a Gaussian random variable. (Tell why.) Plot the PDF for t under each hypothesis, H_0 and H_1. Assume that $A = 2$.

4. Set the threshold τ so that $p_{fa} = 0.01$. What is the corresponding value of p_d assuming $A = 2$? What value of A is necessary so that $p_D \geq 0.95$?

5. For the value of A determined in the last step, plot the ROC for this problem by sweeping over suitable values of the threshold τ.

PART II

Introduction to Random Processes

8

Introduction to Random Processes

This chapter provides a basic introduction to random processes. In previous chapters, the characterization of uncertainty did not concern itself with time. In applications that involve signals, noise, dynamic systems, and so on, one needs to account for the fact that the underlying phenomena *are* dependent on time; thus our probabilistic models need to be augmented to account for this time dependence. In the extension of our studies from random variables to random processes several new concepts and tools are required. Some of these tools are provided in this chapter.

The chapter begins with some basic definitions and examples and moves on to discuss the ensemble concept and some properties and types of random processes. The emphasis in this chapter is to describe random processes from a basic probabilistic point of view without introducing correlation and power spectral density, which are brought up in later chapters.

The chapter also provides a brief discussion of some particular known types of random processes that are important in engineering applications. Both discrete time and continuous time processes are presented, but in an introductory way consistent with our intention. A more extensive treatment of these random processes is reserved for the chapters to follow.

8.1 Introduction

In electrical and computer engineering we frequently encounter phenomena that are random functions of time. Some examples are the power line voltage during an electrical storm, the signal strength in a cellular telephone connection or radio transmission, the number of packets entering a server on the internet, the number of "hits" on a certain web page, and many more. Such random phenomenon are modeled by what is called a *random process* or a *stochastic process*. Even when we do not consider certain phenomena to be truly "random" it may be useful to use a probabilistic model.[1] For example, the stream of bits in a message may have clear meaning but it is treated as a random process because a deterministic model is far too complex to be useful.

8.1.1 The random process model

From the point of view provided by probability theory, the random process model is similar to that of a random variable. A random variable is a mapping of events in the sample space to points on the real line, while a random process is a mapping of the sample space into function space (see Fig. 8.1) where the functions are dependent on time.[2]

---

[1] Webster's Collegiate Dictionary defines "random" as "lacking a definite plan, purpose or pattern," which is not too far away from its use in modern American slang.

[2] The independent variable for a random process is usually considered to be time although it is possible for a random process to be dependent on a spatial parameter or even a set of spatial parameters. Such random processes are usually referred to as "random fields."

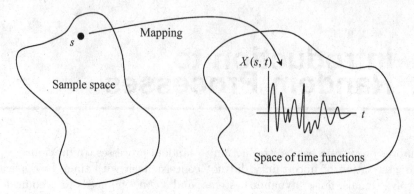

Figure 8.1 The random process model.

Stated another way, a random process $X(s, t)$ is a family of functions for which the underlying mapping takes into account both time t and the outcomes of the random experiment s. A time sample of a random process $X(s, t_1)$ at time $t = t_1$ represents a random variable. Consequently, a random process can be thought of as a continuum of random variables.

8.1.2 Some types of random processes

The values taken on by a random process, like the values taken on by a random variable may be continuous or discrete. In addition, the domain of the time function may be continuous or discrete. This gives rise to four types of random processes. Let us consider an example of each type.

Sound. Natural sound, including speech, music and noise, is a phenomenon caused by pressure vibrations traveling in air or other media. Sounds are a function of a continuous time parameter and correspond to pressure values that are real numbers. Most random processes measured in the physical world, like sound, are functions of *continuous* time and take on *continuous* real values.

Digital audio. Digital audio is at the opposite end of the spectrum of possibilities. A digital audio signal is represented by a sequence of binary numbers that is read out and converted to natural sound by a computer, cell phone, CD player or other device. A digital (N-bit) audio sequence can be modeled as a random process that takes on (2^N) *discrete* values at *discrete* points in time.

Network traffic. Think of packets arriving at a node in a store-and-forward digital network. When the node is busy, arriving packets are placed in a queue and the length of the queue (versus time) is a random process. Since the process may change value at any time due to arrival or departure of a packet, this process is defined for *continuous* time. The values of the process, however, are integers, which are *discrete*.

Time series. Not all discrete random processes derive from electrical or computer engineering applications. Statisticians have studied discrete random processes for years under the topic known as time series analysis. (e.g. [1, 2]). Examples of time series are measures of crop yields on a yearly basis, rainfall data, stock values, or even balances in a bank account. Although these time series frequently take on values that are *continuous*, the processes are inherently discrete in time; measurements that produce the data occur naturally at *discrete* time epochs and measurements between time epochs usually do not exist or are meaningless. Fortunately, the substantial body of work carried out by workers in time series is directly applicable to engineers in signal processing and *vice versa*. Hence the applications in recent years have tended to support each other.

In the literature, it is common practice to drop the variable s and denote the random process $X(s, t)$ as simply $X(t)$. Unless there is a specific reason to show the dependence, that procedure will be followed here as well. Further, when the time variable is discrete, we will use square brackets as in $X[t]$ or $X[k]$ to denote the random process. There will be no specific notation used to distinguish a discrete-valued process from a continuous-valued process; the distinction is usually clear from the application.

8.1.3 *Signals as random processes*

Random processes are used to a large extent in the study of signals and noise. Most signals of interest originate in analog form. For example, sound and light as captured by a microphone or camera are analog in nature. Due to many advantages of digital processing, however, the analog signals are frequently converted into digital form. For light, the conversion is done directly in the (digital) camera. For sound, an analog to digital converter (ADC) such as the one depicted in Fig. 8.2 may be used. Two principal

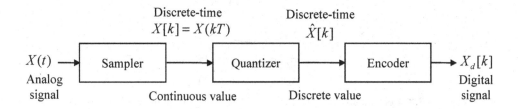

Figure 8.2 Block diagram of an analog-to-digital converter.

blocks of an analog-to-digital converter are the sampler and quantizer. The input $X(t)$ is a continuous-time and continuous-value signal; it has infinite precision in both time and value. The output of the sampler is a discrete-time continuous-value signal. The quantizer takes each of the samples with infinite precision and approximates them to finite values. This step is followed by an encoder that converts each of the quantized samples into a binary sequence, thus creating a discrete-time discrete-value or *digital* signal.

The digital representation that occurs after the quantization is invariably an approximation of the analog signal, which means that a finite amount of error or "noise" is introduced in the process of converting an analog waveform into a digital form. Even if the signal being digitized is completely deterministic, the approximation noise introduced during digitization is a random process. (Further discussion of quantizers and noise can be found in Chapter 7 and Appendix D.)

In the literature, the terms *random process* and *random signal* are often used interchangeably. It may seem appropriate to call an electrical signal exhibiting some form of uncertainty in wave shape, frequency, or phase a random process, but one needs to exercise caution in the use of these terms. Bear in mind that a random process is an abstraction based on a family of functions representing possible outcomes of an experiment. On the other hand, a random signal can be considered as a single realization of the random process.

8.1.4 *Continuous versus discrete*

The remainder of this text discusses both continuous-time and discrete-time random processes. Both types are encountered in topics of interest to electrical and computer engineers. Since the continuous or discrete nature of the value of a random process is usually clear from the application, we will shorten the description and use the term "continuous" to refer to a random process that is continuous in time and "discrete" to one that is discrete in time.

Figure 8.3 illustrates a way to view the relationship between continuous and dis-

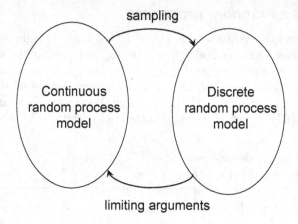

Figure 8.3 Relations between continuous- and discrete-time random processes.

crete random processes. Since many discrete random processes are formed by sampling continuous processes, it seems clear that the tools and models for at least some continuous random processes will lead to corresponding tools and models for the sampled versions. Not all discrete random processes have a continuous counterpart, however. We shall see that it is possible to define certain continuous random processes as limiting cases of a discrete random process as the sampling rate becomes very large. In fact, since the quantum nature of the physical world would tend to argue against any truly continuous model for a real-world phenomenon, one could argue that the discrete random process model is more fundamental. However that may be, it is still useful to treat both cases and develop the appropriate mathematical tools. In aspects of a problem related to measurements in the physical world, a continuous model is typically needed. In aspects related to processing the observations, however, a discrete model best represents the way the data is actually processed using modern methods of digital signal processing.

8.2 *Characterizing a Random Process*

The results and discussion in this section apply to both continuous and discrete random processes. In order not to be repetitive, however, we have chosen to use the notation $X(t)$ pertaining to a continuous process and draw our sketches of random processes as continuous curves. So there will be no confusion, we make a note when a particular result does not apply to the discrete case in an obvious way.

8.2.1 The ensemble concept

In the previous section we defined a random process as a family of functions $X(s,t)$, as a function of both time t and the outcome s of a random experiment. This family of functions is known as an *ensemble*. Members of the ensemble are also referred to as *sample functions* or *realizations*. Ideally, an ensemble may consist of an infinite number of sample functions. Graphically, as shown in Fig. 8.4, an ensemble can be considered as having two directions, one corresponding to time and the other corresponding to the realization. Given a sample function, i.e., for a fixed realization s, one can proceed along the time direction. On the other hand, if time is fixed, we have a collection of time samples corresponding to the various possible values of a random variable.

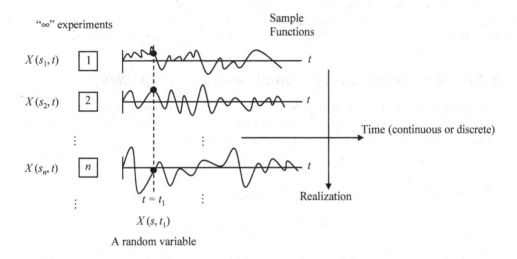

Figure 8.4 Graphical illustration of an ensemble representing a random process.

Since any sample of a random process is a random variable, a random process can be considered a continuum of random variables of which theoretically there may be an infinite number. As shown in previous chapters, a random variable is characterized by a cumulative distribution function (CDF) and a probability density function (PDF) or probability mass function (PMF). This suggests that every time sample of a random process has a CDF and a PDF (or PMF).

Let's speak about our notation for the PDF because notation for the CDF or PMF is similar. If we consider a single sample $X(t_0)$ of a random process $X(t)$, then the PDF of the sample, i.e., the PDF of the random process at the time instant $t = t_0$, is denoted by $f_{X(t_0)}(x_0)$. To shorten notation, we may drop the argument x_0 and simply write $f_{X(t_0)}$. Extending this idea to two samples at times t_0 and t_1, we now write the joint PDF of these samples as $f_{X(t_0)X(t_1)}(x_0, x_1)$, once again dropping arguments x_0 and x_1 if it is desired to simplify notation.

This procedure can be naturally generalized to any number of samples of $X(t)$. Let us denote the sample times as $t_i = t_0 + \tau_i$ for $i = 1, \ldots, n$ where τ_i is the sample time

relative to the first sample. Then the CDF and PDF for $n + 1$ samples are written as
$F_{X(t_0)X(t_0+\tau_1)\cdots X(t_0+\tau_n)}(x_0, x_1, \ldots, x_n)$ and $f_{X(t_0)X(t_0+\tau_1)\cdots X(t_0+\tau_n)}(x_0, x_1, \ldots, x_n)$.

The form of the CDF and PDF brings up an important definition for random processes. It is stated here in terms of the PDF:

A random process is *stationary* (in the strict sense) if the PDF $f_{X(t_0)}$ and the joint PDFs $f_{X(t_0)X(t_0+\tau_1)\cdots X(t_0+\tau_n)}$ of all orders n are independent of the starting time t_0.

An equivalent definition results if the PMF or CDF is substituted for the PDF.

The above definition indicates that the probabilistic description of a stationary random process does not change if the sampling starts at a different point in time (t_0). All that matters is that the relative spacing of the samples, i.e., the spacing *between* samples, remains the same. Stationarity is an important property because without it any algorithms to process the data must depend on the starting point in time.

8.2.2 Time averages vs. ensemble averages: ergodicity

Since time samples of a random process $X(t_0), X(t_1), \ldots$ are random variables, statistical averages based on the expectation operation can be calculated. For example, the mean of a random process at some time t_o is written as

$$m_X(t_o) = \mathcal{E}\{X(t_o)\}$$

The expectation is carried out using the PDF of $X(t)$ at time t_o:

$$\mathcal{E}\{X(t_o)\} = \int_{-\infty}^{\infty} x f_{X(t_o)}(x) dx$$

Since the integral may be regarded as a "summation," the expectation can be interpreted as averaging "down the ensemble." In other words, it is an average over the various sample functions that comprise the ensemble. In general, the mean is a function of where (in time) we have chosen to compute this average.

If it is not possible to perform an ensemble average, it may be possible to base the analysis on averages taken in time over a *single* realization of the random process. This processing is referred to as "time averaging" or "temporal averaging." The time-average mean of a random process $X(t)$ is defined as

$$\langle x(t) \rangle = \lim_{T \to \infty} \frac{1}{2T} \int_{-T}^{T} x(t) dt \tag{8.1}$$

where $x(t)$ is a particular sample function of the random process $X(t)$. The brackets $\langle \rangle$ are used to denote the operation specified on the right side of this equation and can be used to denote other time averages. For example, a time-average correlation between two random processes $X(t)$ and $Y(t)$ would be calculated as

$$\langle x(t)y(t) \rangle = \lim_{T \to \infty} \frac{1}{2T} \int_{-T}^{T} x(t)y(t) dt$$

For a discrete random process $X[k]$, the time-average mean is written as

$$\langle x[k] \rangle = \lim_{K \to \infty} \frac{1}{2K} \sum_{k=-K}^{K-1} x[k]$$

which is analogous to (8.1).

It is helpful to think of the expectation and the time-based averages in the context of the ensemble illustrated in Fig. 8.4. As stated earlier the expectation $\mathcal{E}\{\cdot\}$ (or statistical average) can be thought of as averaging down the ensemble. On the other hand, a time average $\langle \cdot \rangle$ is carried out along the time axis. In this case, the assumption is that only one sample function (not the entire ensemble) is available.

Temporal averages can replace statistical or ensemble averages if the results are in some sense equivalent. This equivalence is a property known as *ergodicity*. Not all random processes are ergodic, however, as shown by the following example.

Example 8.1:   Consider a random process

$$X(t) = A$$

where A is a binary-valued random variable taking on values of ± 1 with equal probabilities. The mean of this process is

$$\mathcal{E}\{A\} = \frac{1}{2}(+1) + \frac{1}{2}(-1) = 0$$

The sample functions of the random process are "d.c." waveforms of two different magnitudes: $+1$ or -1. Each sample function is a constant function of time. Given an ensemble of these sample functions, the statistical mean of this random process (computed above) is zero.

To compute the time-averaged mean of this random process, let us assume that the given sample function is $x(t) = A$. Then the time-averaged mean for this sample function is

$$\langle x(t) \rangle = \lim_{T \to \infty} \frac{1}{2T} \int_{-T}^{T} x(t)dt = A \lim_{T \to \infty} \frac{1}{2T} \int_{-T}^{T} dt = A$$

Since A can only be either $+1$ or -1 but never 0, the time-averaged mean and the statistical mean are not the same. As a result, this is not an ergodic random process.

□

Many random processes of practical importance *do* satisfy ergodicity, at least with respect to some of their moments. Random processes whose statistical and temporal means are the same are called "mean ergodic." This is illustrated in the next example.

Example 8.2:   Consider the random process

$$X(t) = A \sin(\omega t + \Theta)$$

where A and ω are known parameters while Θ is a uniform random variable in the interval $-\pi$ to π. The mean of this random process is given by

$$m_X(t) = \mathcal{E}\{X(t)\} = \int_{-\pi}^{\pi} A \sin(\omega t + \theta) f_\Theta(\theta) d\theta$$

$$= \frac{A}{2\pi} \int_{-\pi}^{\pi} \sin(\omega t + \theta) d\theta = 0$$

where we have used the PDF of Θ, $f_\Theta(\theta) = 1/2\pi$ for $-\pi \leq \theta \leq \pi$ and noticed that the integral over one period of the sinusoid is zero. Thus $X(t)$ is a zero-mean random process.

The time-averaged mean for a sample function $x(t)$ is

$$\bar{m}_X = \langle x(t) \rangle = \lim_{T \to \infty} \frac{1}{2T} \int_{-T}^{T} x(t)dt = \lim_{T \to \infty} \frac{A}{2T} \int_{-T}^{T} \sin(\omega t + \theta)dt = 0$$

The limit on the right is zero because the integral can only take on finite values. Since m_X and $\bar{m}_X$ are identical, the random process is mean ergodic.

□

The preceding examples serve to illustrate the concept of ergodicity by focusing on the mean of the random process. In practice, the ergodicity concept is most useful if, in addition, moments of second and higher order obey the property. Random processes where statistical averages of second moments can be replaced by time averages are said to be *correlation ergodic* while those for which statistical moments of *all* orders can be replaced by time averages are said to be *strictly ergodic*. A particularly important quantity is the *autocorrelation* introduced in the next chapter, whose temporal form is

$$\langle x(t)x(t - \tau) \rangle = \lim_{T \to \infty} \frac{1}{2T} \int_{-T}^{T} x(t)y(t - \tau)dt$$

This quantity measures correlation between samples of the process at time $'t'$ and time $t - \tau$. Notice that since the average is over the time variable, ergodicity does not hold or indeed *make sense* unless the random process is stationary.

8.2.3 Regular and predictable random processes

Section 8.1 described four classes of random processes according to whether the domain and value of the random process was continuous or discrete. Another characterization of random processes that cuts across those previous classifications is discussed here.

Electrical signals in general can be categorized into deterministic and random signals (processes). The parameters of a deterministic signal are completely known to the designer and hence there is no uncertainty about the signal values. Random signals, as realizations of random processes can be further classified as *regular* or *predictable*. To explain the difference between these two it suffices to show an example of the latter.

In a predictable random process, one or more parameters can be random variables but the form of the signal itself is deterministic. For example, consider the signal

$$X(t) = A \sin(2\pi ft + \theta)$$

where f and θ are known constants, and A is a random variable. This signal fits the description of a random process; however, once a few samples of the process are observed, the signal is known precisely for all time. Thus future values of a predictable random process can be estimated with zero error from knowledge of the past values.[3] Figure 8.5 (a) shows two members of the ensemble when amplitude is a random variable while Fig. 8.5 (b) shows two ensemble members when phase is a random variable. If we were to view the complete ensemble for either of these cases we would see a variety of waveforms, all with similar shape, but with differences caused by the values taken on by the random variable (e.g., A or A').

-The waveform of a regular random process has no definite shape and its values cannot be predicted without error from knowledge of its past values. Figure 8.6 depicts

---

[3] A complete discussion of predictable random processes is an advanced topic above the level of this book. The sense in which the process is predictable, however, can be made mathematically precise using concepts of mean-square error or various other probabilistic measures.

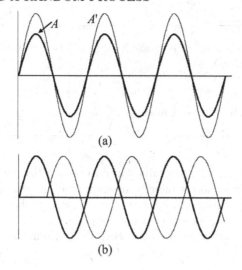

(a)

(b)

Figure 8.5 Examples of predictable random processes. (a) Amplitude is a random variable. (b) Phase is a random variable.

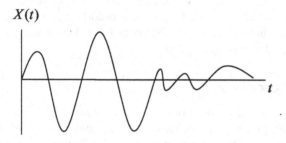

$X(t)$

t

Figure 8.6 Realization of a typical regular random process (non-predictable).

a realization of a regular random process. Although the sample function looks smooth, regular random processes can be very badly behaved from an analytic point of view. The waveform of other realizations in the ensemble would "look" similar but would be different in detail. Electrical noise and some segments of speech are examples of regular random processes.

8.2.4 Periodic random processes

Many random processes exhibit some form of periodicity. This property can be useful, for example, in separating some signals from noise. While there are many ways that periodicity can be exhibited, we will define a random process to be periodic if the distribution is periodic in all of its arguments:

A random process is *strictly periodic* if there exists a value P such that the PDF satisfies the condition

$$f_{X(t_0)X(t_1)\cdots X(t_n)} = f_{X(t_0+k_0P)X(t_1+k_1P)\cdots X(t_n+k_nP)}$$

for all orders n and all sets of integers $k_0, k_1, \ldots k_n$.

Other forms of periodicity are generally special cases of this condition. For example the property known as cyclostationarity in signals occurs when the integers in the above definition satisfy the constraint $k_0 = k_1 = \ldots = k_n = k$.

Observe that a random process need not be stationary in order to be periodic. If the random process is stationary, however, then the periodicity condition only applies to the time variables $\tau_1, \tau_2, \ldots \tau_n$ (see p. 254).

8.3 Some Discrete Random Processes

This section describes a few well-known types of discrete random processes and some of their applications. Many of these are discussed in further detail later in the text.

We have chosen to discuss discrete random processes first because they are easiest to describe at an introductory level. Continuous random processes then follow in the next section. Both of these sections set the stage for a deeper analysis that is carried out in the remaining chapters of the text.

8.3.1 Bernoulli process

One of the simplest types of random processes is represented by a data sequence $X[k]$ with independent samples that takes on only two values, 1 and -1 with probability p and $1 - p$ respectively. This is known as a Bernoulli process, for which a typical realization is shown in Fig. 8.7. A Bernoulli process could be generated by flipping

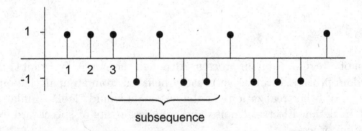

Figure 8.7 Sample function from a Bernoulli random process.

a coin and recording heads as 1 and tails as -1 or vice versa. Although the process is very simple it can be a reasonable model for binary data transmission, errors in a sequence, and other binary events. When the parameter p is equal to $\frac{1}{2}$ the Bernoulli process can be referred to as "binary white noise."

It is fairly easy to show that the Bernoulli process is both stationary and ergodic. To demonstrate stationarity, first notice that the probability of any subsequence of the Bernoulli process can be written easily. For example, consider the subsequence

$\{+1, -1, +1, -1, -1\}$ beginning at some arbitrary point, say $k = 3$. The probability of this subsequence is

$$\Pr\left[+1, -1, +1, -1, -1\right] = p \cdot (1 - p) \cdot p \cdot (1 - p) \cdot (1 - p) = p^2(1 - p)^3$$

and this probability of five such consecutive values is clearly independent of where they are found in the sequence. A moment's thought reveals that the probability of any subsequence can be computed in this way and is independent of the starting point. Therefore the random process is (strictly) stationary.

The ergodicity of the process is a little harder to prove without a proper mathematical definition. Imagine, however, an ensemble of sequences of a Bernoulli process depicted as in Fig. 8.4. Since 1s and -1s occur independently both down the ensemble as well as across the ensemble (in time), both statistical averages and temporal averages produce the same results.

The moments of any sample of the Bernoulli random process are given by

$$\mathcal{E}\{X[k]\} = 2p - 1 \quad \text{and} \quad \mathrm{Var}\,[X[k]] = 4p(1 - p) \tag{8.2}$$

(see Example 9.12 of Chapter IX). For $p = \frac{1}{2}$ the mean is 0 and the variance is $1.$[4]

An interesting use of the Bernoulli process is in what is known as non-parametric detection. In this case only the *sign* ($+$ or $-$) of a random signal is observed and the sequence of signs is modeled as a Bernoulli process. The following example [3] develops this procedure.

Example 8.3: A *counting process* is a random process that counts the number of positive values in a Bernoulli process from some starting time up to a given time k. Taking the starting time as $i = 1$ yields

$$Y[k] = \sum_{i=1}^{k} \tfrac{1}{2}(X[i] + 1)$$

Each time $X[i]$ takes on a value of $+1$ the corresponding term in the sum is equal to one and each time $X[i]$ is -1, the corresponding term is zero. Thus $Y[k]$ represents the number of positive values in the sequence from $i = 1$ to $i = k$.

The counting process can be used to model the evolution of observations in a nonparametric signal detection procedure known as a sign test [4]. In this test a positive-valued signal is observed in zero-mean noise as shown in the following figure and it is desired to determine if the signal is present.

---

[4] Because the Bernoulli random process is defined to have values of 1 and -1 instead of 1 and 0, the mean and variance are different from those of the Bernoulli random variable defined earlier in this text.

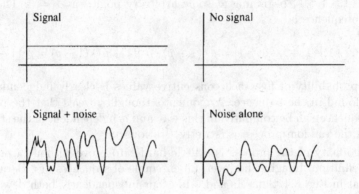

To give the detection procedure robust performance for noise processes with a wide variety of statistical characteristics, only the *sign* of the observations is used in the detector. This is shown in the figure below.

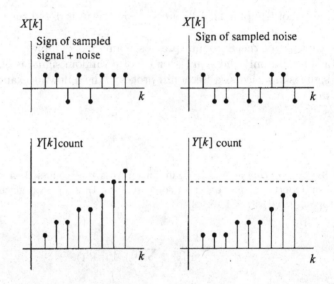

Intuitively, for any fixed number of observations of the received sequence, we would expect to find that there are more positive values if the signal is present, and about equal numbers of positive and negative values if the signal is not present. The number of positive values, which can be modeled as a counting process, is therefore compared to a threshold (shown as a dashed line) to make a detection decision.

Notice that the counting process looks like a set of steps occuring at random times. Observe that at time k, $Y[k]$ can take on any integer value between 0 and k. To compute the probability distribution of this random process suppose that $Y[k]$ is equal to some value n. This means that there are n positive values and $k - n$ negative values between $i = 1$ and $i = k$. These can occur in *any* order. The number of combinations of positive and negative values is given by the binomial coefficient

$$\binom{k}{n} = \frac{k!}{n!(k-n)!}$$

Thus it follows that

$$\Pr\left[Y[k] = n\right] = \begin{cases} \binom{k}{n} p^n (1-p)^{k-n} & n = 0, 1, \ldots, n \\ 0 & \text{otherwise} \end{cases}$$

This is the binomial distribution[5] which can be used to determine the probability that the count exceeds the threshold at any particular time k.

□

8.3.2 Random walk

The Bernoulli process forms the basis for another type of discrete random process known as a random walk. The random walk is defined by

$$X[k] = X[0] + \sum_{i=1}^{k} B[i] \quad k > 0 \tag{8.3}$$

where $X[0]$ is an initial condition and $B[k]$ is a Bernoulli process taking on values of $+1$ and -1 with probability p and $1-p$.

The value of a random walk can be thought of as the position of a point on a line that is subject to a random displacement of $+1$ or -1 from its current position at each epoch of time. If the point started at the coordinate $X[0]$ at time 0 then its coordinate at time k, after k random displacements, would be $X[k]$. Figure 8.8 shows a typical

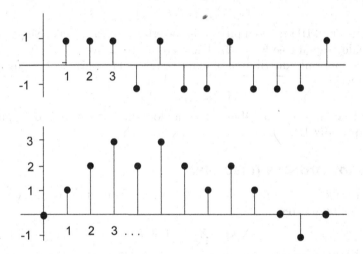

Figure 8.8 Segment of a random walk and the underlying Bernoulli process.

segment of a random walk $X[k]$ and the Bernoulli process $B[k]$ on which it is based.

The mean and the variance of the random walk can be computed most easily by noting that the $B[k]$ are independent and using the previous results (8.2) for a Bernoulli process. If it is assumed that $X[0] = 0$, the mean is given by

$$\mathcal{E}\{X[k]\} = \sum_{i=1}^{k} \mathcal{E}\{B[i]\} = (2p - 1) \cdot k \tag{8.4}$$

---

[5] See Eq. 3.5 of Chapter 3 with k and n interchanged.

while the variance is

$$\text{Var}\,[X[k]] = \sum_{i=1}^{k} \text{Var}\,[B[i]] = 4p(1-p) \cdot k \tag{8.5}$$

For $p \neq \frac{1}{2}$ both the mean and the variance of the random walk are proportional to k. For $p = \frac{1}{2}$ (the binary white noise process) the mean is zero and the variance is equal to k. Since the variance in all cases depends on the time index k the process is *non-stationary*. Nevertheless the random walk is a process fundamental to many areas of statistics, and, as we will see later, to the description of noise in electric and electronic systems.

8.3.3 IID random process

A sequence of random variables that are independent and identically-distributed (IID) forms a discrete random process that is only slightly more general than a Bernoulli process. The process is both stationary and ergodic. If the random process takes on discrete values with PMF $f_X[x]$ then the probability of any sequence of values can be written as a product,

$$\Pr\,[x_1, x_2, x_3 \cdots] = f_X[x_1]f_X[x_2]f_X[x_3] \cdots$$

where the square brackets indicate that the discrete-time process $X[k]$ takes on discrete values. Similarly, if the process takes on continuous values, the *PDF* can be written as a product

$$f_X(x_1)f_X(x_2)f_X(x_3) \cdots$$

An example of an IID random process is the error sequence that results when quantizing an analog signal (see Fig. 8.2). The quantization error $X[k] - \hat{X}[k]$ is generally modeled as a uniformly-distributed random variable. The output of the quantizer can therefore be written as

$$\hat{X}[k] = X[k] + W[k]$$

where $W[k]$ is a uniformly-distributed IID random process referred to as quantization noise (see Appendix D).

8.3.4 Markov process (discrete)

The random walk is a random process that can be thought of as "state-based." For $k > 1$ (8.3) can be written as

$$X[k] = X[k-1] + B[k] \tag{8.6}$$

Let us call $X[k]$ the *state* of the process at time k. Then the equation says that the state of the process at time k is equal to the state at time $k-1$ plus a step (± 1).

If (8.6) is generalized slightly, by allowing $B[k]$ to be an IID random process with some arbitrary distribution, then we have a very powerful model for a random process known as a *Markov process*.

Markov processes have the property

$$\Pr\,[X[k]|X[[k-1]X[k-2] \cdots X[0]] = \Pr\,[X[k]|X[[k-1]] \tag{8.7}$$

In other words the probability of the current value $X[k]$ given all the past history of the process is the same as the probability of $X[k]$ given just the *previous* value of the process. The previous value summarizes all of the past information; that's why it is called the "state."

To see that (8.6) has the Markov property, it is only necessary to notice that $X[k-2]$ and the earlier values of X have no influence on $X[k]$ if $X[k-1]$ is known. Further, the conditional distribution for X can be written as

$$F_X(x_k|x_{k-1}) = F_B(x_k - x_{k-1})$$

where F_B is the distribution for B. Thus a complete statistical description of the process is possible.

State-based models occur in all kinds of engineering applications from satellite tracking systems to linguistic models in speech. Claude Shannon introduced Markov models in communication theory and gave this interesting example in terms of words of the English language [5, 6]. Words are chosen independently but with probabilities equal to their relative frequency of occurance in English, and concatenated to form a "sentence." The result is:

> REPRESENTING AND SPEEDILY IS AN GOOD APT OR COME CAN DIFFERENT NATURAL HERE HE THE A IN CAME THE TO OF TO EXPERT GRAY COME TO FURNISHES THE LINE MESSAGE HAD BE THESE.

This sequence of words not only lacks meaning, but also seems strange because of peculiar juxtapositions of words such as "THE" and "TO" that would not occur in the natural language. On the other hand, if words are chosen with probabilities that depend on the previous word as in a Markov process, the following "sentence" occurs:

> THE HEAD AND IN FRONTAL ATTACK ON AN ENGLISH WRITER THAT THE CHARACTER OF THIS POINT IS THEREFORE ANOTHER METHOD FOR THE LETTERS THAT THE TIME OF WHO EVER TOLD THE PROBLEM FOR AN UNEXPECTED.

While the result is still nonsensical, it sounds a good deal more like an English sentence. The studies here are not concerned with generating sequences of words to form sentences. However the sequences of numbers that form a discrete signal can likewise be made to resemble many real world measured signals if they are modeled by a Markov random process. That is the point in this example.

8.4 Some Continuous Random Processes

Having spent some time discussing random processes that are discrete in time, let us now turn to the discussion of some important continuous time random processes. It will be seen that these continuous random processes involve limiting arguments for discrete processes over small increments of time that shrink to zero.

8.4.1 Wiener process

A type of noise occurring in electronic systems results from electron movement at the molecular level. A brief discussion of this process from a physical point of view is provided in Appendix D. Early in his career Norbert Wiener studied Brownian motion, the random motion that small particles suspended on the surface of a liquid undergo due to their continued bombardment by molecules of the liquid. This led to the formulation of a random process now known as the Wiener process[6] and its role in a model for noise. The Wiener process is defined to be the limit of a random walk as the time interval between samples approaches zero.

Let us begin the development of the Wiener process by considering a random walk (say representing motion of a particle in one dimension) with a step size s and parameter $p = \frac{1}{2}$. That is, at each time epoch the random process increases or decreases its value by $\pm s$ with probability $p = \frac{1}{2}$. Assume that the process is represented on a continuous time scale with microscopic time increments Δ such that $t = k\Delta$. Then the process is

$$X(t) = X(k\Delta) = \sum_{i=1}^{k} s\,\zeta[i]\ ; \qquad k > 0 \tag{8.8}$$

where, for simplicity, the process is assumed to start at $t = k = 0$. From the previous analysis of the random walk the mean of this random process is zero and the variance is given by

$$\text{Var}\,[X(k\Delta)] = s^2 k = \frac{s^2 t}{\Delta} \tag{8.9}$$

Now suppose that the interval Δ approaches zero and the step size s also approaches zero in such a way that the ratio s^2/Δ approaches a constant ν_o; then

$$\text{Var}\,[X(t)] = \lim_{\Delta \to 0} \frac{s^2 t}{\Delta} = \nu_o t \tag{8.10}$$

The resulting process in this limiting form is the *Wiener process*. Since $X(t)$ is the sum of a large number of IID random variables, the Central Limit Theorem states that $X(t)$ has a Gaussian density function.

$$f_{X(t)}(x) = \frac{1}{\sqrt{2\pi\nu_o t}} e^{-x^2/2\nu_o t} \tag{8.11}$$

It is also fairly easy to see that any number of samples of this process are jointly Gaussian. For example, if there are two samples, say $X(t_1)$ and $X(t_2)$, with $t_1 < t_2$, then form the random variable

$$v = c_1 X(t_1) + c_2 X(t_2)$$

where c_1 and c_2 are any real numbers. The new random variable can be written using (8.8) as

$$
\begin{aligned}
v &= c_1 \sum_{i=1}^{k_1} s\,\zeta[i] + c_2 \sum_{i=1}^{k_2} s\,\zeta[i] \\
&= (c_1 + c_2) \sum_{i=1}^{k_1} s\,\zeta[i] + c_2 \sum_{i=k_1+1}^{k_2} s\,\zeta[i]
\end{aligned}
$$

where $k_1 = t_1/\Delta$ and $k_2 = t_2/\Delta$. The terms in the first sum are independent of the terms in the second sum. Therefore, each of these two sums is independent and Gaussian by the Central Limit Theorem. Because v is the sum of two independent

---

[6] Also called a Wiener-Lévy process, a Brownian motion process, or a diffusion process.

Gaussian random variables, it too is Gaussian. Since the argument holds for *any* choice of the constants c_1 and c_2, it follows that that $X(t_1)$ and $X(t_2)$ are jointly Gaussian [7, p. 218]. A similar argument shows that *any* *number* of samples of the Wiener process are jointly Gaussian. Such a process is called a *Gaussian random process*. Since the variance of the Wiener process depends on t however, the Wiener process is *not* stationary.

8.4.2 *Gaussian white noise*

The Wiener process by itself does not find much use in communications and signal processing. The process obtained by taking the derivative with respect to time, however, is ubiquitous in these and other areas of electrical engineering. This process is known as *Gaussian white noise*. Gaussian white noise, while derived from the Wiener process, is an idealized model for signals that may appear in the real world. Although the term "noise" would imply that this process is undesirable, it appears as an input in linear system models for many other types of other random signals. White noise, as it occurs in nature, depends on temperature; hence it is also referred to as thermal noise and it has been studied extensively [8]. A brief discussion of white noise from this physical point of view is provided in Appendix D.

Returning to our model, the assumed Gaussian nature of the white noise derives from the Gaussian property of the Wiener process. Linear operations such as the derivative preserve the Gaussian property. The discussion of Gaussian white noise is continued in the next two chapters dealing with analysis of random signals, where its importance becomes abundantly clear.

8.4.3 *Other Gaussian processes*

While the steps involved in the derivation of the Wiener process lead to a process that is non-stationary, the argument can be modified to have the impulsive process applied to a stable linear filter. Since the response time of any real filter is long compared to the time at which electrical charges arrive, the effect of filtering is like integration. The output of the filter, which is the sum of a large number of random variables, therefore has a jointly Gaussian distribution. This argument can be made rigorously (see [3]) and so provides an additional argument for the modeling of signals as Gaussian random processes.

8.4.4 *Poisson process*

The Gaussian random processes described above can take on different values for different time samples no matter how close these time samples may be. In fact, these random processes appear to be evolving continuously in time.

A different type of continuous random process having a stair-step-like appearance is the *Poisson* process. A typical sample function for the Poisson process is depicted in Fig. 8.9. Notice that $X(t)$ may change its value by +1 at any set of points on the time axis but otherwise remains constant. The value of the random process, at any point in time, is called its "state."

The Poisson random process is widely used to model the arrival of random events such as telephone calls, messages at an email server, printer requests and many other occurances. The process $X(t)$ counts the number of such events from some given starting time; it is similar to the discrete counting process described in Example 8.3.

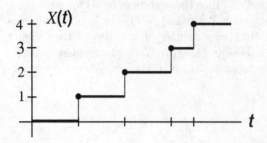

Figure 8.9 Sample function (realization) of a Poisson counting process.

The difference of course, is that this process may change value at *any* point along the time axis and not at just integer values of time. It will be seen in Chapter 11, where the process is described in detail, that the probability of an "arrival" is based on a single parameter λ. The time between arrivals, the number of events arriving in a time interval, and many other properties can be computed by well-known procedures.

The model can be turned around and used to describe the length of "service times" at a service facility such as a cashier, a computer printer, mail delivery system or other service. It is shown later that combining these models leads to a basic form of queueing system suitable for describing traffic on a computer network.

Figure 8.10 shows an example of a related type of process known as a "random

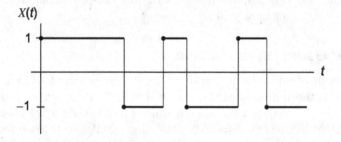

Figure 8.10 Random telegraph wave.

telegraph wave" that has two possible states. The process turns "on" (+1) and "off" (−1) at random times along the continuous time axis as determined by a pair of Poisson models. Although the name of this process alludes to communication in former days, this process has many applications in modern information and electronic systems.

8.4.5 Markov chain (continuous)

A Markov *chain* is defined as a Markov process with a countable set of states. Transitions in the process from one state to another are defined by a set of state transition probabilities. Although Markov chains exist as both discrete time and continuous time random processes, the models used for computer networks are based on queueing theory where the associated random processes are continuous.

In continuous time Markov chains, a state transition may occur at any point along the time axis, after which the probabilities of transition are defined by the new state. The state transitions are typically defined by a set of Poisson processes. The random telegraph wave described above is an example of a simple continuous time Markov process with just two states. Figure 8.11 depicts a more general process, although

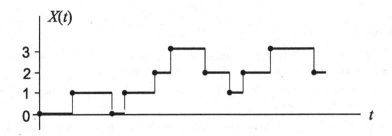

Figure 8.11 Sample function of a Markov chain with Poisson transitions.

transitions that occur in a Poisson model can only increase or decrease the state by $+1$ or -1.

In the general analysis of queueing systems Markov and Poisson random processes are combined in a complementary manner to describe statistical and dynamical properties of the system. Some particular measures of performance include average length of queues, waiting times, number of requests in the system and service times. An introduction to this important topic is provided in Chapter 11.

8.5 Summary

This chapter provides an introduction to the theory of random processes. While random variables represent mappings from the sample space to the real line, random processes represent mappings to a space of functions collectively known as the ensemble. Random processes may be defined in continuous or discrete time; in either case they are described from a probabilistic point of view by joint distributions of their samples.

Many types of electrical signals are best characterized as random processes. Random processes are said to be stationary (in the strict sense) if their statistical description does not change over time. This provides a certain consistency that allows for their exploitation. For example, the motion of an object in flight may not be deterministic but has consistent average characteristics to allow another system to track the object.

Stationary random processes may also be ergodic. If so, statistical averages can be replaced by averages in time. Since it is often difficult to acquire independent sample functions approximating the ensemble, random processes that exhibit ergodicity are important in many field experiments. Random processes may also exhibit periodic characteristics and even be entirely predictable.

A number of continuous and discrete random processes are described in this chapter and discussed in an introductory manner. Among these are processes that lead to a model for noise. Sources of noise are found in most real-world systems. Noise may arise both from thermal agitation in analog electronic systems as well as quantization error in digital systems. Without noise we could communicate freely anywhere in the universe no matter how distant. However, it is as hard to imagine a world without noise as it is to imagine a world without friction.

A final topic in this chapter that is treated later in the text has to deal with Markov and Poisson processes. These processes are important in the analysis of computer systems—especially computer networks. These models have become even more important in recent years because of the worldwide propensity of digital systems and digital communication.

References

[1] George E. P. Box and Gwilym M. Jenkins. *Time Series Analysis: Forecasting and Control.* Holden-Day, Oakland, California, revised edition, 1976.

[2] David R. Brillinger. *Time Series: Data Analysis and Theory.* Holden-Day, Oakland, California, expanded edition, 1981.

[3] Charles W. Therrien. *Discrete Random Signals and Statistical Signal Processing.* Prentice Hall, Inc., Upper Saddle River, New Jersey, 1992.

[4] J. D. Gibson and J. L. Melsa. *Introduction to Nonparametric Detection with Applications.* Academic Press, New York, 1975.

[5] Claude E. Shannon. A mathematical theory of communication. *Bell System Technical Journal*, 27(3):379–422, July 1948. (See also [6].).

[6] Claude E. Shannon and Warren Weaver. *The Mathematical Theory of Communication.* University of Illinois Press, Urbana, IL, 1963.

[7] Athanasios Papoulis and S. Unnikrishna Pillai. *Probability, Random Variables, and Stochastic Processes.* McGraw-Hill, New York, fourth edition, 2002.

[8] Wilbur B. Davenport, Jr. and William L. Root. *An Introduction to the Theory of Random Signals and Noise.* McGraw-Hill, New York, 1958.

Problems

Random Processes

8.1 Below are two waveforms corresponding to voiced speed and unvoiced speech. By examination, comment on whether or not each of these waveforms could be a random process. Support your answer.

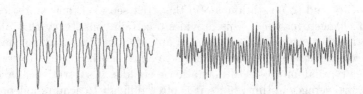

8.2 Repeat Prob. 8.1 for the following waveforms.

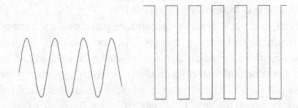

8.3 If certain digital audio packets are being transported over a store-and-forward packet switched network, describe the ways in which the quality of the audio can be affected by the queue length.

8.4 Consider the scenario of packets arriving at a network node. Explain how the packet inter-arrival time can be a random process. How does it affect the queue length in the node?

8.5 Suppose that a random experiment consists of observing vehicle traffic at a traffic light.

(a) Describe how the length of the lines formed at the traffic light can be a random process. Also describe the nature of this random process in terms of time and magnitude.

(b) What other quantities in this experiment can be random processes? Briefly explain.

8.6 A random process (signal) $X(t)$ is sampled and then approximated by a 4-step quantizer with output values of 0, 1, 2, and 3. The PDF of a sample $X_0 = X[k_0] = X(k_0 T)$ is given by

$$f_{X_0}(x_0) = \begin{cases} \frac{1}{2}(2-x), & 0 \le x \le 2 \\ 0, & \text{otherwise.} \end{cases}$$

Determine and plot the PMF of the discrete output sample $\hat{X}[k_0]$.

8.7 A random process $X(t)$ is digitized by an analog to digital converter system having an input dynamic range from -1.6 volts to 1.6 volts with an eight-step quantizer. Quantization noise in the system $W_q[k]$ is characterized by

$$f_{W_q}(w_q) = \begin{cases} \frac{4}{\Delta^2}\left(w_q + \frac{\Delta}{2}\right), & -\frac{\Delta}{2} \le w_q \le 0 \\ \frac{4}{\Delta^2}\left(\frac{\Delta}{2} - w_q\right), & 0 \le w_q \le \frac{\Delta}{2} \\ 0, & \text{otherwise.} \end{cases}$$

(a) For a given sample of the random process, what is the probability that its quantization error exceeds $\frac{\Delta}{4}$?

(b) Determine the mean and the variance of the quantization noise.

Characterizing a Random Process

8.8 Consider a random process $X(t)$ defined by a finite sized ensemble consisting of four equally probable sample functions given by

$$\begin{aligned} x_1(t) &= -2u(t); \\ x_2(t) &= \cos 5\pi t u(t); \\ x_3(t) &= u(t); \\ x_4(t) &= 2\sin(6\pi t + 0.2)u(t) \end{aligned}$$

(a) Draw the ensemble.

(b) For $t = 0.2$, determine the sample space.

8.9 Consider three samples $X(t_0)$, $X(t_1)$ and $X(t_2)$ of a random process $X(t)$. Assume that all three samples are independent and are uniformly distributed in the range $[-1, 3]$. Based on this information, what is the PDF of this process?

8.10 In a complex random process given by

$$X(t) = Ae^{-j\omega t}$$

the magnitude A is a random variable having the following PDF:

$$f_A(a) = \begin{cases} 1+a & -1 \le a \le 0 \\ 1-a & 0 \le a \le 1 \end{cases}$$

Is this a mean ergodic random process?

8.11 A random process $X(t)$ is defined as

$$X(t) = A$$

where A is a three-valued random variable taking on values of -1, 0, and $+1$ with equal probabilities.

(a) Draw two representative sample functions of this random process.

(b) Is this a mean ergodic random process?

8.12 Is the random process defined in Prob. 8.11 a regular or predictable random process? Explain.

8.13 Consider a random process $X(t) = A\cos(2\pi Ft)$ where F is a uniform random variable in the range 1200 to 1203 Hz. Is this a regular or predictable random process? Explain your answer.

Discrete random processes

8.14 Consider a Bernoulli-like random process defined as

$$X[k] = \begin{cases} +2 & \text{with probability } p \\ -1 & \text{with probability } 1-p \end{cases}$$

(a) Sketch a typical sample function of $X[k]$.

(b) Compute the mean and variance. Are they independent of k?

(c) Is the process stationary?

(d) Is the process mean ergodic?

8.15 A discrete-time random process $X[k]$ is defined by

$$Y[k] = \sum_{i=0}^{k} X[i]$$

where $X[k]$ is the process defined in Prob. 8.14.

(a) Sketch some typical realizations of $X[k]$.

(b) Sketch the corresponding realizations of $Y[k]$.

(c) Determine the mean and variance of $Y[k]$ for $p = 0.48$ and $k = 4$.

8.16 A discrete-time random process $X[k]$ is defined by

$$X[k] = \sum_{i=0}^{k} B[i]$$

where $B[k]$ is binary white noise (Bernoulli process with $p = \frac{1}{2}$). The process $X[k]$ is called a *discrete Wiener process*.

(a) Sketch a typical realization of binary white noise.

(b) Sketch the corresponding realization of the discrete Wiener process.

(c) Determine the mean and variance of the discrete Wiener process for $k = 5$.

8.17 A discrete-time random process $Y[k]$ is obtained as a sum of successive samples of another random process as follows

$$Y[k] = X[k] + X[k-1]$$

where $X[k]$ is the random process defined in Prob. 8.14.

(a) Sketch a typical sample function of $Y[k]$.

(b) Determine the probability mass function of a sample of $Y[k]$.

(c) Find the mean and the variance of $Y[k]$.

(d) Is $Y[k]$ mean ergodic?

8.18 A discrete random process is defined by

$$X[k] = A \cos(\omega_0 k + \phi)$$

where ϕ is a uniform random variable in the range of $-\pi$ to π.

(a) Sketch a typical sample function of $X[k]$.

(b) Are its mean and variance constants (i.e., independent of k)?

(c) Is $X[k]$ stationary?

(d) Is it mean ergodic?

8.19 Given a discrete random process

$$X[k] = A \cos(\omega_0 k)$$

where A is a Gaussian random variable with mean m_A and variance σ_A^2

(a) Sketch a typical sample function of $X[k]$.

(b) Are its mean and variance constants (independent of k)?

(c) Is it stationary?

(d) Is it mean ergodic?

8.20 At a certain time k_0 a sample from a Bernoulli process is observed to have a value of $+1$. What is the distribution for the waiting time j until *two* more $+1$s occur in the sequence?

8.21 A common way to characterize the frequency content of a random process is in terms of the number of "zero-crossings" in a given interval of time. One can define "zero-crossing" for a discrete-time signal or sequence as whenever the sequence changes from a positive to a negative value or *vice-versa*.

Consider a Bernoulli sequence over some time interval $[0, N-1]$. What are the probabilities for 0 zero-crossings and 1 zero-crossing in this interval? Write also a recursive formula that expresses the probability of k zero-crossings in terms of the probability of $k-1$ zero-crossings in some interval of length L.

Continuous random processes

8.22 A Wiener process has a PDF given by

$$f_{X(t)}(x) = \frac{1}{\sqrt{8\pi t}} \exp\left(-\frac{x^2}{8t}\right)$$

   (a) Given two samples of this random process $X(1.2)$ and $X(1.6)$, determine their joint PDF.

   (b) Write the correlation matrix (see Section 7.2 of Chapter 7) based on these two samples.

8.23 Add the two random variables in Prob. 22 to obtain a new random variable. Determine and write the PDF of the new random variable.

8.24 At a traffic light, automobiles arrive at a rate of 10 vehicles per minute along the North-South road and at a rate of 5 vehicles per minute along the East-West road. Treating the experiment as a counting process:

   (a) Sketch a typical realization of the counting process for each of these cases.

   (b) In which direction would you leave the green light on longer? Explain.

8.25 In the experiment in Prob. 24, consider the traffic coming from and leaving for all directions from the light. Sketch a realization of the corresponding Markov process.

8.26 At a copy center, when a typical color copier is used by customers, the center makes a profit at a rate of $1 per minute while the center loses at a rate of $0.50 per minute when the copier is idle. On average, the probability that a color copier is busy being used is 0.40.

   (a) Sketch a typical realization of this two-state random process.

   (b) Over a period of a 15-hour day, how much profit per hour is the copy center likely to make?

8.27 For the Wiener process in Prob. 8.22, define a new random variable as follows

$$U = c_2 X(t_2) - c_1 X(t_1)$$

where $t_1 = 1.2$ seconds and $t_2 = 1.6$ seconds; c_1 and c_2 are any real numbers.

   (a) Is U a Gaussian random variable?

   (b) Find the mean and variance of U.

Noise Sources (see Appendix D)

8.28 Noise power of a thermal noise source is given to be -120 dBm. What is the noise power in Watts?

8.29 In a power amplifier operating at $17°C$ with a bandwidth of 10 kHz:

   (a) What is the thermal noise power in dB?

   (b) Find the rms noise voltage given that the internal resistance is 100 ohms and a load resistance of 100 ohms.

   (c) If the signal being amplified is a perfect sinusoid with a magnitude of 3 mV, what is the SNR as the signal leaves the amplifier?

8.30 Consider an N-step quantizer. The input to the quantizer is a sinusoidal waveform with a peak magnitude of A. The quantization error due to rounding is
characterized by the following PDF

$$f_E(e) = \left\{ \frac{2}{\Delta}\left(1 - \frac{2}{\Delta}|e|\right) \qquad -\frac{\Delta}{2} \le e \le \frac{\Delta}{2},\right.$$

(a) Determine the mean and the variance of the quantization error.

(b) Obtain the expression for the signal to quantization noise in dB.

(c) If the SQNR desired is at least 80 dB, how many (binary) bits are required
to achieve this?

Computer Projects

Project 8.1

In this project, you will generate sample functions of different random processes.

1. Generate three sample functions of each of the following types of random
   processes; the length of each sample function should be 500 or fewer points.

   (a) A Bernoulli random process. Let the values taken by the Bernoulli (binary)
       random process be -1 and $+1$. The probability that the value is a $+1$ is
       given to be 0.52.

   (b) A binomial random process. Let the number of Bernoulli trials be $n = 12$.
       The values of the binomial random variable are defined as the number of
       1s in n trials of a Bernoulli random variable that takes values of 0 and 1.
       Let the parameter p in the Bernoulli trials be $p = 0.49$.

   (c) A uniform random process. Let the range of values taken by the random
       process be -1 to $+2$.

   (d) A Gaussian random variable with mean $m = 1.2$ and variance $\sigma^2 = 1.6$.

2. Plot the random process as an ensemble of three sample functions. Comment
   on the results obtained.

Project 8.2

(a) Generate and plot some typical sample functions (sequences) for a discrete
    Wiener process (random walk with $p = \frac{1}{2}$). Assume the process starts at
    $k = 0$ and that the initial values $x[0] = 1$, $x[0] = 0$, and $x[0] = -1$ are equally
    likely. Plot several members of the ensemble using at least 50 time samples.
    Observe that the temporal means computed along these sequences may be
    significantly different from zero due to long runs of positive or negative
    values.

(b) Now generate 100 realizations of a discrete Wiener process but do not plot
    them. Estimate the mean $m[k] = \mathcal{E}\{X[k]\}$ by averaging "down the ensem-
    ble" and plot this function. Is the ensemble mean close to zero?

(c) Repeat part (b) for a random walk with $p = 0.8$.

Project 8.3

A random sequence $x[k]$ is generated according to the difference equation

$$x[k] = \alpha\, x[k-1] + b[k]$$

where α is a constant and $b[k]$ is a binary white noise sequence taking on values -1 and $+1$ with equal probabilities.

Generate and plot 50 samples of the random sequence for $\alpha = .95, .70,$ and $-.95$. What differences do you observe in these three different random sequences? What happens if $|\alpha| > 1$?

9 Random Signals in the Time Domain

A given random process can be characterized by an n-dimensional joint PDF, where n is the number of samples taken. Such a characterization is not always practical and may not be necessary for solving many engineering problems. The first and second moments of a random process, on the other hand, provide a partial characterization of a random process which for most engineering problems is adequate.

This chapter focuses on first and second moment analysis of random signals in the time domain. The various second moment quantities for random signals are defined and and the concept of wide sense stationarity (stationarity of the moments) is described. Complex random signals are introduced along with their usual method of description.

With these topics established, a parallel set of second moment quantities and their properties are described for discrete-time random signals. While the discrete forms of these quantities bring in no fundamentally new concepts, it is useful to describe them explicitly because virtually all modern electronic systems employ digital signal processing.

The chapter moves on to a discussion of linear systems and their effect on random signals in the time domain. Both continuous- and discrete-time signals and systems are treated. The chapter ends with some applications of the theory to the problems of system identification and optimal filtering.

9.1 First and Second Moments of a Random Process

Characterizing random signals using first and second moments can go a long way. In fact, for Gaussian random processes, those moments are all that are ever needed. The moments of a random process are defined and discussed below. The first moment is the mean of the random process, introduced in the previous chapter. The second moments appear in many different related forms as functions that express correlation between the samples.

9.1.1 Mean and variance

The *mean* of a random process $X(t)$ is denoted by $m_X(t)$ and defined by

$$m_X(t) = \mathcal{E}\{X(t)\} \tag{9.1}$$

where the expectation is defined in terms of the PDF of the random variable $X(t)$:

$$\mathcal{E}\{X(t)\} = \int_{-\infty}^{\infty} x f_{X(t)}(x)dx$$

Note that the mean is not necessarily constant but may be a function of time. This follows because $f_{X(t)}$ itself may be a function of time. As discussed in Section 8.2.2 of the previous chapter, $m_X(t)$ can be considered to be the result of averaging *down* the ensemble at a particular point in time (t).

The *variance* of a random process is defined in a similar manner as

$$\sigma_X^2(t) = \mathcal{E}\left\{\left((X(t) - m_X(t))^2\right)\right\} \tag{9.2}$$

where the expectation is again defined defined by the PDF of $X(t)$. The mean and variance are related (as for any random variable) by

$$\sigma_X^2(t) = \mathcal{E}\left\{X^2(t)\right\} - m_X^2(t) \tag{9.3}$$

where $\mathcal{E}\left\{X^2(t)\right\}$ is the second moment of the random process.

9.1.2 Autocorrelation and autocovariance functions

When more than one sample of a random process appears in a problem, we need to deal with the correlation or covariance of the samples. Consider the case of two samples $X(t_1)$ and $X(t_0)$. The correlation of these samples is given by

$$\mathcal{E}\left\{X(t_1)X(t_0)\right\} = \int_{-\infty}^{\infty}\int_{-\infty}^{\infty} x_1 x_0 f_{X(t_1)X(t_0)}(x_1, x_0)dx_1 dx_0$$

where the expectation is defined using the *joint* PDF of the samples. Let us take t_1 to be an arbitrary point t on the time axis and t_0 equal to $t - \tau$ where τ is the interval between the samples (see Fig. 9.1). Then, the correlation expressed by the previous

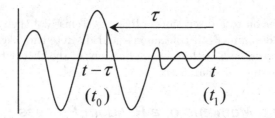

Figure 9.1 Relation of variables t and τ in autocorrelation function.

equation, can be written as

$$R_X(t, \tau) \overset{\text{def}}{=} \mathcal{E}\left\{X(t)X(t - \tau)\right\} \tag{9.4}$$

$R_X(t, \tau)$ is called the *autocorrelation function* for the random process. The variable τ is called the *lag* and (9.4) is referred to as the *time-lag* form of the autocorrelation function[1].

The autocorrelation function is a measure of how closely two points in a random process track each other statistically. When the autocorrelation is high, points separated by a distance τ tend to have similar values; when the autocorrelation is low, these points exhibit little similarity. When the lag τ is zero, the two selected points are identical and the autocorrelation function measures the *average power* of the signal (at time t).

$$R_X(t, 0) = \mathcal{E}\left\{X^2(t)\right\} \tag{9.5}$$

The *autocovariance function* of a random process $X(t)$ is defined as the joint central

---

[1] It is also common to write the autocorrelation function as $R_X(t_1, t_0)$. The time-lag form is more convenient for our purposes.

moment of $X(t)$ and $X(t-\tau)$:

$$C_X(t,\tau)) = \mathcal{E}\{(X(t)-m_X(t))(X(t-\tau)-m_X(t-\tau))\} \tag{9.6}$$

where $m_X(t)$ and $m_X(t-\tau)$ are the mean values of the random process at times t and $t-\tau$. When $\tau = 0$, the autocovariance function is equal to the variance of the random process

$$\sigma_X^2(t) = C_X(t,0) \tag{9.7}$$

Note that the variance may, in general, depend on the sampling time t.

It follows from (5.28) of Chapter 5 that the autocorrelation function and autocovariance function are related to each other as

$$\boxed{C_X(t,\tau) = R_X(t,\tau) - m_X(t)m_X(t-\tau)} \tag{9.8}$$

Since both the autocorrelation function and the autocovariance function can take on arbitrarily large (or small) values, it may be difficult in certain instances to determine whether the correlation for a random process is "high" or "low." Following along the lines of (5.29) in Chapter 5, the correlation coefficient $\rho_X(t,\tau)$ of a random process is defined as the normalized covariance between the two samples $X(t)$ and $X(t-\tau)$:

$$\rho_X(t,\tau) = \frac{C_X(t,\tau)}{\sigma_X(t)\sigma_X(t-\tau)} \tag{9.9}$$

The range of values taken by the correlation coefficient so defined is limited to

$$-1 \le \rho_X(t,\tau) \le 1$$

Therefore the correlation coefficient may be more appropriate to use in comparative analyses.

The following example illustrates computation of the first and second moment quantities of a typical random process.

Example 9.1: Consider the random process of Example 8.2 (Chapter 8)

$$X(t) = A\sin(\omega t + \Theta)$$

where A and ω are known parameters while Θ is a uniform random variable in the range $-\pi$ to π.

As previously shown, the mean of this random process is given by

$$m_X(t) = \mathcal{E}\{X(t)\} = \int_{-\pi}^{\pi} A\sin(\omega t + \theta)f_\Theta(\theta)d\theta$$

$$= \frac{A}{2\pi}\int_{-\pi}^{\pi}\sin(\omega t + \theta)d\theta = 0$$

The autocorrelation function of $X(t)$ is given by[2]

$$R_X(t,\tau) = \mathcal{E}\{X(t)X(t-\tau)\} = \int_{-\infty}^{\infty} A\sin(\omega t + \theta)A\sin(\omega(t-\tau)+\theta)f_\Theta(\theta)d\theta$$

$$= A^2\int_{-\pi}^{\pi}\sin(\omega t + \theta)\sin(\omega(t-\tau)+\theta)\frac{1}{2\pi}d\theta$$

---

[2] A trigonometric identity is used in the third line. Also remember that the integration is over the variable θ.

$$R_X(t,\tau) \quad = \quad \frac{A^2}{2\pi} \int_{-\pi}^{\pi} \frac{1}{2}[\cos\omega\tau - \cos(\omega(2t-\tau)+2\theta)]d\theta$$

$$= \quad \frac{A^2}{2}\cos\omega\tau - \underbrace{\frac{A^2}{4\pi}\int_{-\pi}^{\pi}\cos(\omega(2t-\tau)+2\theta)d\theta}_{0} = \frac{A^2}{2}\cos\omega\tau$$

Since the final result does not depend on t, the autocorrelation function is independent of the starting point and it can be written simply as $R_X(\tau)$. The significance of this is discussed in the section below.

Since the mean is zero, the autocovariance function is the same as the autocorrelation function (see (9.8)). The average power and the variance is given by

$$\sigma_X^2 = R_X(\tau)|_{\tau=0} = \frac{A^2}{2}$$

The correlation coefficient $\rho_X(\tau)$ is computed from (9.9) as

$$\rho_X(\tau) = \frac{C_X(\tau)}{\sigma_X \cdot \sigma_X} = \frac{(A^2/2)\cos\omega\tau}{A/\sqrt{2} \cdot A/\sqrt{2}} = \cos\omega\tau$$

This clearly satisfies the condition $-1 \le \rho_X(\tau) \le 1$.

□

9.1.3 Wide sense stationarity

A random process is said to be stationary if its statistical properties do not vary over time. This is important for many practical engineering problems. Strict sense stationarity, introduced in Chapter 8, is a powerful property that requires knowledge of the density or distribution functions. Recall that a random process is stationary in the strict sense if the distribution or density functions of all orders do not depend on time.

In most problems of practical interest in engineering, however, we work with first and second order moments (the mean and the autocorrelation function) of a random process. Therefore, it is useful from a practical standpoint if just these moments are independent of time. This property is known as wide sense stationarity (abbreviated wss). A workable definition is given below.

A random process is *wide sense stationary* (wss) if the mean is constant, $m_X(t) = m_X$, and the autocorrelation function is only a function of lag, $R_X(t,\tau) = R_X(\tau)$.

Since these quantities do not depend on time, it is possible to start anywhere in the process and find the same values of mean and correlation. A second order strict sense stationary is also wss, but the reverse is not true.

Since the autocovariance function and the correlation coefficient of a random process that satisfies the above definition are also independent of time, we can drop the time variable t and summarize the various equations in Table 9.1.

To determine if a random process is wss, proceed as in Example 9.1. First compute the mean. In Example 9.1 the mean turns out to be zero; therefore the process passes

quantity	definition
Mean	$m_X = \mathcal{E}\{X(t)\}$
Autocorrelation Function	$R_X(\tau) = \mathcal{E}\{X(t)X(t-\tau)\}$
Autocovariance Function	$C_X(\tau) = \mathcal{E}\{(X(t)-m_X)(X(t-\tau)-m_X)\}$
Interrelation	$C_X(\tau) = R_X(\tau) - m_X^2$
Correlation Coefficient	$\rho_X(\tau) = \dfrac{C_X(\tau)}{C_X(0)}$

Table 9.1 Definitions and interrelations for a wide sense stationary random process.

the first test for wide sense stationarity. (Note that the wss condition does not require the mean to be zero; it only requires that the mean be a constant.) Next compute the autocorrelation function to see if there is a dependence on the time variable t. Since in the example the autocorrelation function turns out to depend only on the lag τ, the random process passes both tests and therefore is wide sense stationary.

Let us consider another example.

Example 9.2: A random process is defined as

$$X(t) = A \cos \omega t$$

where A is a Gaussian random variable with mean zero and variance σ^2 and ω is a known parameter.

The mean of this random process is

$$m_X(t) = \mathcal{E}\{A \cos \omega t\} = \mathcal{E}\{A\} \cos \omega t = 0$$

Since the mean is constant, $X(t)$ passes the first test for stationarity.

The autocorrelation is defined by

$$
\begin{aligned}
R_X(t,\tau) &= \mathcal{E}\{X(t)X(t-\tau)\} = \mathcal{E}\{(A \cos \omega t)(A \cos \omega(t-\tau))\} \\
&= \mathcal{E}\{A^2\}(\cos \omega t \cos \omega(t-\tau)) = (\sigma^2/2)(\cos \omega(2t-\tau) + \cos \omega\tau)
\end{aligned}
$$

Since the autocorrelation function depends explicitly on both t and τ the process fails the second test and so is not wss.

□

The random process in Example 9.1 has a uniform random phase and turns out to be wss. Let us see what happens if a uniform random phase is added to the process in Example 9.2.

Example 9.3: A random process is defined as

$$X(t) = A \cos(\omega t + \Theta)$$

where A is a Gaussian random variable with mean zero and variance σ^2 and Θ is an independent random variable uniformly distributed on the interval $-\pi$ to π.

Since A and Θ are independent random variables, we can compute the mean as follows:

$$m_X(t) = \mathcal{E}\{A\cos(\omega t + \Theta)\} = \mathcal{E}\{A\}\mathcal{E}\{\cos(\omega t + \Theta)\} = 0$$

Thus $X(t)$ passes the first test for stationarity.

The autocorrelation is defined by

$$
\begin{aligned}
R_X(t,\tau) &= \mathcal{E}\{X(t)X(t-\tau)\} = \mathcal{E}\{(A\cos(\omega t + \Theta))\,(A\cos(\omega(t-\tau)+\Theta)\} \\
&= \mathcal{E}\{A^2\}\frac{1}{2}\mathcal{E}\{\cos\omega\tau + \cos(2\omega t - \omega\tau + 2\Theta)\} \\
&= \frac{\sigma^2}{2}\left[\cos\omega\tau + \int_{-\pi}^{\pi}\cos(2\omega t - \omega\tau + 2\theta)\frac{1}{2\pi}d\theta\right] = \frac{\sigma^2}{2}\cos\omega\tau
\end{aligned}
$$

Since the autocorrelation function depends only on the lag τ the random process is wss.

□

These last three examples illustrate that a uniform random phase is important for stationarity where sinusoids are involved.

It is not an implicit requirement that random processes in general be wss. For example, speech signals are not wss. Many random processes, such as speech, may be considered wss over short periods of time, however, during which analysis techniques for wss processes can be applied repeatedly. For the remainder of this chapter it can be assumed that that the random processes dealt with are wss. If the assumption is wrong, this will generally show up in the analysis as an (unexpected) dependence on t.

9.1.4 Properties of autocorrelation functions

The autocorrelation function $R_X(\tau)$ and the autocovariance function $C_X(\tau)$ of a wss process $X(t)$ are defined in Table 9.1. Both of these functions exhibit properties that are important in engineering applications. Some of these properties are listed below without proofs. Further, although these properties are stated in terms of $R_X(\tau)$, they are equally applicable to $C_X(\tau)$.

1. The point $\tau = 0$ is considered the origin of the autocorrelation function and R_X is strictly positive at the origin:

$$R_X(0) > 0 \tag{9.10}$$

$R_X(0)$ represents the average power in the signal (see (9.5)). Likewise, $C_X(0)$ is strictly positive and represents the variance (see (9.7)).

2. The autocorrelation function has its highest value at $\tau = 0$ and its magnitude is bounded by $R_X(0)$ as follows

$$|R_X(\tau)| \leq R_X(0) \tag{9.11}$$

3. The autocorrelation function has even symmetry about the origin

$$R_X(\tau) = R_X(-\tau) \tag{9.12}$$

4. The autocorrelation function is a positive semidefinite function, mathematically stated as

$$\int_{-\infty}^{\infty} \int_{-\infty}^{\infty} a(t_1) R_X(t_1 - t_0) a(t_0) dt_1 dt_0 \geq 0 \qquad (9.13)$$

where $a(t)$ is any arbitrary real-valued function.

Although they are listed last, properties 3 and 4 are the most fundamental. In particular, property 4 *implies* property 2 which in turn implies property 1. (The proof of this is explored in Prob. 9.7.) Properties 3 and 4 are *necessary and sufficient* properties of an autocorrelation function. If a function satisfies properties 2 and 3 it isn't certain to be a valid autocorrelation function unless it also satisfies property 4. On the other hand, if it fails to satisfy *any* of the four properties, then the proposed function is not valid. Property 4 is given an interpretation later in this chapter where it is related to power of the random process.

The following example shows how to test for these properties.

Example 9.4: The following functions are claimed to be valid autocorrelation functions.

· (a) $R_X(\tau) = A \sin \omega \tau$ (b) $R_X(\tau) = A \cos \omega \tau$ (c) $R_X(\tau) = A e^{|\tau|}$ (d) $R_X(\tau) = A\delta(\tau)$

Let us check by testing their properties. It is assumed that A and ω are positive real parameters and δ is the unit impulse.

(a) This function fails the even symmetry test (9.12); it also fails (9.10) and (9.11). Therefore it is not a legitimate autocorrelation function.

(b) This function satisfies the even symmetry requirement (9.12). It also satisfies (9.10) and (9.11) but these are necessary and not sufficient conditions. We must show that the function satisfies the required condition (9.13):

$$\int_{-\infty}^{\infty} \int_{-\infty}^{\infty} a(t_1) A \cos \omega(t_1 - t_0) a(t_0) dt_1 dt_0 \geq 0$$

To evaluate this integral let us first use a trigonometric identity to write

$$\cos \omega(t_1 - t_0) = \cos \omega t_1 \cos \omega t_0 - \sin \omega t_1 \sin \omega t_0$$

Taking the first term on the right and substituting it in the double integral produces

$$\int_{-\infty}^{\infty} \int_{-\infty}^{\infty} a(t_1) A \cos \omega t_1 \cos \omega t_0 \, a(t_0) \, dt_1 dt_0 =$$

$$A \left(\int_{-\infty}^{\infty} a(t_1) \cos \omega t_1 \, dt_1 \right) \left(\int_{-\infty}^{\infty} a(t_0) \cos \omega t_0 \, dt_0 \right) = A \left(\int_{-\infty}^{\infty} a(t_1) \cos \omega t_1 \, dt_1 \right)^2$$

which is a positive quantity. We can likewise substitute the sine terms in the double integral and arrive at the same conclusion. Since the overall integral is the sum of these two positive terms, the positive semidefinite property is satisfied and case (b) is a legitimate autocorrelation function.

(c) This function fails to satisfy the second property (9.11). Therefore it is not a valid autocorrelation function.

(d) The unit impulse satisfies the symmetry requirement (9.12) and also satisfies (9.10) and (9.11). To see if it satisfies the positive semidefinite condition (9.13) we substitute in and proceed directly:

$$\int_{-\infty}^{\infty} \int_{-\infty}^{\infty} a(t_1) A\delta(t_1 - t_0)\, a(t_0) dt_1 dt_0 =$$

$$A \int_{-\infty}^{\infty} a(t_0) \left(\int_{-\infty}^{\infty} a(t_1)\delta(t_1 - t_0) dt_1 \right) dt_0 = A \int_{-\infty}^{\infty} a^2(t_0) dt_0 > 0$$

Since the condition is satisfied, the impulse is a valid autocorrelation function.

□

9.1.5 The autocorrelation function of white noise

The autocorrelation function $A\delta(l)$ introduced in the last example belongs to a very important process, namely Gaussian white noise. It is shown here that the process introduced in Chapter 8 (Section 8.4.2) as the formal derivative of the Wiener process has this special autocorrelation function.

To begin, let us first look at the correlation between two points in a Wiener process $X(t)$ with parameter ν_0. If two time points t_1 and t_2 are chosen with $t_1 < t_2$ then the correlation can be written as

$$\mathcal{E}\{X(t_1)X(t_2)\} = \mathcal{E}\{X(t_1)X(t_1)\} + \mathcal{E}\{X(t_1)[X(t_2) - X(t_1)]\}$$

The first term here is the variance of the process at time t_1, i.e., $\nu_0 t_1$ (see (8.10) of Chapter 8). The second term is zero because the microscopic events occurring up to time t_1 are independent of the events occurring in the interval $t_2 - t_1$. Thus when $t_1 < t_2$ we find $\mathcal{E}\{X(t_1)X(t_2)\} = \nu_0 t_1$ and by a similar argument when $t_2 < t_1$ we find $\mathcal{E}\{X(t_1)X(t_2)\} = \nu_0 t_2$. In summary, we have the concise relation

$$\boxed{\mathcal{E}\{X(t_1)X(t_2)\} = \nu_0 \min(t_1, t_2) \quad \text{(Wiener process)}} \tag{9.14}$$

Next, consider a process $W'(t)$ which is an approximation to the derivative of the Wiener process

$$W'(t) \stackrel{\text{def}}{=} \frac{X(t + \Delta t) - X(t)}{\Delta t} \tag{9.15}$$

where Δt is any very small but *finite* positive change in t. It will be shown that the limiting form of this process is the white noise process.

The autocorrelation of this new process is defined by

$$\mathcal{E}\{W'(t)W'(t - \tau)\} = \mathcal{E}\left\{ \frac{[X(t+\Delta t) - X(t)][X(t-\tau+\Delta t) - X(t-\tau)]}{(\Delta t)^2} \right\} =$$

$$\frac{\mathcal{E}\{X(t+\Delta t)X(t-\tau+\Delta t) - X(t)X(t-\tau+\Delta t) - X(t+\Delta t)X(t-\tau) + X(t)X(t-\tau)\}}{(\Delta t)^2}$$

The four terms in the numerator are evaluated using (9.14). To aid in this evaluation the four terms and their sum are listed in the table following for three different conditions on the lag variable τ. Results for negative values of τ are obtained in a similar manner.

condition	term 1	term 2	term 3	term 4	result
$\tau = 0$	$\frac{\nu_o(t+\Delta t)}{\Delta t^2}$	$-\frac{\nu_o t}{\Delta t^2}$	$-\frac{\nu_o t}{\Delta t^2}$	$\frac{\nu_o t}{\Delta t^2}$	$\frac{\nu_o}{\Delta t}$
$0 \leq \tau \leq \Delta t$	$\frac{\nu_o(t+\Delta t)}{\Delta t^2}$	$-\frac{\nu_o t}{\Delta t^2}$	$-\frac{\nu_o(t-\tau)}{\Delta t^2}$	$\frac{\nu_o t}{\Delta t^2}$	$\frac{\nu_o}{\Delta t}\left(1-\frac{\tau}{\Delta t}\right)$
$\tau > \Delta t$	$\frac{\nu_o(t-\tau+\Delta t)}{\Delta t^2}$	$-\frac{\nu_o(t-\tau+\Delta t)}{\Delta t^2}$	$-\frac{\nu_o(t-\tau)}{\Delta t^2}$	$\frac{\nu_o(t-\tau)}{\Delta t^2}$	0

It is conventional in engineering practice to denote the parameter ν_o as $\nu_o = N_o/2$ for later convenience. The resulting autocorrelation function can then be written as

$$R_{W'}(\tau) = \begin{cases} \frac{N_o}{2\Delta t}\left(1 - \frac{|\tau|}{\Delta t}\right) & |\tau| \leq \Delta t \\ 0 & \text{otherwise} \end{cases}$$

and is depicted in Fig. 9.2) (a). In the limit as Δt approaches 0, $R_{W'}(\tau)$ becomes the

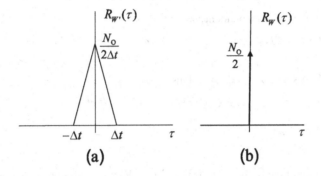

(a)   **(b)**

Figure 9.2 Autocorrelation function of white noise. (a) Approximation for small Δt. (b) Limiting form $(N_o/2)\delta(\tau)$.

autocorrelation function of the white noise process

$$R_W(\tau) = \frac{N_o}{2}\delta(\tau) \tag{9.16}$$

which is depicted in Fig. 9.2 (b).

Notice that this autocorrelation function depends only upon the lag variable τ and not upon the time variable t. Further, the process has *zero mean* because the underlying Wiener process has zero mean. Therefore the Gaussian white noise process is a *wide-sense stationary* random process. Note also that the variance of the white noise process $R_W(0)$ is *infinite*! In other words the ideal white noise process has infinite power. More is said about this fact in the next chapter discussing frequency domain characterization of random processes.

9.2 Cross Correlation

The analysis of two or more related random processes involves the concept of cross-correlation. Starting with the most general case, we define the cross-correlation function for random processes $X(t)$ and $Y(t)$ as

$$R_{XY}(t, \tau) = \mathcal{E}\{X(t)Y(t-\tau)\}$$

where the expectation is taken over both $X(t)$ and $Y(t)$. The cross-correlation function compares a point in random process X at time "t" to a point in random process Y at an earlier time $t - \tau$ (later time if τ is negative). The following example illustrates the procedure.

Example 9.5: Consider two random processes

$$X(t) = A\sin(\omega t + \Theta) + B\cos(\omega t)$$
$$Y(t) = B\cos(\omega t)$$

where A and ω are known parameters while Θ is a uniform random variable in the interval $-\pi$ to π] and B is a Gaussian random variable with mean 2 and variance 3. Random variables Θ and B are assumed to be independent. The cross-correlation function is given by

$$R_{XY}(t,\tau) = \mathcal{E}\{X(t)Y(t-\tau)\} = \mathcal{E}\{([A\sin(\omega t + \Theta) + B\cos(\omega t)] B\cos(\omega(t-\tau)))\}$$

$$= \mathcal{E}\{A\sin(\omega t + \Theta)B\cos\omega(t-\tau) + B^2\cos\omega t\cos\omega(t-\tau)\}$$

Since Θ and B are independent, we can write

$$R_{XY}(t,\tau) = A\mathcal{E}\{\sin(\omega t + \Theta)\}\mathcal{E}\{B\}\cos\omega(t-\tau) + \mathcal{E}\{B^2\}\cos\omega t\cos\omega(t-\tau)$$

The first term is 0 because $\mathcal{E}\{\sin(\omega t + \Theta)\} = 0$ and the second moment of B is given by $\mathcal{E}\{B^2\} = \text{Var}[B] + m_B^2 = 7$. This yields

$$R_{XY}(t,\tau) = 7\cos(\omega t)\cos(\omega t - \tau)$$

□

As with autocorrelation, it is useful if R_{XY} does not depend on absolute time t. (This is not the case in the preceding example.) We make the following definition.

Two random processes are said to be *jointly wide sense stationary* if,

1. Each of $X(t)$ and $Y(t)$ is wide sense stationary.
2. The cross-correlation is a function only of lag: $R_{XY}(t,\tau) = R_{XY}(\tau)$.

The following examples illustrate how to check for jointly wss random processes.

Example 9.6: Consider two random processes

$$X(t) = A\sin\omega t + B\cos\omega t$$
$$Y(t) = A\sin\omega t - B\cos\omega t$$

where ω is a constant and A and B are uncorrelated random variables with zero mean and equal variance σ^2. We want to examine $X(t)$ and $Y(t)$ to determine if they are jointly wide sense stationary.

Since the mean values are zero, we can proceed to check if the autocorrelation functions to see if they are only a function of τ.

The autocorrelation function of $X(t)$ is given by

$$R_X(t,\tau) = \mathcal{E}\{X(t)X(t-\tau)\} = \mathcal{E}\{A^2\}\sin\omega t\sin\omega(t-\tau) + \mathcal{E}\{AB\}(\ \cdots\ )$$
$$+ \mathcal{E}\{AB\}(\ \cdots\ ) + \mathcal{E}\{B^2\}\cos\omega t\cos\omega(t-\tau)$$

Since $E\{AB\} = E\{A\}E\{B\} = 0$, the cross-terms drop out leaving

$$
\begin{aligned}
R_X(t,\tau) &= \sigma^2 \sin \omega t \sin \omega(t - \tau) + \sigma^2 \cos \omega t \cos \omega(t - \tau) \\
&= \sigma^2 \cos(\omega t - \omega(t - \tau)) = \sigma^2 \cos \omega \tau
\end{aligned}
$$

Since the autocorrelation function depends only on τ, $X(t)$ is wss with

$$
R_X(\tau) = \sigma^2 \cos \omega \tau
$$

A similar set of steps shows that $Y(t)$ is wss with the same form

$$
R_Y(\tau) = \sigma^2 \cos \omega \tau
$$

Since each process is individually wss, we proceed to compute the cross-correlation function as

$$
\begin{aligned}
R_{XY}(t,\tau) = E\{X(t)Y(t-\tau)\} &= E\{A^2\} \sin \omega t \sin \omega(t - \tau) - E\{AB\}(\; \cdots \;) \\
&\quad + E\{AB\}(\; \cdots \;) - E\{B^2\} \cos \omega t \cos \omega(t - \tau) \\
&= \sigma^2(\sin \omega t \sin \omega(t - \tau) - \cos \omega t \cos \omega(t - \tau)) \\
&= -\sigma^2 \cos(\omega t + \omega(t - \tau)) = -\sigma^2 \cos \omega(2t - \tau))
\end{aligned}
$$

Since this depends on *both* t and τ, the two random processes are *not* jointly wide sense stationary.

□

The previous example illustrates that two processes can be individually wss but not necessarily jointly wss. Let us consider one more example.

Example 9.7: By modifying the random processes given in Example 9.6, let us form two new random processes

$$
\begin{aligned}
X(t) &= A \sin \omega t + B \cos(\omega t) \\
Y(t) &= B \sin \omega t - A \cos(\omega t)
\end{aligned}
$$

where ω is a constant, and A and B are uncorrelated random variables with zero mean and equal variance σ^2. The autocorrelation functions of $X(t)$ and $Y(t)$ are found as in the previous example to be

$$
R_X(\tau) = \sigma^2 \cos \omega \tau \quad \text{and} \quad R_Y(\tau) = \sigma^2 \cos \omega \tau
$$

so the processes are individually wss.

The cross-correlation function is given by

$$
\begin{aligned}
R_{XY}(t,\tau) &= E\{X(t)Y(t-\tau)\} \\
&= E\{AB\}(\; \cdots \;) - E\{A^2\} \sin \omega t \cos \omega(t - \tau) \\
&\quad + E\{B^2\} \cos \omega t \sin \omega(t - \tau) - E\{AB\}(\; \cdots \;) \\
&= -\sigma^2(\sin \omega t \cos \omega(t - \tau) - \cos \omega t \sin \omega(t - \tau)) = -\sigma^2 \sin \omega \tau
\end{aligned}
$$

where the product terms $E\{AB\}$ drop out. Since the random processes are each wss and the cross-correlation function depends only on τ, X and Y are jointly wss.

□

When $X(t)$ and $Y(t)$ are jointly wss, the cross-correlation function can be *defined* as

$$
\boxed{R_{XY}(\tau) = E\{X(t)Y(t - \tau)\}} \tag{9.17}
$$

In addition, the *cross-covariance function* for the random processes can be defined as

$$C_{XY}(\tau) = \mathcal{E}\{(X(t) - m_X)(Y(t - \tau) - m_Y)\} \tag{9.18}$$

and the cross-correlation and cross-covariance functions are related as

$$C_{XY}(\tau) = R_{XY}(\tau) - m_X m_Y \tag{9.19}$$

Notice that in dealing with cross-correlation, the order of subscripts in the definition *is important!* Changing the roles of $X(t)$ and $Y(t)$ in (9.17) produces

$$R_{YX}(\tau) = \mathcal{E}\{Y(t)X(t - \tau)\}$$

which is *not* the same as $R_{XY}(\tau)$. By virtue of their definitions, however, the two cross-correlation functions are related as

$$R_{YX}(\tau) = R_{XY}(-\tau) \tag{9.20}$$

A few further facts about cross correlation are worth mentioning here. First, let us define two common terms.

- Random processes $X(t)$ and $Y(t)$ are said to be *orthogonal* if $R_{XY}(\tau) = 0$.
- Random processes $X(t)$ and $Y(t)$ are said to be *uncorrelated* if $C_{XY}(\tau) = 0$.

Notice that in view of (9.19), when X and Y are uncorrelated R_{XY} is equal to the product of the means. In fact, "uncorrelated" could be defined by the condition $R_{XY}(\tau) = m_X m_Y$. On the other hand, random processes X and Y are *independent* if and only if their joint density functions of all orders are equal to the product of their marginal densities. This condition also implies that R_{XY} is the product of the means. Hence *independent processes are uncorrelated* but uncorrelated processes are not necessarily independent.[3]

Finally, let us cite two common bounds for the cross-correlation function:

$$\begin{aligned} |R_{XY}(\tau)| &\leq \sqrt{R_X(0)R_Y(0)} & (a) \\ |R_{XY}(\tau)| &\leq \tfrac{1}{2}(R_X(0) + R_Y(0)) & (b) \end{aligned} \tag{9.21}$$

The top equation (a) is a form of the well-known Cauchy-Schwartz inequality (see Section 6.1.4 of Chapter 6.) The proof of the lower equation (b) is left as a problem (Prob. 9.12). These bounds are demonstrated in the example below.

Example 9.8: Consider the random processes described in Example 9.7. The correlation functions were found to be

$$R_X(\tau) = \sigma^2 \cos \omega\tau, \quad R_Y(\tau) = \sigma^2 \cos \omega\tau, \quad R_{XY}(\tau) = -\sigma^2 \sin \omega\tau$$

Thus $R_X(0) = R_Y(0) = \sigma^2$ and $|R_{XY}(\tau)| = \sigma^2 \sin \omega\tau$

Let us now check if $R_{XY}(\tau)$ satisfies (9.21).

Substituting in (9.21)(a) on both sides yields $\sigma^2 |\sin \omega\tau| \leq \sigma^2$, which reduces to $|\sin \omega\tau| \leq 1$.

---

[3] Except in the Gaussian case.

Substituting in (9.21)(b) yields $\sigma^2|R_{XY}(\tau)| \leq \frac{1}{2}(2\sigma^2)$ or $|\sin\omega\tau| \leq 1$.

Thus the cross-correlation function satisfies both bounds.

□

9.3 Complex Random Processes

Electrical engineers involved in signal processing and communication generally formulate problems using what is called "complex envelope notation." It is not necessary to delve into the origins of this method; however, it is sufficient to say that narrowband communications receivers deal with two parts of a signal known as the "in-phase" (with respect to the carrier) and "quadrature" components. In carrying out the analysis these components are represented as real and imaginary parts of a complex signal. The complex random signal is written as

$$X(t) = X_r(t) + \jmath X_i(t)$$

where $\jmath = \sqrt{-1}$ and X_r and X_i are the real and imaginary parts.

The complete analysis of complex random signals involves the joint distribution and density functions for all of the real and imaginary parts. Fortunately, it is sufficient for most practical problems to deal with just the first and second moments.

9.3.1 Autocorrelation for complex processes

When a random signal is complex, it is necessary to modify the definitions of correlation slightly. The autocorrelation function of a wss complex signal is defined as

$$R_X(\tau) = \mathcal{E}\{X(t)X^*(t-\tau)\} \tag{9.22}$$

where $*$ denotes complex conjugate.

The reason for the conjugate in the definition is the following. The autocorrelation function (9.22) is a complex quantity which upon expansion can be written as

$$R_X(\tau) = (R_{X_r}(\tau) + R_{X_i}(\tau)) + \jmath (R_{X_iX_r}(\tau) - R_{X_rX_i}(\tau))$$

where the terms on the right are real-valued autocorrelation and cross-correlation functions for the components of the random signal. The signals found in most communications systems have a certain symmetry property such that $R_{X_r}(\tau) = R_{X_i}(\tau)$ and $R_{X_iX_r}(\tau) = -R_{X_rX_i}(\tau)$. Therefore all of the needed correlation terms can be extracted from the complex autocorrelation (9.22) by observing that $R_r = R_i = \frac{1}{2}\text{Re}\,R_X$ and $R_{ri} = -R_{ir} = \frac{1}{2}\text{Im}\,R_X$. In fact, if the conjugate is left out of the definition in (9.22), then under these symmetry conditions, the result is identically zero.[4] (Check it out!)

The autocorrelation of a complex random process is illustrated in the following example.

Example 9.9: Consider a complex random process

$$X(t) = Ae^{j\omega t}$$

where A is a Gaussian random variable with mean μ and variance σ^2 while ω is a deterministic parameter.

---

[4] There are several papers in the signal processing literature concerning cases where the symmetry does not hold. This, however, goes way beyond our discussion here and does not invalidate the use of (9.22) in the numerous cases where it applies.

The mean of this random process is given by

$$m_X(t) = \mathcal{E}\{X(t)\} = \mathcal{E}\{A\}e^{-j\omega t} = \mu e^{j\omega t}$$

Note that the mean retains the dependence on $e^{j\omega t}$ and therefore the random process is not wss unless $\mu = 0$. Let us make that assumption and continue.

The autocorrelation function is

$$\begin{aligned} R_X(t,\tau) &= \mathcal{E}\{X(t)X^*(t-\tau)\} = \mathcal{E}\{Ae^{j\omega t}Ae^{-j\omega(t-\tau)}\} \\ &= \mathcal{E}\{A^2\}e^{j\omega\tau} = \sigma^2 e^{j\omega\tau} \end{aligned}$$

where we have assumed $\mu = 0$ so $\mathcal{E}\{A^2\} = \sigma^2$. Note that the autocorrelation function is complex with both real and imaginary parts:

$$R_X(\tau) = \sigma^2 \cos\omega\tau + j\sigma^2 \sin\omega\tau$$

(This follows from the Euler identity $e^{j\phi} = \cos\phi + j\sin\phi$.) Since the mean is zero and the autocorrelation function depends only on lag τ, this random process is wss.

□

In this example the mean was an important factor in determining stationarity. In practice, we can often remove a time-varying mean and carry out the analysis or design with a modified random signal that has zero mean and is wss.

The next two examples further illustrate the role of symmetry in complex random processes.

Example 9.10: Consider a complex random process

$$Z(t) = X(t) - jY(t)$$

where $X(t) = A\cos\omega t$, $Y(t) = B\sin\omega t$, and A and B are random variables of unknown distribution.

The mean of this random process is

$$m_Z(t) = \mathcal{E}\{Z(t)\} = \mathcal{E}\{A\}\cos\omega t - j\mathcal{E}\{B\}\sin\omega t$$

which is clearly a function of time.

The autocorrelation function is given by

$$\begin{aligned} R_Z(t,\tau) &= \mathcal{E}\{Z(t)Z^*(t-\tau)\} = \mathcal{E}\{[X(t) - jY(t)][X(t-\tau) + jY(t-\tau)]\} \\ &= R_X(t,\tau) + jR_{XY}(t,\tau) - jR_{YX}(t,\tau) + R_Y(t,\tau) \end{aligned}$$

where the terms on the right are real-valued auto- and cross-correlation functions for X and Y given by:

$$R_X(t,\tau) = \mathcal{E}\{X(t)X(t-\tau)\} = \mathcal{E}\{A^2\}\cos\omega t \cos\omega(t-\tau)$$
$$R_{XY}(t,\tau) = \mathcal{E}\{X(t)Y(t-\tau\} = \mathcal{E}\{AB\}\cos\omega t \sin\omega(t-\tau)$$
$$R_{YX}(t,\tau) = \mathcal{E}\{Y(t)X(t-\tau)\} = \mathcal{E}\{AB\}\sin\omega t \cos\omega(t-\tau)$$
$$R_Y(t,\tau) = \mathcal{E}\{Y(t)Y(t-\tau)\} = \mathcal{E}\{B^2\}\sin\omega t \sin\omega(t-\tau)$$

Substituting these into the above equation for R_Z and collecting terms yields

$$\begin{aligned} R_Z(t,t-\tau) = \ & \mathcal{E}\{A^2\}\cos\omega t \cos\omega(t-\tau) + \mathcal{E}\{B^2\}\sin\omega t \sin\omega(t-\tau) \\ & + j\left(\mathcal{E}\{AB\}\cos\omega t \sin\omega(t-\tau) - \mathcal{E}\{AB\}\sin\omega t \cos\omega(t-\tau)\right) \end{aligned}$$

The autocorrelation function, like the mean, is also a function of time t which again indicates that the process is not stationary.

□

Although the general random process described in the last example is not wss, $Z(t)$ can be made wide sense stationary if the random variables meet certain criteria. If random variables A and B have zero mean, then the mean of $Z(t)$ is zero, but this is not enough. The conditions on the second moments are illustrated in the example below.

Example 9.11:   Consider the complex random process introduced in Example 9.10

$$Z(t) = X(t) - jY(t)$$

where $X(t) = A\cos\omega t$, $Y(t) = B\sin\omega t$, and A and B are random variables of unknown distribution. Let us require, however, that the random variables A and B each have mean zero and have equal power $\mathcal{E}\{A^2\} = \mathcal{E}\{B^2\} = P_o$.

Under these conditions, the autocorrelation function computed in Example 9.10 becomes

$$
\begin{aligned}
R_Z(t, t-\tau) &= P_o\left[\cos\omega t\cos\omega(t-\tau) + \sin\omega t\sin\omega(t-\tau)\right] \\
&\quad + j\mathcal{E}\{AB\}\left[\cos\omega t\sin\omega(t-\tau) - \sin\omega t\cos\omega(t-\tau)\right] \\
&= P_o(\cos(\omega t - \omega(t-\tau))) + j\mathcal{E}\{AB\}(\cos(\omega t - \omega(t-\tau)))
\end{aligned}
$$

where trigonometric identities have been used in the second line. The imaginary term drops out if and only if the random variables are *uncorrelated*, $\mathcal{E}\{AB\} = 0$ and we are left with

$$R_Z(\tau) = P_o\cos\omega\tau$$

Summarizing the above, the random process $Z(t)$ is wide sense stationary if A and B are zero-mean uncorrelated random variables with identical second moments.

□

9.3.2 Cross-correlation, covariance and properties

The cross-correlation for two jointly stationary complex random processes is defined in a way analogous to (9.22).

$$R_{XY}(\tau) = \mathcal{E}\{X(t)Y^*(t-\tau)\} \tag{9.23}$$

The cross-covariance has a similar definition with the mean of X and Y removed.

Some quantities and properties related to complex random processes are listed in Table 9.2. The equations supersede previous equations in the sense that they can be applied to real-valued random processes by dropping the complex conjugate.

9.4 Discrete Random Processes

Discrete random processes may arise from sampling continuous random signals or from measurements taken at regular intervals that are inherently discrete. Several examples of discrete random processes are cited in Chapter 8. Again, first and second moments of these random processes are usually the primary method of analysis.

The mean of a discrete random process $X[k]$ is simply the expectation

$$m_X[k] = \mathcal{E}\{X[k]\} \tag{9.24}$$

and is in general a function of the time index k. The discrete autocorrelation function compares a sample of the random process at time k to a sample at time $k - l$ and is defined for a complex random process as

$$R_X[k, l] = \mathcal{E}\{X[k]X^*[k-l]\} \tag{9.25}$$

quantity	definition
Mean	$m_X = \mathcal{E}\{X(t)\}$
Autocorrelation function	$R_X(\tau) = \mathcal{E}\{X(t)X^*(t-\tau)\}$
Cross-correlation function	$R_{XY}(\tau) = \mathcal{E}\{X(t)Y^*(t-\tau)\}$
Autocovariance function	$C_X(\tau) = \mathcal{E}\{(X(t)-m_X)(X(t-\tau)-m_X)^*\}$
Cross-covariance function	$C_{XY}(\tau) = \mathcal{E}\{(X(t)-m_X)(Y(t-\tau)-m_Y)^*\}$
Interrelations	$C_X(\tau) = R_X(\tau) - \lvert m_X\rvert^2$
	$C_X(\tau) = R_{XY}(\tau) - m_X m_Y^*$
Correlation Coefficient	$\rho_X(\tau) = \dfrac{C_X(\tau)}{C_X(0)}$ with $\lvert\rho(\tau)\rvert \le 1$

property	
Symmetry †	$R_X(\tau) = R_X^*(-\tau)$
	$R_{XY}(\tau) = R_{YX}^*(-\tau)$
Average power	$R_X(0) \ge 0$
Variance	$\sigma_X^2 = C_X(0) \ge 0$
Bound on magnitude †	$\lvert R_X(\tau)\rvert \le R_X(0)$
Positive semidefiniteness †	$\int_{-\infty}^{\infty}\int_{-\infty}^{\infty} a(t_1)R_X(t_1-t_0)a^*(t_0)dt_1 dt_0 \ge 0$
	for any arbitrary $a(t)$

† also pertains to the covariance.

Table 9.2 Definitions and properties for complex wss random processes.

where l represents the discrete lag value. (The conjugate is dropped if the random process is real.)

A discrete random process is wss if and only if the moments do not depend on the time variable k. In this case, the mean is written as a constant m_X and the autocorrelation function is written with a single argument: $R_X[l]$.

To see how these moments are computed for a simple random process, consider the Bernoulli process introduced in Section 8.3.1.

Example 9.12: A Bernoulli process $X[k]$ takes on a value of $+1$ with probability p and -1 with probability $1-p$.

The mean of the Bernoulli process is

$$m_X[k] = \mathcal{E}\{X[k]\} = +1 \cdot p + -1 \cdot (1-p) = 2p - 1$$

As long as $l \neq 0$ the autocorrelation function is given by

$$
\begin{aligned}
R_X[k,l] = \mathcal{E}\{X[k]X^*[k-l]\} &= (+1)(+1) \cdot p^2 + (+1)(-1) \cdot p(1-p) \\
&+ (-1)(+1) \cdot (1-p)p + (-1)(-1) \cdot (1-p)^2 \\
&= (2p-1)^2
\end{aligned}
$$

For $l = 0$, $X[k]$ and $X[k-l]$ are the same point; thus

$$
R_X[k,0] = \mathcal{E}\{X^2[k]\} = (+1)^2 \cdot p + (-1)^2 \cdot (1-p) = 1
$$

We see that the autocorrelation function depends on the lag l but not the time k. Therefore the Bernoulli process is wss.

The autocorrelation function is depicted below. Notice for $l \neq 0$ $R_X[l]$ is equal to the mean squared. Notice further that when $p = 0.5$ the mean is zero and $R_X[l] = \delta[l]$.

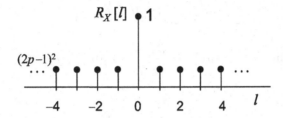

Thus the autocorrelation functions consists of a single value at the origin and zeros elsewhere. A zero-mean random process with this form of autocorrelation function is known as a discrete white noise process. The Bernoulli process with $p = 0.5$ is sometimes referred to as "binary white noise."

$\square$

A number of other discrete second moment quantities and relations can be defined which are analogous to the quantities defined for continuous random processes. For example, the autocovariance function for a discrete wss random process is defined by

$$
C_X[l] = \mathcal{E}\{(X[k] - m_X)(X[k-l] - m_X)^*\}
\tag{9.26}
$$

and satisfies

$$
C_X[l] = R_X[l] - m_X m_X^*
\tag{9.27}
$$

If the last formula is applied to the Bernoulli process of Example 9.12, the covariance function is seen to be zero everywhere except at the origin where it has a value of

$$
C_X[0] = 1 - (2p-1)^2 = 4p(1-p)
$$

This is the variance of the Bernoulli process.

The Bernoulli process is an example of a discrete random process that does not originate as a continuous time signal. When a discrete time signal *is* formed by taking samples of a continuous time process, however, the correlation and covariance functions turn out to be *samples* of the continuous time functions. In other words, if $X(t)$ is a random process with autocorrelation function $R_X(\tau)$ and $X(t)$ is sampled at time intervals T to obtain the discrete time sequence $X[k] \overset{\text{def}}{=} X(kT)$, then $X[k]$ has the discrete autocorrelation function $R_X[l] = R_X(lT)$.

The counterpart to Table 9.2 for discrete random processes is shown on the next page. A number of points can be made concerning this table.

quantity	definition
Mean	$m_X = \mathcal{E}\{X[k]\}$
Autocorrelation function	$R_X[l] = \mathcal{E}\{X[k]\,X^*[k-l]\}$
Cross-correlation function	$R_{XY}[\tau] = \mathcal{E}\{X[k]\,Y^*[k-l]\}$
Autocovariance function	$C_X[l] = \mathcal{E}\{[X[k]-m_X]\,[X[k-l]-m_X]^*\}$
Cross-covariance function	$C_{XY}[l] = \mathcal{E}\{[X[k]-m_X]\,[Y[k-l]-m_Y]^*\}$
Interrelations	$C_X[l] = R_X[l] - |m_X|^2$
	$C_X[l] = R_{XY}[l] - m_X m_Y^*$
Correlation Coefficient	$\rho_X[l] = \dfrac{C_X[l]}{C_X[0]}$ with $|\rho[l]| \leq 1$

property	
Symmetry †	$R_X[l] = R_X^*[-l]$
	$R_{XY}[l] = R_{YX}^*[-l]$
Average power	$R_X[0] \geq 0$
Variance	$\sigma_X^2 = C_X[0] \geq 0$
Bound on magnitude †	$|R_X[l]| \leq R_X[0]$
Positive semidefiniteness †	$\displaystyle\sum_{k_1=-\infty}^{\infty}\sum_{k_0=-\infty}^{\infty} a[k_1]R_X[k_1-k_0]a^*[k_0] \geq 0$
	for any arbitrary $a[k]$

† also pertains to the covariance.

Table 9.3 Discrete-time wss random processes: definitions and properties.

- Square brackets are used to indicate that all of the functions are discrete-time sequences.

- Wide sense stationarity is assumed. This means that first moments are constant and correlation functions depend only on *lag (l)*.

- All functions are complex in general. Equivalent formulas for real-valued random processes are obtained by dropping the conjugate.

- Certain parameters, such as average power $R_X[0]$ and variance $C_X[0]$ are always real and positive.

- The positive semidefinite property is expressed as a double summation instead of a double integral. The arbitrary sequence a should be real if R_X is real.

A simple model for a discrete random process has an exponential covariance function. Some of the items in Table 9.3 are illustrated with an example employing this type of random process.

Example 9.13:   A wss discrete-time random process $X[k]$ has a mean and autocorrelation function

$$m_X = 3 \qquad R_X[l] = 9 + 4e^{-0.2|l|}$$

The autocovariance function is computed using (9.27):

$$C_X[l] = R_X[l] - m_X^2 = 9 + 4e^{-0.2|l|} - 3^2 = 4e^{-0.2|l|}$$

Plots of the autocorrelation and autocovariance function are provided below. The

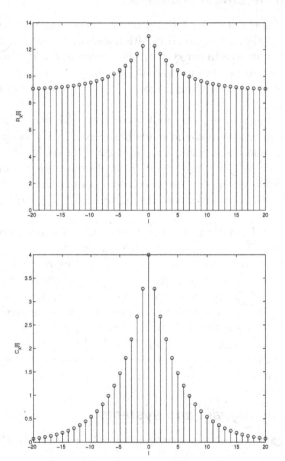

plots display the even symmetry of the two functions and illustrate that the function is positive at the origin and obtains its largest value for $l = 0$.

Let us now consider two points in time, $k = 5$ and $k = 8$ and compute some parameters for the two samples $X[5]$ and $X[8]$.

Since the mean is not a function of time, the mean of both samples is the same, namely $m_X = 3$. The variance is also the same for the two samples and is given by $C_X[0] = 4$ (see Table 9.3).

The correlation coefficient is given by

$$\rho_X[l] = \frac{C_X[l]}{C_X[0]} = \frac{e^{-0.2|l|}}{4} = e^{-0.2|l|}$$

The lag for two time samples at 8 and 5 is $l = 8 - 5 = 3$. Therefore the correlation coefficient for the two samples is

$$\rho_X[3] = e^{(-0.2)(3)} = 0.5488$$

□

9.5 Transformation by Linear Systems

The discussion in this chapter has focused on describing random signals by first and second moment quantities, i.e., mean, correlation and covariance functions. When these random signals are input to a linear system, the output of the system is also a random signal. It is important to be able to describe the first and second moment quantities for the output in terms of the first and second moment quantities of the input and the linear system parameters.

Some continuous linear transformations occur naturally in real world experiments. For example, in ocean studies a sound made in one part of the ocean propagates to a hydrophone array at another location and is recorded for study. Frequently the sound at the source, such as marine life or some form of ocean noise, can be described as a random process. The ocean can be described as a continuous linear system that changes the statistical characteristics of the input and it is important to understand this effect.

Discrete linear transformations are most often used in the *processing* of signals. Therefore it is essential to understand the effect of linear transformations on discrete random signals in order to optimize the processing and extract information.

This section discusses the effect of both continuous and discrete-time linear systems on random processes of the corresponding type. For now, the analysis is conducted in the time domain. Many important techniques can be formulated in the frequency domain, however, but that is the topic of the next chapter.

In all of the discussion here, the system is assumed to be linear and time-invariant (LTI) while the input is assumed to be a wss random process. Under these conditions the output is also a wss process.

9.5.1 Continuous signals and systems

A linear time-invariant system is depicted in Fig. 9.3 where $X(t)$ is the input and $Y(t)$

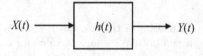

$X(t) \longrightarrow \boxed{h(t)} \longrightarrow Y(t)$

Figure 9.3 A linear time-invariant system driven by a random process.

is the output. The function $h(t)$ printed on the box is called the impulse response and completely describes the linear system. Like the random signals, the impulse response may be complex, *but it is not random!*

The output of the linear system is given by the integral

$$Y(t) = \int_{-\infty}^{\infty} h(\lambda)X(t-\lambda)\,d\lambda \tag{9.28}$$

which is known as the convolution integral[5]. The computation itself is called *convolution* and is conventionally written as

$$Y(t) = h(t) \circledast X(t)$$

The term "impulse response" is used because $h(t)$ represents the output of the system if the input is a unit impulse $\delta(t)$.

9.5.2 Transformation of moments

Output mean

The mean value of the system output can be easily obtained by applying the expectation to (9.28). In particular

$$m_Y = \mathcal{E}\{Y(t)\} = \mathcal{E}\left\{\int_{-\infty}^{\infty} h(\lambda)X(t-\lambda)\,d\lambda\right\} = \int_{-\infty}^{\infty} h(\lambda)\mathcal{E}\{X(t-\lambda)\}\,d\lambda$$

or

$$\boxed{m_Y = m_X \int_{-\infty}^{\infty} h(\lambda)\,d\lambda} \tag{9.29}$$

which is a constant. This equation implies an important fact for an LTI system: *if the input has zero mean, then the output has zero mean.*

The following example illustrates computation of the output mean.

Example 9.14: A linear system with the following impulse response[6]

$$h(t) = e^{-2t}u(t)$$

is driven by the random process $X(t) = 2 + 5\sin(\omega t + \Theta)$. Here Θ is a uniform random variable over the interval $-\pi$ to π and ω is a known parameter.

The mean of the input random process is

$$m_X = \mathcal{E}\{2 + 5\sin(\omega_0 t + \theta)\} = 2 + 5\mathcal{E}\{\sin(\omega_0 t + \theta)\} = 2$$

where we observe that the expectation involving the sinusoid is 0 (see Example 9.1).

The mean of the output random process, from (9.29), is

$$m_Y = 2\int_0^{\infty} e^{-2\lambda}\,d\lambda = 2\frac{e^{-2\lambda}}{-2}\bigg|_0^{\infty} = 1$$

□

Input-output cross-correlation functions

The cross-correlation function of the system output and input is defined by

$$R_{YX}(\tau) = \mathcal{E}\{Y(t)X^*(t-\tau)\}$$

---

[5] Convolution is encountered in another context in Chapter 5, see Section 5.5.2.
[6] This function is sketched in Example 9.15.

Substituting for $Y(t)$ from (9.28) results in

$$R_{YX}(\tau) = \mathcal{E}\left\{\left(\int_{-\infty}^{\infty} h(\lambda)X(t-\lambda)\,d\lambda\right)X^*(t-\tau)\right\}$$

$$= \int_{-\infty}^{\infty} h(\lambda)\mathcal{E}\{X(t-\lambda)X^*(t-\tau)\}\,d\lambda$$

Now observe that for any two points in a wss random process we can write[7]

$$\mathcal{E}\{X(t_1)X^*(t_0)\} = R_X(t_1 - t_0)$$

Applying this in the above equation yields

$$R_{YX}(\tau) = \int_{-\infty}^{\infty} h(\lambda)R_X(\tau-\lambda)\,d\lambda$$

This latter result can be recognized as the convolution of the system impulse response with the input autocorrelation function.

$$\boxed{R_{YX}(\tau) = h(\tau) \circledast R_X(\tau)} \qquad (9.30)$$

The following example determines both cross-correlation functions when the system input is white noise.

Example 9.15: Consider the LTI system of Example 9.14. Suppose that the input is white noise with an autocorrelation function given by

$$R_X(\tau) = \frac{N_0}{2}\delta(\tau)$$

The system impulse response and the input autocorrelation function are illustrated in the figure below.

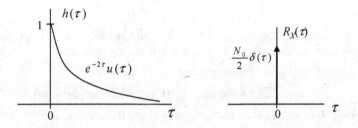

From (9.30), the cross-correlation function $R_{YX}(\tau)$ is[8]

$$R_{YX}(\tau) = h(\tau) \circledast R_X(\tau) = \left(e^{-2\tau}u(\tau)\right) \circledast \left(\frac{N_0}{2}\delta(\tau)\right) = \frac{N_0}{2}e^{-2\tau}u(\tau)$$

and from (9.20), the cross-correlation function $R_{XY}(\tau)$ is

$$R_{XY}(\tau) = R_{YX}(-\tau) = \frac{N_0}{2}e^{2\tau}u(-\tau)$$

The figure following shows plots of these two cross-correlation functions.

---

[7] This follows from the earlier definition of the autocorrelation function in terms of t and τ.
[8] Convolution with an impulse leaves a function unchanged.

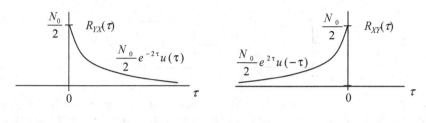

□

In the last example, observe that $R_{YX}(\tau)$ turns out to be just a scaled version of the system impulse response. This is an important result that can be used to determine the impulse response of an unknown system. The general precedure is considered in Section 9.6.1.

Output autocorrelation function

The autocorrelation function of the system output $R_Y(\tau)$ is

$$R_Y(\tau) = \mathcal{E}\{Y(t)Y(t-\tau)\}$$

Substituting (9.28) for the first $Y(t)$ results in

$$
\begin{aligned}
R_Y(\tau) &= \mathcal{E}\left\{\left(\int_{-\infty}^{\infty} h(\lambda)X(t-\lambda)\,d\lambda\right)Y^*(t-\tau)\right\} \\
&= \int_{-\infty}^{\infty} h(\lambda)\mathcal{E}\{X(t-\lambda)X^*(t-\tau)\}\,d\lambda = \int_{-\infty}^{\infty} h(\lambda)R_X(\tau-\lambda)\,d\lambda
\end{aligned}
$$

This is recognized as the convolution

$$R_Y(\tau) = h(\tau) \circledast R_{XY}(\tau) \tag{9.31}$$

Let us now combine (9.31) with (9.30) by first writing

$$R_{XY}(\tau) = R_{YX}^*(-\tau) = h^*(-\tau) \circledast R_X^*(-\tau)$$

Then substituting this result in (9.31) and recognizing that $R_X^*(-\tau) = R_X(\tau)$ produces the desired formula

$$R_Y(\tau) = h(\tau) \circledast h(-\tau) \circledast R_X(\tau) \tag{9.32}$$

This result can also be written in integral form as

$$R_Y(\tau) = \int_{-\infty}^{\infty}\int_{-\infty}^{\infty} h(\lambda)h(\mu)R_X(\tau-\lambda+\mu)\,d\lambda\,d\mu$$

The double convolution of (9.32) seems more intuitive, however, and that is the method we emphasize. This operation is depicted in Figure 9.4. The purpose of the figure is not to suggest a way of computing the output autocorrelation function but rather to illustrate the concept.

The expressions for the cross-covariance functions and the ouput autocovariance function are identical to those for the correlation functions. (After all, a covariance function *is* a correlation function for the random process with the mean removed.) A number of formulas useful for the covariance are listed in Table 9.4.

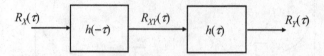

Figure 9.4 Conceptual schematic representation of output autocorrelation function (power spectral density) in terms of the system and the input autocorrelation (power spectral density).

quantity	definition
Cross-covariance Function	$C_{YX}(\tau) = h(\tau) \circledast C_X(\tau)$
	$C_{XY}(\tau) = h^*(-\tau) \circledast C_X(\tau)$
Output Autocovariance Function	$C_Y(\tau) = h(\tau) \circledast C_{XY}(\tau)$
	$C_Y(\tau) = h(\tau) \circledast h^*(-\tau) \circledast C_X(\tau)$
Interrelationships	$C_{YX}(\tau) = R_{YX}(\tau) - m_Y m_X^*$
	$C_{XY}(\tau) = R_{XY}(\tau) - m_X m_Y^*$
	$C_Y(\tau) = R_Y(\tau) - |m_Y|^2$

Table 9.4 Covariance functions pertaining to an LTI system.

This section ends with an example that illustrates the calculations for the methods developed here.

Example 9.16: Consider the linear system of Example 9.14 having impulse response

$$h(t) = e^{-2t}u(t)$$

The system input is a white noise process with added mean $m_X = \sqrt{2}$ and autocorrelation function

$$R_X(\tau) = 2 + 3\delta(\tau)$$

We will determine the output mean, the cross-correlation functions, the output autocorrelation function, and the output power spectral density. The impulse response and the input autocorrelation function are shown below.

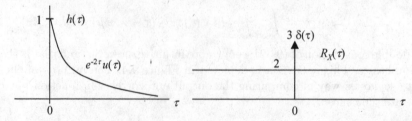

From the autocorrelation function, the input mean $m_X = \sqrt{2}$ (see Table 9.4). The output mean is

$$m_Y = \sqrt{2} \int_0^\infty e^{-2t} dt = \frac{1}{\sqrt{2}}$$

The input autocovariance function is given by

$$C_X(\tau) = R_X(\tau) - m_X^2 = 3\,\delta(\tau)$$

The cross-covariance function is thus

$$C_{YX}(\tau) = h(\tau) \circledast C_X(\tau) = \left(e^{-2\tau}u(\tau)\right) \circledast 3\,\delta(\tau) = 3\,e^{-2\tau}u(\tau)$$

and the corresponding cross-correlation function becomes

$$R_{YX}(\tau) = C_{YX}(\tau) + m_Y m_X = 3e^{-2\tau}u(\tau) + 1$$

which is shown in the figure below.

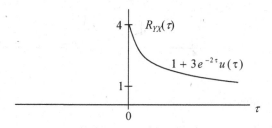

We then have

$$C_{XY}(\tau) = C_{YX}(-\tau) = 3\,e^{2\tau}u(-\tau)$$

$$R_{XY}(\tau) = C_{XY}(\tau) + m_X m_Y = 3e^{2\tau}u(-\tau) + 1$$

and a plot of the cross-correlation function is shown below.

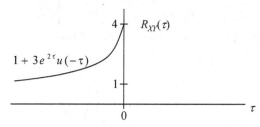

Now turn to computing the second moment quantities associated with the output $Y(t)$. For ease of computation, let us first determine the output autocovariance function

$$C_Y(\tau) = h(\tau) \circledast C_{XY}(\tau) = 3e^{-2\tau}u(\tau) \circledast e^{2\tau}u(-\tau)$$

The above convolution is carried out for two cases:

Case 1: $\tau \geq 0$

The figure below provides a graphical illustration of this case. Note how the second function is "turned around in time."

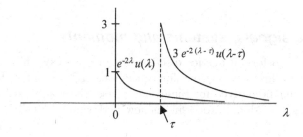

The integral is evaluated as follows:

$$C_Y(\tau) = 3 \int_\tau^\infty e^{-2\lambda} e^{-2(-\tau+\lambda)} d\lambda = 3e^{2\tau} \frac{e^{-4\lambda}}{-4}\bigg|_\tau^\infty = \frac{3}{4} e^{-2\tau}$$

Case 2: $\tau \leq 0$

The figure below provides a graphical illustration of this case. The corresponding

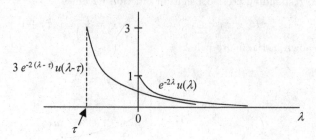

integral is evaluated as follows:

$$C_Y(\tau) = 3 \int_0^\infty e^{-2\lambda} e^{-2(-\tau+\lambda)} d\lambda = 3e^{2\tau} \frac{e^{-4\lambda}}{-4}\bigg|_0^\infty = \frac{3}{4} e^{2\tau}$$

Combination of these two results yields the output autocovariance function:

$$C_Y(\tau) = \frac{3}{4} e^{-2|\tau|}$$

The output autocorrelation function is then given by

$$R_Y(\tau) = C_Y(\tau) + m_Y^2 = \frac{3}{4} e^{-2|\tau|} + \frac{1}{2}$$

Finally, a plot of the output autocorrelation function is shown in the figure below.

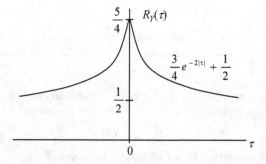

Clearly, the output autocorrelation function exhibits the required even symmetry and satisfies the necessary condition $|R_Y(\tau)| \leq R_Y(0)$.

□

9.5.3 Discrete signals, systems and moments

The discussion of transformations for discrete random processes follows closely upon the discussion in the last few sections for continuous signals and systems. Although many results will not be derived for the discrete case, the results are *at least* equally important since most of signal processing is now performed digitally.

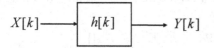

$$X[k] \longrightarrow \boxed{h[k]} \longrightarrow Y[k]$$

Figure 9.5 Discrete linear time-invariant system with random input and output sequences.

Figure 9.5 depicts a discrete linear time-invariant system[9] with random input $X[k]$ and output $Y[k]$. The box is labeled with the discrete impulse response sequence $h[k]$. The system output is given by

$$Y[k] = \sum_{i=-\infty}^{\infty} h[i]X[k-i] \tag{9.33}$$

which is the discrete form of convolution, also written as

$$Y[k] = h[k] \circledast X[k] \tag{9.34}$$

Output mean

The output mean of a discrete system is easily computed by taking the expectation on both sides of (9.33):

$$\mathcal{E}\{Y[k]\} = \sum_{i=-\infty}^{\infty} h[i]\mathcal{E}\{X[k-i]\}$$

which yields

$$\boxed{m_Y = m_X \sum_{i=-\infty}^{\infty} h[i]} \tag{9.35}$$

As for continuous linear systems, whenever the input mean is zero, the output mean is zero.

The following example illustrates the output mean computation.

Example 9.17:   A finite impulse response (FIR) linear system is described by the following equation:

$$Y[k] = 0.5X[k] + 0.2X[k-1]$$

The input to the system has mean $m_X = 2$. Let us compute the output mean.

The impulse response of this system is

$$h[k] = \begin{cases} 0.5 & k = 0 \\ 0.2 & k = 1 \\ 0 & \text{otherwise} \end{cases}$$

or equivalently, $h[k] = 0.5\delta[k] + 0.2\delta[k-1]$. From (9.35), the mean of the output is therefore

$$m_Y = 2\,(0.5 + 0.2) = 2 \times 0.7 = 1.4$$

$\square$

---

[9] Also called a linear *shift*-invariant system.

Note that in this example the output mean could also be computed by applying the expectation directly to the difference equation:

$$\mathcal{E}\{Y[k]\} = 0.5\mathcal{E}\{X[k]\} + 0.2\mathcal{E}\{X[k-1]\}$$

which leads to the same result.

Correlation functions

Let us now consider calculation of correlation functions. Figure 9.6 shows a linear

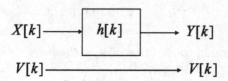

Figure 9.6 A linear system driven in the presence of a third random process.

system with input $X[k]$ and output $Y[k]$ in the presence of another random process $V[k]$. Beginning with the input-output relation (9.34), it is easy to show that

$$R_{YV}[l] = h[l] \circledast R_{XV}[l] \qquad (9.36)$$

which holds for any random process $V[k]$ (see Prob. 9.26). In particular, taking $V = X$ as a special case yields

$$R_{YX}[l] = h[l] \circledast R_X[l] \qquad (9.37)$$

or taking $V = Y$ yields

$$R_Y[l] = h[l] \circledast R_{XY}[l] \qquad (9.38)$$

Combining these last two equations with the relation $R_{XY}[l] = R_{XY}^*[-l]$ produces

$$R_Y[l] = h[l] \circledast h[-l] \circledast R_X[l] \qquad (9.39)$$

Equation (9.39) is a compact formula; but bear in mind that the convolutions can be computed in any order. Moreover, it may be easier to start with (9.37), reverse the result, and then apply (9.38). The following example illustrates the procedure.

Example 9.18:   Consider the system in Example 9.17 having impulse response

$$h[k] = \begin{cases} 0.5 & k = 0 \\ 0.2 & k = 1 \\ 0 & \text{otherwise} \end{cases}$$

Assume that this system is driven by a discrete random process having an autocorrelation function

$$R_X[l] = \delta[l+1] + 2\delta[l] + \delta[l-1]$$

These two functions are illustrated graphically opposite.

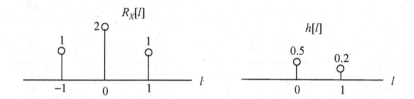

The cross-correlation function $R_{YX}[l]$ is first computed from (9.37). This is done by directly convolving impulses with impulses.

$$R_{YX}[l] = h[l] \circledast R_X[l] = \left(0.5\delta[k] + 0.2\delta[k-1]\right) \circledast \left(\delta[l+1] + 2\delta[l] + \delta[l-1]\right)$$

$$= 0.5\delta[l+1] + 1.2\delta[l] + 0.9\delta[l-1] + 0.2\delta[l-2]$$

A plot of the result is shown below.

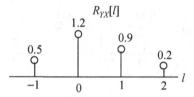

The other cross-correlation function $R_{XY}[l]$ is then computed as

$$R_{XY}[l] = R_{YX}[-l] = 0.2\delta[l+2] + 0.9\delta[l+1] + 1.2\delta[l] + 0.5\delta[l-1]$$

and is shown in the plot below.

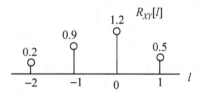

Finally, the output autocorrelation function is computed by convolving this function with the impulse response.

$$R_Y[l] = h[l] \circledast R_{XY}[l] = \sum_{i=-2}^{1} R_{XY}[i]h[l-i]$$

A graphical approach is used here. The figure on the next page illustrates the two functions in the sum for $l = -2$ and $l = 1$ and the corresponding output autocorrelation values.

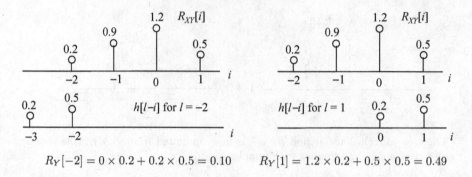

$$R_Y[-2] = 0 \times 0.2 + 0.2 \times 0.5 = 0.10 \qquad R_Y[1] = 1.2 \times 0.2 + 0.5 \times 0.5 = 0.49$$

The remaining output autocorrelation values are computed similarly.

A plot of the output autocorrelation function is shown below.

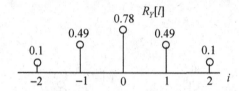

By observation, $R_Y[l]$ is even symmetric and $|R_Y[l]| \leq R_Y[0]$ for all values of l.

□

Covariance functions

The discrete covariance functions obey rules essentially identical to the discrete correlation functions [1, 2]. A summary of expressions involving the cross-covariance functions and the output autocovariance function is provided in Table 9.5.

quantity	definition
Cross-covariance Function	$C_{YX}[l] = h[l] \circledast C_X([l])$
	$C_{XY}[l] = h^*(-l) \circledast C_X[l]$
Output Autocovariance Function	$C_Y[l] = h[l] \circledast C_{XY}[l]$
	$C_Y[l] = h[l] \circledast h^*[-l] \circledast C_X[l]$
Interrelations	$C_{YX}[l] = R_{YX}[l] - m_Y m_X^*$
	$C_{XY}[l] = R_{XY}[l] - m_X m_Y^*$
	$C_Y[l] = R_Y[l] - |m_Y|^2$

Table 9.5 Discrete covariance functions pertaining to an LTI sysytem.

The first four entries in the table can also serve as reminders of the expressions pertaining to the correlation functions.

9.6 Some Applications

Two examples of applications of the tools developed in this chapter are provided here. The system identification problem is formulated in terms of continuous signals and systems. The optimal filtering problem is formulated in discrete time.

9.6.1 System identification

System identification is a method of determining the impulse response $h(t)$ of an unknown system by use of the output-input cross-correlation function discussed in Section 9.5.2. Figure 9.7 shows a block diagram of a system identification scheme. The

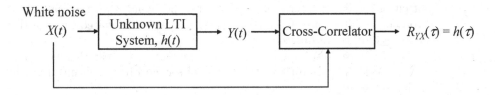

Figure 9.7 Schematic representation of system identification of an unknown system.

unknown system is driven by a white noise having $N_0/2 = 1$, that is,

$$R_X(\tau) = \delta(\tau)$$

The cross-correlation function between the output $Y(t)$ and the input $X(t)$ is determined from (9.30) as

$$R_{YX}(\tau) = h(\tau) \circledast R_X(\tau) = h(\tau) \circledast \delta(\tau) = h(\tau) \tag{9.40}$$

In Fig. 9.7 a cross-correlation device is used to compute or estimate $R_{YX}(\tau)$, which is equal to the system impulse response.

   This is a useful and a more practical method for determining the impulse response than driving a system with an impulse directly. Ideally, an impulse has infinite magnitude and 0 width, which makes generating an impulse impractical. For the sake of this argument, suppose that it is possible to generate a waveform having a narrow pulse width and a large magnitude in order to closely resemble an impulse. Such a waveform may cause undue stress on the system when applied at the input. On the other hand, using white noise as in the above procedure alleviates this problem since white noise is a much gentler input.

9.6.2 Optimal filtering

The optimal filtering problem[10] is depicted graphically in Figure 9.8. A random signal $D[k]$ is generated in the environment and arrives at the receiver. Along the path there is noise and interference of various sorts so that the signal $X[k]$ at the receiver is

---

[10] Also called the Wiener filtering problem after Norbert Wiener [1864–1964].

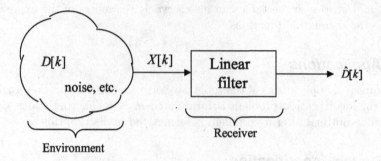

Figure 9.8 Depiction of optimal filtering.

not precisely the same as what was transmitted. The receiver decides to process $X[k]$ using a linear filter in an attempt to produce a result that is closer to the original $D[k]$. The output of the linear filter is an *estimate* for the signal, which we call $\hat{D}[k]$. It is desired to find the filter, optimal in the sense that the mean squared value of the error $\mathcal{E}[k] = D[k] - \hat{D}[k]$ is minimized.

The filter is an FIR linear system with a set of parameters a_i. To keep the derivation as simple as possible let's assume that the filter is of second order; specifically, it is described by the equation

$$\hat{D}[k] = a_0 X[k] + a_1 X[k-1]$$

The error is therefore given by

$$\mathcal{E}[k] = D[k] - \hat{D}[k] = D[k] - a_0 X[k] - a_1 X[k-1] \tag{9.41}$$

while the quantity to be minimized is $\mathcal{E}\left\{\mathcal{E}^2[k]\right\}$.

The coefficients to minimize the mean squared error can be found by taking derivatives; we require

$$\frac{\partial}{\partial a_i}\left(\mathcal{E}\left\{\mathcal{E}^2[k]\right\}\right) = \mathcal{E}\left\{2\mathcal{E}[k]\frac{\partial \mathcal{E}[k]}{\partial a_i}\right\} = -2\mathcal{E}\left\{\mathcal{E}[k]X[k-i]\right\} = 0 \qquad i = 0, 1 \tag{9.42}$$

where the partial derivative and expectation were interchanged and we noticed from (9.41) that $\dfrac{\partial \mathcal{E}[k]}{\partial a_i} = -X[k-i]$. The last part of (9.42) is known as the "orthogonality principal." It states that in order to minimize the mean-squared error, the error $\mathcal{E}[k]$ must be *orthogonal* to the observations[11]. Now substituting (9.41) in (9.42) leads to the condition

$$\mathcal{E}\left\{\left(D[k] - a_0 X[k] - a_1 X[k-1]\right) X[k-i]\right\} = 0 \qquad i = 0, 1$$

and taking the expectation and rearranging terms yields

$$R_X[i]a_0 - R_X[i-1]a_1 = R_{DX}[i] \qquad i = 0, 1 \tag{9.43}$$

When evaluated at $i = 0$ and $i = 1$, (9.43) represents a pair of equations that can be solved for the two filter parameters a_0 and a_1:

$$R_X[0]a_0 \;+\; R_X[-1]a_1 \;=\; R_{DX}[0]$$

$$R_X[1]a_0 \;+\; R_X[0]a_1 \;=\; R_{DX}[1]$$

---

[11] See discussion of orthogonality in Section 5.3.1 and Table 5.3 of Chapter 5.

Clearly, this procedure can be generalized to FIR filters with any number of parameters and the resulting equations can be solved by matrix inversion.

Let us now derive an expression for the mean-squared error produced by the optimal filter. By substituting (9.41) for one instance of $\mathcal{E}[k]$ we can write

$$MSE = \mathcal{E}\left\{\mathcal{E}^2[k]\right\} = \mathcal{E}\left\{\mathcal{E}[k]\left(D[k] - a_0 X[k] - a_1 X[k-1]\right)\right\} = \mathcal{E}\left\{\mathcal{E}[k]D[k]\right\}$$

where the terms involving X vanish because of (9.42) (orthogonality).

Once again substituting (9.41) into this equation and taking the expectation leads to

$$MSE = R_D[0] - a_0 R_X[0] - a_1 R_X[1] \tag{9.44}$$

This is the mean squared error produced by the optimal filter and serves as its figure of merit.

Let us consider two examples to show how the optimal filter is applied. In the examples a filter of slightly higher order is used and the above formulas are generalized accordingly.

Example 9.19: A real-valued random signal $S[k]$ is observed in white noise which is uncorrelated with the signal. The observed sequence is given by

$$X[k] = S[k] + N[k]$$

where

$$R_S[l] = 2(0.8)^{|l|} \quad \text{and} \quad R_N[l] = 2\delta[l]$$

It is desired to estimate $S[k]$ using the present and previous two observations.

For this problem the "desired" signal $D[k]$ is equal to $S[k]$, and the optimal filter is of the form

$$\hat{D}[k] = a_0 X[k] + a_1 X[k-1] + a_2 X[k-2]$$

Following the development above we can find the equations for the optimal filter and its mean squared error.

$$\begin{bmatrix} R_X[0] & R_X[-1] & R_X[-2] \\ R_X[1] & R_X[0] & R_X[-1] \\ R_X[2] & R_X[1] & R_X[0] \end{bmatrix} \begin{bmatrix} a_0 \\ a_1 \\ a_2 \end{bmatrix} = \begin{bmatrix} R_{DX}[0] \\ R_{DX}[1] \\ R_{DX}[2] \end{bmatrix} \qquad MSE = R_D[0] - \sum_{i=0}^{2} a_i R_X[i]$$

The necessary correlation functions are found as follows:

$$R_D[l] = R_S[l] = 2(0.8)^{|l|}$$

$$R_{DX}[l] = \mathcal{E}\left\{S[k]X[k-l]\right\} = \mathcal{E}\left\{S[k]\left(S[k-l] + N[k-l]\right)\right\} = R_S[l] = 2(0.8)^{|l|}$$

$$R_X[l] = \mathcal{E}\left\{\left(S[k] + N[k]\right)\left(S[k-l] + N[k-l]\right)\right\} = R_S[l] + R_N[l] = 2(0.8)^{|l|} + 2\delta[l]$$

where the terms which are products of S and N vanish because the signal and noise are uncorrelated.

The needed numerical values are then

$$R_D[0] = 2$$

$$R_X[0] = 4.00 \quad R_X[1] = R_X[-1] = 1.60 \quad R_X[2] = R_X[-2] = 1.28$$
$$R_{DX}[0] = 2.00 \quad R_{DX}[1] = 1.60 \quad R_{DX}[2] = 1.28$$

The equation for the filter parameters now becomes

$$\begin{bmatrix} 4.00 & 1.60 & 1.28 \\ 1.60 & 4.00 & 1.60 \\ 1.28 & 1.60 & 4.00 \end{bmatrix} \begin{bmatrix} a_0 \\ a_1 \\ a_2 \end{bmatrix} = \begin{bmatrix} 2.00 \\ 1.60 \\ 1.28 \end{bmatrix}$$

This can be solved to find

$$a_0 = 0.3824 \qquad a_1 = 0.2000 \qquad a_2 = 0.1176$$

The mean-square error is then

$$
\begin{aligned}
MSE &= R_D[0] - \sum_{i=0}^{2} a_i R_X[i] \\
&= 2.000 - 0.3824 \cdot 4.00 - 0.2000 \cdot 1.60 - 0.1176 \cdot 1.28 \\
&= 0.7647
\end{aligned}
$$

□

A slight modification of this example illustrates the problem of *prediction* for a random signal in noise.

Example 9.20: Let us consider the problem of estimating the sequence in Example 9.19 two points ahead, using the same three observations. For this problem $D[k] = S[k+2]$ and

$$
\begin{aligned}
R_{DX}[l] &= \mathcal{E}\{S[k+2]X[k-l]\} = \mathcal{E}\{S[k+2](S[k-l] + N[k-l])\} \\
&= R_S[l+2] = 2(0.8)^{|l+2|}
\end{aligned}
$$

which produces the values

$$R_{DX}[0] = 1.280 \qquad R_{DX}[1] = 1.024 \qquad R_{DX}[2] = 0.8192$$

$R_D[l]$ and $R_X[l]$ are found to be the same as before.

The equation for the filter parameters becomes

$$
\begin{bmatrix} 4.00 & 1.60 & 1.28 \\ 1.60 & 4.00 & 1.60 \\ 1.28 & 1.60 & 4.00 \end{bmatrix}
\begin{bmatrix} a_0 \\ a_1 \\ a_2 \end{bmatrix} =
\begin{bmatrix} 1.280 \\ 1.024 \\ 0.8192 \end{bmatrix}
$$

which is solved to find

$$a_0 = 0.2447 \qquad a_1 = 0.1280 \qquad a_2 = 0.07529$$

The mean-square error in this case is

$$
\begin{aligned}
MSE &= R_D[0] - \sum_{i=0}^{2} a_i R_X[i] \\
&= 2.000 - 0.2447 \cdot 1.280 - 0.1280 \cdot 1.024 - 0.07529 \cdot 0.8192 \\
&= 1.494
\end{aligned}
$$

This is larger than the mean-squared error of the previous example by about a factor of 2. This indicates that prediction of the signal two points ahead is considerably less accurate than estimation of its current value.

□

9.7 Summary

This chapter focuses on the description of random signals and other random processes in the time domain. The discussion follows the usual procedure of analysis using first and second moment quantities. The chapter begins by defining mean, correlation and covariance functions for a single random process. The important notion of wide sense stationarity is then developed and stated as a time invariance of mean and correlation. Properties of the autocorrelation and autocovariance functions are then cited. Following these developments the discussion is broadened to the second moment description of two random processes through the cross-correlation and cross-covariance functions.

The chapter then introduces complex-valued random processes and the special conditions that apply to their analysis. This is followed by a parallel discussion for random processes in discrete time.

As a final topic we consider how random processes are affected by linear systems. This is an important topic because linear systems are used to extract information and otherwise transform random processes to suit our needs. The chapter ends with two applications that illustrate the use of random processes and suitable processing in two important engineering applications.

References

[1] Charles W. Therrien. *Discrete Random Signals and Statistical Signal Processing*. Prentice Hall, Inc., Upper Saddle River, New Jersey, 1992.

[2] Henry Stark and John W. Woods. *Probability, Random Processes, and Estimation Theory for Engineers*. Prentice Hall, Inc., Upper Saddle River, New Jersey, third edition, 2002.

Problems

First and second moments of a random process

9.1 Given the following random signal $X(t) = A + \cos(\omega t + \Phi)$, where Φ is a uniform random variable in the range $-\pi$ to π and A is a constant, determine the following:

  (a) The mean of $X(t)$.

  (b) The autocorrelation function of $X(t)$.

  (c) The autocovariance function of $X(t)$.

9.2 A certain signal is modeled by a continuous Gaussian random process $X(t)$ with mean $\mathcal{E}\{X(t)\} = 2$ and autocorrelation function $R(t, \tau) = 4 + e^{-0.5|\tau|}$.

  (a) Compute the covariance function $C(t, \tau)$.

  (b) Compute the variance of $X(t)$.

  (c) Write the PDF for a single sample $X(t_0)$.

  (d) Write the joint PDF for two samples $X(t_1)$ and $X(t_0)$ for $t_1 = 7$ and $t_0 = 3$.

9.3 Let $X(t) = A \cos \omega t + B$ where A is a Gaussian random variable with mean zero and variance 2, B is a Gaussian random variable with mean 4 and variance 3, A and B are independent, and $\omega = 0.10\pi$.

  (a) Find the mean of $X(t)$.

  (b) Determine the autocorrelation function of $X(t)$.

  (c) Compute the first moment and the second moment of $X(2)$.

9.4 The mean and the autocovariance function of a random process are given by

$$m(t) = 0; \quad C_X(t, \tau) = \cos(\pi \tau).$$

(a) Find the lag τ for two points $t_1 = 3.1$ and $t_0 = 1.8$.

(b) Compute the values of mean and correlation.

(c) Compute the correlation coefficient.

(d) Determine the average total power.

(e) Find the rms value of $X(t)$.

9.5 Given the random process $X(t) = A\cos(\omega t + \phi)$, where A is a random variable uniformly distributed between 2 and 5, and ϕ and ω are constants, determine

(a) The mean of $X(t)$.

(b) The autocorrelation function of $X(t)$.

(c) Is $X(t)$ wide sense stationary? Provide a brief explanation.

9.6 Tell if the following two processes are (1) wide-sense stationary and (2) mean ergodic. A and Φ are independent random variables where A has a Rayleigh distribution and Φ has a uniform distribution over the interval $[-\pi, \pi]$.

(a) $X(t) = A\cos(\omega t + \Phi)$.

(b) $X(t) = A\cos(\omega t + 0.2\pi)$.

9.7 Show that the positive semidefinite condition (9.13) implies the property (9.11) $|R_X(\tau)| \leq R_X(0)$. Hint: Choose $a(t) = \delta(t) \pm \delta(t - \tau)$.

9.8 The mean and autocorrelation function of a random process $X(t)$ are given by $m(t) = 1$ and $R_X(t, \tau) = 1 + 3\cos(2\pi\tau)$.

(a) Given two time points $t_1 = 1.6$ and $t_0 = 1.0$, compute the correlation coefficient.

(b) Write the 2×2 covariance matrix of this random process corresponding to the vector

$$\mathbf{X} = \begin{bmatrix} X(1.0) \\ X(1.6) \end{bmatrix}$$

All entries in the matrix must be numerical values.

9.9 A certain signal is modeled by a continuous Gaussian random process $X(t)$ with mean $\mathcal{E}\{X(t)\} = 0$ and autocorrelation function $R(t, \tau) = e^{-500|\tau|}$.

(a) Is the signal wide sense stationary?

(b) Find $\mathrm{Var}[X(t)]$. Is it a function of t?

(c) Write the probability density function for a single sample $X(t_0)$.

Cross Correlation

9.10 Two random processes $X(t)$ and $Y(t)$ are independent and jointly wide sense stationary.

(a) If we define, $Z(t) = X(t) + Y(t)$, is $Z(t)$ wide sense stationary?

(b) If we define, $Z(t) = X(t)Y(t)$, is $Z(t)$ wss?

9.11 Show that cross-covariance function $C_{YX}(\tau)$ is given by

$$C_{YX}(\tau) = R_{YX}(\tau) - m_Y m_X$$

as listed in Table 9.4.

9.12 By forming a random process $Z(t) = X(t) \pm Y(t-\tau)$ and computing $R_Z(0)$, prove the bound (9.21)(b).

Complex random processes

9.13 A *complex*-valued continuous random signal is defined by

$$X(t) = Ae^{\jmath(\omega t + \Phi)}$$

where A and ω are constant parameters (i.e., not random variables) and Φ is a random variable uniformly distributed between $-\pi$ and π.

    (a) What is the mean, $E\{X(t)\}$?

    (b) If the covariance function for a complex random process is defined as $C(t, \tau) = \text{Cov}\,[X(t), X^*(t-\tau)]$, what is the autocovariance function of this process?

    (c) What is $\text{Cov}\,[X(t), X(t-\tau)]$ (*without* the conjugate)?

9.14 A *complex* continuous random process is defined by

$$X(t) = Ae^{\jmath(\omega_o t + \Phi)} + W(t)$$

where A and Φ are real parameters, ω_o is a known radian frequency, and $W(t)$ is a stationary zero-mean white noise process with autocorrelation function $R_W(\tau) = \frac{1}{2}\delta(\tau)$. Assume that $W(t)$ is independent of the parameters A and Φ. Find the autocorrelation function for $X(t)$ under the following conditions:

    (a) $\Phi = 0$ and A is a Rayleigh random variable with parameter $\alpha = \frac{1}{2}$.

    (b) Φ is uniformly distributed between $-\pi$ and π and $A = 1/\sqrt{2}$.

    (c) Φ is uniformly distributed between $-\pi$ and π and A is a Rayleigh random variable with $\alpha = \frac{1}{2}$. A and Φ are independent.

    (d) Same conditions as Part (b) but $W(t)$ has a autocorrelation function $R_W(\tau) = \frac{1}{2}e^{-|\tau|}$.

Discrete random processes

9.15 A *Bernoulli process* is defined by

$$X[k] = \begin{cases} +1 & \text{with probability } p \\ -1 & \text{with probability } 1 - p \end{cases}$$

Successive values of the process $X[k], X[k+1], X[k+2]\ldots$ are *independent*.

    (a) What are the mean and variance of the process, $E\{X[k]\}$ and $\text{Var}\,[X[k]]$? Do they depend on k?

    (b) What are the autocorrelation function $R[k, l]$ and autocovariance function $C[k, l]$ of the process? Is the random process stationary?

9.16 $X[k]$ is a binary white noise process (Bernoulli process with $p = \frac{1}{2}$).

    (a) What is the mean of this random process?

    (b) What is the autocorrelation function?

(c) What is the covariance function?

(d) If a new random process is defined as the difference

$$Y[k] = X[k] - X[k-1]$$

what is the mean of the process $Y[k]$?

(e) What is the autocorrelation function for the process $Y[k]$?

9.17 Consider the random process defined in Prob. 8.15 of Chapter 8.

(a) Compute the mean and variance of the process $Y[k]$. Do they depend on the time index k?

(b) For the special case $p = \frac{1}{2}$, do the mean and variance depend on k?

9.18 The autocorrelation functions for two discrete-time random processes $X[k]$ and $Y[k]$ are sketched below:

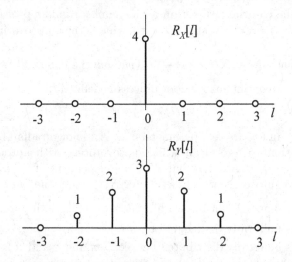

(a) Are both of these autocorrelation functions legitimate?

(b) Which process has higher average power?

9.19 A discrete-time random process $X[k]$ is defined by

$$X[k] = 2W[k] - 3W[k-1] + W[k-2]$$

where $W[k]$ is a wide-sense stationary white noise process with mean 0 and autocorrelation function $R_W[l] = \delta[l]$.

(a) Find the mean and autocorrelation function for the process $X[k]$. Is it stationary? If so, express the autocorrelation function in terms of just the lag variable l, and plot $R_X[l]$ versus l.

(b) Another random process is defined by

$$Y[k] = W[k] - 2W[k-1]$$

Find the autocorrelation function for $Y[k]$ and the cross-correlation function for X and Y. Are the two random processes *jointly stationary*? If so, express all correlation functions in terms of the lag variable l.

9.20 $X[k]$ is the input to a discrete-time linear shift-invariant system. $X[k]$ has mean 0 and autocorrelation function $R_X[l] = \delta[l]$. A zero-mean process with this form of autocorrelation function is referred to as discrete-time white noise.

The linear system has an impulse response given by

$$h[k] = 2\delta[k] - 3\delta[k-1] + \delta[k-2]$$

The output of the system is a random process $Y[k]$.

(a) Find the cross-correlation function between input and output $R_{XY}[l]$.

(b) Find the autocorrelation function of the output $R_Y[l]$.

(c) If the mean of the input is not 0, but $m_X = \mathcal{E}\{X[k]\} = 2.5$, then what is the mean of the output m_Y?

9.21 Given that the cross-covariance function $C_{XY}[l]$ is

$$C_{XY}[l] = R_{XY}[l] - m_X m_Y$$

show that the output autocovariance function $C_Y[l]$ is given by

$$C_Y[l]) = R_Y[l] - m_Y^2$$

as listed in Table 9.5.

9.22 An estimate for the autocorrelation function of a discrete-time random process $X[k]$ is given by

$$\hat{R}_X^{(N)}[l] = \frac{1}{N-l} \sum_{k=0}^{N-l-1} X[k]X[k+l]$$

where N is the number of time samples used for the estimate. (Since we already have been using a capital letter for the autocorrelation function, we use a "hat" here for the estimate and a superscript to indicate its dependence on N.)

(a) Is this estimate unbiased, asymptotically unbiased, or neither?

(b) What about the following estimate?

$$\hat{R}_X^{(N)'}[l] = \frac{1}{N} \sum_{k=0}^{N-l-1} X[k]X[k+l]$$

Response of linear systems

9.23 The impulse response $h(\tau)$ of an LTI system and the autocorrelation function of a random process $X(t)$ at the input of this system, respectively, are

$$
\begin{aligned}
h(\tau) &= e^{-2t}u(t) \\
R_X(\tau) &= \begin{cases} 2 - |\tau| & \text{for } -2 \le \tau \le 2 \\ 0 & \text{otherwise} \end{cases}
\end{aligned}
$$

Determine and sketch the cross-correlation function $R_{YX}(\tau)$.

9.24 The input to a continuous-time linear system is a white noise process with autocorrelation function $R_X(\tau) = 5\delta(\tau)$. The impulse response of the system is

$$h(t) = e^{-at}u(t)$$

where a is a fixed positive deterministic parameter and $u(t)$ represents the unit step function.

(a) Find the output autocorrelation function under the assumption that the mean of the input is 0. Be sure to specify it for $-\infty < \tau < \infty$.

(b) If the expected value of the input is 2, what is the expected value (mean) of the output?

9.25 A continuous white noise process is applied to the continuous-time system shown below. The input $W(t)$ has 0 mean and autocorrelation function given by $R_W(\tau) = \delta(\tau)$.

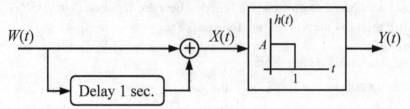

(a) What is $X(t)$ and what is the autocorrelation function $R_X(t_1, t_0)$? Is $X(t)$ stationary?

(b) What is the autocorrelation function of the output signal $Y(t)$?

9.26 With respect to the situation shown in Fig. 9.6, derive the result (9.36), $R_{YV}[l] = h[l] \circledast R_{XV}[l]$.

9.27 The input random process $X[k]$ to a discrete-time linear shift-invariant system has mean 0 and autocorrelation function $R_X[l] = 2\delta[l]$. The output of the system is $Y[k]$, and the impulse response of the system is given by

$$h[k] = \delta[k] - 2\delta[k-1]$$

(a) Find and plot the autocorrelation function of the output $R_Y[l]$.

(b) Find the variance of the system output.

9.28 Refer to Prob. 9.18. What would be the impulse response $h[k]$ of a causal time-invariant linear system such that if $X[k]$ were the input, $Y[k]$ would be the output of this system?

Computer Projects

Project 9.1

The data sets s00.dat, s01.dat, s02.dat, and s03.dat from the data package represent realizations of four different random signals. In this project you will examine the autocorrelation function for these signals.

1. Plot each of the signals versus time. It is best to use two plots per page and orient the plot in landscape mode. Tell whether you think each random signal exhibits high correlation, low correlation, or negative correlation.

2. Estimate the mean for each process; it should be close to 0.

3. Estimate the autocorrelation function $R[l]$ for each process and plot it over the interval $-128 \le l \le 127$. Compare the autocorrelation of each of the signals based on the estimated autocorrelation function. How well does this compare to your guesses about correlation in Step 1?

MATLAB programming notes:

Use the MATLAB function "getdata" from the software package to retrieve the data. The MATLAB function "xcorr" may be used to estimate the autocorrelation function.

Project 9.2

In this project you can generate and analyze some random processes of your own choice.

1. Generate some realizations of different types of random processes on your own. (Do at least two different random processes that are wide sense stationary.) These random processes must necessarily be discrete-time random processes, but they can be discrete-valued or continuous-valued.
2. For each random process that you generate, write a brief analysis (one page or less) describing the random process and showing what its theoretical mean and autocorrelation function are.
3. Now estimate the mean and autocorrelation and/or autocovariance function for your data and show how it compares to the theoretical results. Provide a brief discussion.

MATLAB programming notes:

The MATLAB functions 'rand' and 'randn' can be used to generate uncorrelated sequences of uniform or normally-distributed (Gaussian) random variables. The MATLAB function 'xcorr' may be used to estimate the autocorrelation function.

Project 9.3

This project explores the identification of an unknown linear time-invariant system using the cross-correlation approach. The discrete-time system to be identified has the transfer function

$$H(z) = \frac{0.50 - 1.32z^{-1} + 0.82z^{-2}}{1 + 1.75z^{-1} + 0.89z^{-2}}$$

1. Compute and plot the impulse response $h[k]$ of the above system for $0 \le k \le 127$.
2. Generate 10,000 samples of the input $X[k]$ as a unit-variance white Gaussian sequence. Plot the first 100 samples of $X[k]$.
3. Compute the autocorrelation function of the input sequence and plot the autocorrelation function $R_X[l]$ for $-128 \le l \le 127$. Verify that the input sequence has the autocorrelation function you expect for white noise.
4. With $X[k]$ as input, generate the output of the "unknown" system $Y[k]$. Compute the cross-correlation function $R_{YX}[l]$ between $Y[k]$ and $X[k]$.
5. Plot the cross-correlation function $R_{YX}[l]$ for $0 \le l \le 127$ and compare it to the previously computed impulse response of the system.
6. Repeat Steps 2 through 5 for input sequences of lengths 1,000 and 100,000.

MATLAB programming notes:

You may use the MATLAB functions 'filter' (signal processing toolbox) to compute the impulse response in step 1. The MATLAB function 'xcorr' can be used to estimate both the autocorrelation function and the cross-correlation function.

Project 9.4

The sequence s00.dat in the data package is a sample function of Gaussian white noise. The sequence s01.dat is a sample function of a random signal.

Consider an optimal filtering problem where the observed signal is of the form

$$X[k] = S[k] + \sigma_{\mathrm{o}} N[k]$$

where $S[k]$ is the "desired" signal (i.e., $D[k] = S[k]$) and $N[k]$ is the noise. The parameter σ_{o} is determined from the signal-to-noise ratio

$$\mathrm{SNR} = 10 \log_{10} \frac{R_S[0]}{\sigma_{\mathrm{o}}^2}$$

and S and N are assumed to be statistically independent.

1. Estimate, print and plot the autocorrelation function for the two data sets up to a lag value of $l = 15$.
2. Determine the value of σ_{o} corresponding to a SNR of 0 dB.
3. Following the procedure in Example 9.19, find and solve the equations for the optimal filters with 2, 4, 8, and 16 coefficients. Also find the mean-squared error for each case. List this information in a table.
4. For each of the four cases, filter the sequence $X[k]$ to produce the estimate $\hat{D}[k]$. For example, with two coefficients the filtering equation would be

$$\hat{D}[k] = a_0 X[k] + a_1 X[k-1]$$

(Assume $X[0]$ is the first time sample and $X[-1]$, $X[-2]$, etc. are zero.)
5. Plot the original signal $S[k]$, the signal $X[k]$ with added noise, and the estimated signal $\hat{D}[k]$ for each of the four filters. Comment on the performance of the filters.

MATLAB programming notes:

Use the MATLAB function 'getdata' from the software package to retrieve the data. The MATLAB function 'xcorr' may be used to estimate the autocorrelation function. Also, the MATLAB function 'filter' (signal processing toolbox) may be used to filter the data in Step 4.

10

Random Signals in the Frequency Domain

Up to now, the discussion of random processes has focused on the random process as a function of time. For example, the mean and the autocorrelation function provide a time domain characterization of the random process. For well established reasons, deterministic signals, as well as the systems that process them, are characterized in both the time domain and the *frequency* domain. The purpose of this chapter is to complete the description of random signals by developing tools to be used in the frequency domain.

The chapter begins with a definition of the power spectral density function that serves as a fundamental descriptor of a random signal in the frequency domain. Properties of the power spectral density function then are then derived from properties of the random process in the time domain. Cross spectral density functions are also introduced for the joint characterization of two random processes.

The discussion then moves on to white noise and bandlimited white noise as frequency domain concepts. It is seen that white noise has a very useful but uncomplicated description in the frequency domain, unlike its description in the time domain.

This topic is followed by a discussion of linear transformations for random signals in the frequency domain. The formulas that describe the transformations are easily derived from corresponding time domain formulas and properties of the Fourier transform.

The chapter moves on to a formulation of the above concepts for discrete-time random processes and systems. The formulas differ in detail from the continuous-time case, but they are important because most modern devices use digital signal processing.

The last section of the chapter provides a demonstration of the frequency domain methods in two important application areas. A digital communication application is provided which uses some of the analog or continuous time ideas. Analog spectral analysis is relevant here because although the information transmitted is digital, the transmitter is an analog device. This application is followed by a development of optimal filtering in the "digital" spectral domain. The latter is compared to an optimal filtering application presented in Chapter 9 that is carried out in the time domain.

10.1 Power Spectral Density Function

Analysis of random signals, like the analysis of deterministic signals in the frequency domain is based on the Fourier transform and its many important properties. It is assumed here that the reader is familiar with the Fourier transform and its properties on the level of Ref [1].

10.1.1 Definition

Consider a wss random process $X(t)$; i.e., its mean m_X isx constant t and its autocorrelation function $R_X(\tau)$ is a function only of the time lag τ. The *power spectral density* function (PSD) for $X(t)$ is then defined as the Fourier transform of the autocorrelation function:

$$S_X(f) = \mathcal{F}\{R_X(\tau)\} = \int_{-\infty}^{\infty} R_X(\tau)e^{-j2\pi f\tau}d\tau \qquad (10.1)$$

where $\mathcal{F}\{\}$ denotes the Fourier transform operation and f is the frequency domain variable.

The autocorrelation function can be recovered from the power spectral density by the inverse Fourier transform $\mathcal{F}^{-1}\{\}$:

$$R_X(\tau) = \mathcal{F}^{-1}\{S_X(f)\} = \int_{-\infty}^{\infty} S_X(f)e^{j2\pi f\tau}df \qquad (10.2)$$

Thus the autocorrelation function and the power spectral density form a Fourier transform pair. Note that for deterministic signals, it is customary to take the Fourier transform of the given signal and then determine the "amplitude" spectral density. On the contrary, for random signals, the "power" spectral density is based on the autocorrelation function, which is a second moment. Consequently, the power spectral density can be considered a second moment quantity.

Since the autocorrelation function and autocovariance function are related as

$$R_X(\tau) = C_X(\tau) + |m_X|^2$$

(see Table 9.2 of Chapter 9) taking the Fourier transform produces

$$S_X(f) = \mathcal{F}\{C_X(\tau) + m_X^2\} = \mathcal{F}\{C_X(\tau)\} + |m_X|^2\delta(f)$$

where the impulse or delta function in frequency is a result of taking the Fourier transform of a constant. The power spectral density thus separates into two parts; the first part is due to the covariance while the second represents the power at zero frequency (the "d.c." component of the signal). When the random signal has a d.c. component (nonzero mean) it is frequently removed when performing linear operations and added back in later.

10.1.2 Properties of the power spectral density

The properties of the autocorrelation function (see Table 9.2) induce a corresponding set of properties for the PSD. Some of these properties are presented here with a brief discussion.

1. Recall that the average power of a random process is equal to $\mathcal{E}\{|X(t)|^2\} = R_X(0)$. Thus evaluating (10.2) at $\tau = 0$ produces

$$R_X(0) = \mathcal{E}\{|X(t)|^2\} = \int_{-\infty}^{\infty} S_X(f)df \qquad (10.3)$$

   which indicates that the average power of the random process is obtained as the area under the power spectral density. While the average power can be obtained from a single value of the autocorrelation function in the time domain, it is spread over all frequencies in the frequency domain. This is why $S_X(f)$ is called *power spectral density*. The units of the PSD are *watts/hertz*.

2. The autocorrelation function of a *real-valued* random process has even symmetry $(R_X(\tau) = R_X(-\tau))$. Correspondingly, the power spectral density is a real-valued *even* function.[1]

---

[1] The real property occurs because R_X is even; the even property occurs because R_X is real.

To show this let us write (10.1) as

$$S_X(f) = \int_{-\infty}^{\infty} R_X(\tau)(\cos 2\pi f\tau - j\sin 2\pi f\tau)\, d\tau$$

$$= \int_{-\infty}^{\infty} R_X(\tau)\cos 2\pi f\tau\, d\tau \tag{10.4}$$

where the imaginary term drops out because of odd symmetry. From (10.4), it is easy to see that when R_X is real the power spectral density is even, i.e.,

$$S_X(f) = S_X(-f)$$

3. From the positive semidefinite property of the autocorrelation function (Table 9.2), it can be shown that the corresponding power spectral density is nonnegative.

$$\boxed{S_X(f) \geq 0} \tag{10.5}$$

The argument to prove this is provided later in this chapter.

The properties described above can be illustrated by two examples.

Example 10.1:   The real random process of Example 9.1 (Chapter 9) has a mean $m_X = 0$ and autocorrelation function

$$R_X(\tau) = \frac{A^2}{2}\cos \omega_0 \tau$$

where the constant $\omega_0 = 2\pi f_0$ has been introduced and a subscript has been added to distinguish f_0 from the Fourier transform variable f.

The power spectral density of this random process is

$$S_X(f) = \mathcal{F}\{R_X(\tau)\} = \frac{A^2}{2}\mathcal{F}\{\cos 2\pi f_0\tau\}$$

$$= \frac{A^2}{4}\left(\mathcal{F}\{e^{-j2\pi(f-f_0)\tau}\} + \mathcal{F}\{e^{-j2\pi(f+f_0)\tau}\}\right) = \frac{A^2}{4}\delta(f-f_0) + \frac{A^2}{4}\delta(f+f_0)$$

The power spectral density is clearly even $(S_X(f) = S_X(-f))$, and nonnegative $(S_X(f) \geq 0$ for all values of f).

Let us compute the mean squared value (average power) from the time and the frequency domain representations and compare the results.

From the autocorrelation function, $E\{X^2(t)\} = R_X(0) = A^2/2$.

From (10.3), by integrating the spectral density:

$$\int_{-\infty}^{\infty} S_X(f)df = \frac{A^2}{4}\underbrace{\int_{-\infty}^{\infty}\delta(f-f_0)df}_{1} + \frac{A^2}{4}\underbrace{\int_{-\infty}^{\infty}\delta(f-f_0)df}_{1} = \frac{A^2}{2}$$

□

Here is another example which does not involve any singularities. This form often serves as a simple model for a family of random processes.

Example 10.2: The autocorrelation function for a random process has the exponential form

$$R_X(\tau) = 5e^{-\alpha|\tau|}$$

where α is a real parameter with $\alpha > 0$. The function is sketched below.

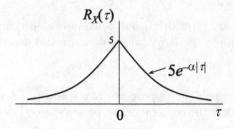

Let us determine its power spectral density. From (10.1), we have

$$
\begin{aligned}
S_X(f) &= \mathcal{F}\{R_X(\tau)\} = 5 \int_{-\infty}^{\infty} e^{-\alpha|\tau|} e^{-j2\pi f\tau} d\tau \\
&= 5 \int_{-\infty}^{0} e^{\alpha\tau} e^{-j2\pi f\tau} d\tau + 5 \int_{0}^{\infty} e^{-\alpha\tau} e^{-j2\pi f\tau} d\tau \\
&= 5 \int_{-\infty}^{0} e^{(\alpha-j2\pi f)\tau} d\tau + 5 \int_{0}^{\infty} e^{-(\alpha+j2\pi f)\tau} d\tau \\
&= \frac{5}{\alpha - j2\pi f} + \frac{5}{\alpha + j2\pi f} = \frac{10\alpha}{\alpha^2 + 4\pi^2 f^2}
\end{aligned}
$$

which is a second order rational polynomial in f. The figure following

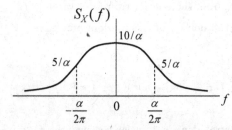

shows a sketch of the power spectral density function.

Again we can verify that $S_X(f)$ has the three properties cited above.

□

The last example illustrates some further points about this random process. The PSD shows clearly that the power in the random process is distributed over frequency and that the power density is higher at lower frequencies. This is in contrast to the previous example where the power is concentrated at only two points in frequency corresponding to the positive and negative frequencies of the sinusoid.

In the time domain the parameter α of this example controls the correlation of the process; when α is large the autocorrelation function falls off rapidly. This implies that points in the process at some fixed separation τ will have much lower correlation that when α is small. A larger value for α also means that the process "wiggles around" more in time and therefore has more high frequency components. The PSD confirms this assessment. If α is large the two half power points at $f = \pm\alpha/2\pi$ move out to increase the bandwidth.

10.1.3 Cross-spectral density

The cross-spectral density function of two jointly wide-sense stationary random processes $X(t)$ and $Y(t)$ is defined as

$$S_{XY}(f) = \mathcal{F}\{R_{XY}(\tau)\} = \int_{-\infty}^{\infty} R_{XY}(\tau)e^{-j2\pi f \tau}d\tau \tag{10.6}$$

The cross-spectral density $S_{YX}(f)$ is defined similarly with X and Y interchanged and satisfies

$$S_{YX}(f) = S_{XY}^*(f) \tag{10.7}$$

The two cross-power spectral densities are, in general, complex-valued functions even if $R_{XY}(\tau)$ is real (see Example 10.3 below). The magnitude of the cross-spectral density shows how the "cross correlation" between two random processes is distributed over frequency. If the magnitude of the cross spectrum is high in some frequency regions and low in others there may be some common phenomenon that is responsible for the regions of high spectral magnitude.

Some facts pertaining to the cross-spectral density of jointly wss random processes are as follows.

1. If two random processes are *orthogonal*, then the cross-correlation function vanishes and

$$S_{XY}(f) = \mathcal{F}\{R_{XY}(\tau)\} = 0$$

2. If the random processes having mean values of m_X and m_Y are *uncorrelated*, then the cross-covariance function vanishes leaving

$$S_{XY}(f) = \mathcal{F}\{R_{XY}(\tau)\} = m_X m_Y \, \delta(f) \tag{10.8}$$

This follows since $R_{XY}(\tau) = m_X m_Y$ and the Fourier transform of a constant is a delta function.

Example 10.3: In Example 9.7 of Chapter 9 we computed the cross-correlation function for a certain pair of real random processes to be

$$R_{XY}(\tau) = -\sigma^2 \sin \omega_0 \tau$$

where $\omega_0 = 2\pi f_0$. From (10.6), the cross-power spectral density is

$$
\begin{aligned}
S_{XY}(f) &= \mathcal{F}\{R_{XY}(\tau)\} = -\sigma^2 \mathcal{F}\{\sin \omega_0 \tau\} \\
&= j\frac{\sigma^2}{2}\mathcal{F}\{e^{-2\pi(f-f_0)\tau}\} - j\frac{\sigma^2}{2}\mathcal{F}\{e^{-2\pi(f+f_0)\tau}\} \\
&= j\frac{\sigma^2}{2}\delta(f-f_0) - j\frac{\sigma^2}{2}\delta(f+f_0)
\end{aligned}
$$

Note that $S_{XY}(f)$ in this case is *purely imaginary*.

We can compute the other cross-power spectral density function as well. Since the random processes are real we have $R_{YX}(\tau) = R_{XY}(-\tau) = \sigma^2 \sin \omega_0 \tau$. The cross-power spectral density $S_{YX}(f)$ is then computed as

$$
\begin{aligned}
S_{YX}(f) &= \mathcal{F}\{R_{YX}(\tau)\} = \sigma^2 \mathcal{F}\{\sin \omega_0 \tau\} \\
&= -j\frac{\sigma^2}{2}\delta(f-f_0) + j\frac{\sigma^2}{2}\delta(f+f_0)
\end{aligned}
$$

Observe that S_{XY} and S_{YX} satisfy the conjugate relation (10.7) above. That is, $S_{XY}(f)$ is the complex conjugate of $S_{YX}(f)$.

□

10.2 White Noise

10.2.1 Spectral representation of Gaussian white noise

The concept of a continuous time white noise process is discussed in several places in this text from a time domain perspective. Recall from Chapter 8 that the Wiener process is used as a model for the random motion of electrons in an electronic system. It is argued that the Wiener process is a zero-mean Gaussian random process whose time derivative is used as a mathematical model for noise. The origin of Gaussian white noise is further discussed from a physical point of view in Appendix D.

In Chapter 9 the autocorrelation function of a white noise process based on the Wiener model is derived and found to have the form

$$R_W(\tau) = \frac{N_o}{2}\delta(\tau) \tag{10.9}$$

where the factor of 2 is included with the parameter N_o for later convenience. The power spectral density is obtained by taking the Fourier transform and is therefore a constant.

$$S_W(f) = \frac{N_o}{2} \qquad \text{for all values of } f \tag{10.10}$$

The autocorrelation function and corresponding PSD are depicted in Fig. 10.1.

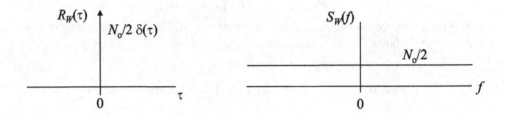

Figure 10.1 The autocorrelation function and the power spectral density of white noise.

Conceptually, white noise has a flat power spectral density that exists for the entire frequency spectrum. This is analogous to physical white light, which contains all the natural colors in more-or-less equal proportions. Random processes that do not have a flat spectrum are typically referred to as "colored." The power spectral density in Example 10.2 is that of a colored random process.

It has already been noted that the units of the power density spectrum are watts/hertz; therefore the white noise parameter N_o has these same units. Observe that since the integral of the PSD in Fig. 10.1 is infinite, white noise has *infinite power*. This is in correspondence with the fact that power from a time domain perspective is $R_W(0) = \infty$. (The impulse is infinite at the origin.) Also, since two time samples of white noise, no matter how close, are uncorrelated, one has to conclude that white noise has power at all possible frequencies.

One of the widely encountered application examples is the case of a sinusoidal signal in additive white Gaussian noise, e.g., communications signals and radar waveforms observed at a receiver. Typically, sinusoid is the desired signal while the additive noise is undesired. The following two examples deal with the cases of sinusoidal signals in noise. The objective here is to extract the desired signal from a signal "corrupted" by noise.

Example 10.4: Consider the random process consisting of a sinusoid with random phase in additive white noise,

$$X(t) = A\sin(\omega_0 t + \Theta) + W(t)$$

where A and $\omega_0 = 2\pi f_0$ are constants while Θ is a uniform random variable in the interval $-\pi$ to π, $W(t)$ is assumed to be white noise with power spectral density $N_0/2$, and Θ and $W(t)$ are uncorrelated, i.e.,

$$\mathcal{E}\{A\sin(\omega_0 t + \Theta)W(t)\} = \mathcal{E}\{A\sin(\omega_0 t + \Theta)\}\,\mathcal{E}\{W(t)\} = 0$$

(Notice that both terms have zero mean.)

The autocorrelation function of $X(t)$ is computed as

$$
\begin{aligned}
R_X(\tau) &= \mathcal{E}\{X(t)X(t-\tau)\} \\
&= \mathcal{E}\{[A\sin(\omega_0 t + \Theta) + W(t)]\,[A\sin(\omega_0(t-\tau) + \Theta) + W(t-\tau)]\} \\
&= A^2\mathcal{E}\{\sin(\omega_0 t + \Theta)\sin(\omega_0(t-\tau) + \Theta)\} + \mathcal{E}\{W(t)W(t-\tau)\}
\end{aligned}
$$

where the cross terms are 0 because Θ and $W(t)$ are uncorrelated. Following Example 10.1, the first term simplifies to

$$A^2\mathcal{E}\{\sin(\omega_0 t + \Theta) \cdot \sin(\omega_0(t-\tau) + \Theta)\} = \frac{A^2}{2}\cos\omega\tau$$

while the second term is recognized as the autocorrelation function of white noise.

$$\mathcal{E}\{W(t)W(t-\tau)\} = \frac{N_0}{2}\delta(\tau)$$

Substituting these into the expression for $R_X(\tau)$ yields

$$R_X(\tau) = \frac{A^2}{2}\cos\omega\tau + \frac{N_0}{2}\delta(\tau)$$

The power spectral density of $X(t)$ is then obtained as

$$
\begin{aligned}
S_X(f) &= \mathcal{F}\{R_X(\tau)\} = \frac{A^2}{2}\mathcal{F}\{\cos\omega\tau\} + \frac{N_0}{2}\mathcal{F}\{\delta(\tau)\} \\
&= \frac{A^2}{4}\delta(f - f_0) + \frac{A^2}{4}\delta(f + f_0) + \frac{N_0}{2}
\end{aligned}
$$

The figure below illustrates the power spectral density function of $X(t)$. The flat line with a height of $N_0/2$ represents what is called the *noise floor*. As seen, the two impulses representing the sinusoid exceed the noise, but only in an infinitely narrow band around $\pm f_0$. Unless some filtering is performed to reduce the noise, a receiver will be unable to detect the signal because the signal power is finite while the total noise power is infinite.[2]

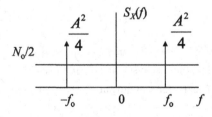

□

---

[2] In practice, a narrowband filter centered around $\pm f_0$ would be used to increase the signal-to-noise ratio to some desired level.

Another approach to detecting the signal is illustrated in the following example.

Example 10.5: Frequently, in certain applications such as digital communication, the receiver has knowledge of the transmitted signal waveforms. What is not known is which of the transmitted waveforms from a predefined set is transmitted during a given symbol period. This scenario can be summarized in the following figure. The received signal is corrupted by additive noise. The reference signal indicates the receiver's knowledge of transmitted waveforms. In order to extract the desired sinusoid from the received signal, the cross-correlation function between the received signal and the reference signal is computed.

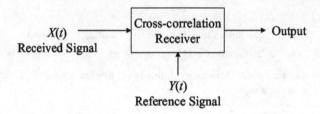

To be more specific, assume that the transmitter sends one of two signals representing a logical 1 and a logical 0:[3]

$$\text{Logical 1} \quad \Rightarrow \quad A\sin(\omega_0 t + \Theta)$$
$$\text{Logical 0} \quad \Rightarrow \quad 0 \quad \text{(nothing is transmitted)}$$

Now, let us consider two cases. First consider the case of no sinusoid present (logical 0 sent). At the receiver we then have the following random processes:

$$X(t) \quad = \quad W(t)$$
$$Y(t) \quad = \quad A\sin(\omega_0 t + \Theta)$$

The corresponding cross-correlation function is

$$R_{XY}(\tau) = A\mathcal{E}\left\{W(t)\sin(\omega_0(t - \tau) + \Theta)\right\} = 0$$

because Θ and $W(t)$ are uncorrelated.

In the second case a logical 1 is transmitted. In this case the signals at the receiver are given by

$$X(t) \quad = \quad A\sin(\omega_0 t + \Theta) + W(t)$$
$$Y(t) \quad = \quad A\sin(\omega_0 t + \Theta)$$

The cross-correlation function is then

$$R_{XY}(\tau) \quad = \quad A^2\mathcal{E}\left\{\sin(\omega_0 t + \Theta)\sin(\omega_0(t - \tau) + \Theta)\right\} + A\mathcal{E}\left\{\sin(\omega_0 t + \Theta)W(t - \tau)\right\}$$
$$= \quad \frac{A^2}{2}\cos\omega\tau$$

(The second term drops out because the signal and the noise are uncorrelated.)

The power spectral density for the output in this case is given by

$$S_X(f) \quad = \quad \mathcal{F}\left\{R_X(\tau)\right\} = \frac{A^2}{2}\mathcal{F}\left\{\cos\omega\tau\right\}$$
$$= \quad \frac{A^2}{4}\delta(f - f_0) + \frac{A^2}{4}\delta(f + f_0)$$

---

[3] A model of this type can also be used to represent target detection by a radar (see Chapter 7).

Notice that the noise has been totally eliminated, i.e., the resulting SNR is infinite!

The figure below illustrates the power spectral density function of $X(t)$. The signal extraction using the cross-correlation approach is clearly superior to that using the autocorrelation approach (Example 10.4). The reason for better performance is due to the availability of the reference signal.

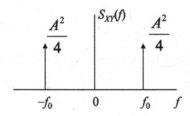

□

10.2.2 Bandlimited white noise

Although white noise is a theoretically important concept, it is not realizable in practice. Figure 10.1 and equation (10.3) show that white noise has infinite power.

$$\int_{-\infty}^{\infty} S_X(f)df = \infty$$

This alone would preclude any physical realization. In practice white noise needs only to be defined over a finite band of frequencies

$$S_W(f) = \begin{cases} \dfrac{N_0}{2} & \text{for } -B \le f \le B \\ 0 & \text{otherwise} \end{cases} \tag{10.11}$$

This modified version of white noise is called *bandlimited* white noise and B is typically referred to as the *absolute* bandwidth. Figure 10.2 shows the power spectral density of bandlimited white noise. Since almost all physical systems of practical interest have finite bandwidth, and our interest is usually limited to the system bandwidth, bandlimited white noise can be considered to be white noise for analysis purposes.

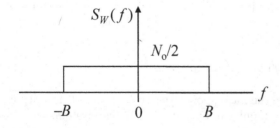

Figure 10.2 The power spectral density of bandlimited white noise.

The autocorrelation function corresponding to bandlimited white noise is given by

$$R_W(\tau) = \mathcal{F}^{-1}\{S_W(f)\} = \frac{N_0}{2}\int_{-B}^{B} e^{j2\pi f\tau}df$$

$$= \frac{N_0}{2\pi\tau}\frac{e^{j2\pi B\tau} - e^{-j2\pi B\tau}}{2j} = N_0 B\frac{\sin 2\pi B\tau}{2\pi B\tau} \tag{10.12}$$

Figure 10.3 shows a plot of the autocorrelation function of bandlimited white noise.

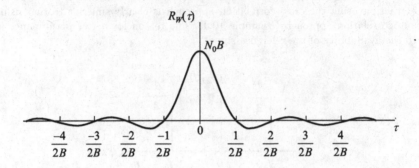

Figure 10.3 The autocorrelation function of bandlimited white noise.

The autocorrelation function is maximum at $\tau = 0$ where $R_W(0) = N_0 B$, and is zero for $\tau = k/2B$ for all integers k. Since the mean of $W(t)$ is zero and

$$R_W(\tau) = \mathcal{E}\{W(t)W(t-\tau)\} = 0 \quad \text{for } \tau = \tfrac{k}{2B},$$

the samples of bandlimited white noise are uncorrelated only for specific values of τ. Compare this to the case of *theoretical* white noise with infinite bandwidth whose samples are uncorrelated for all values of τ (see Fig. 10.1). As $B \to \infty$ the $(\sin x)/x$ shape of the autocorrelation function in Fig. 10.3 approaches an impulse. In other words, bandlimited white noise approaches theoretical white noise in the limit as the absolute bandwidth B approaches infinity.

10.3 Transformation by Linear Systems

Transformations of signals by linear time-invariant systems are described in the frequency domain by the frequency response of the system. It is not surprising therefore that the mean and spectral densities involving the output of these linear systems can be expressed in terms of the frequency response. Fortunately the relations are not difficult to derive given the relations among first and second moment quantities in the time domain.

The frequency response of a linear time-invariant system measures the response of the system to a complex exponential input ($e^{j2\pi f}$) and is the Fourier transform of the system impulse response. In particular, for a system with impulse response $h(t)$ the frequency response $H(f)$ is defined by

$$H(f) = \int_{-\infty}^{\infty} h(t)e^{-2\pi ft}dt \tag{10.13}$$

With this definition, we can derive the mean, power spectral density of the output and the cross-spectral density between input and output of the linear system.

10.3.1 Output mean

In the preceding chapter it is shown that the output mean of a LTI system with input $X(t)$ and output $Y(t)$ can be expressed as

$$m_Y = m_X \int_{-\infty}^{\infty} h(t)dt \quad \text{(see Section 9.5.2)}$$

Since the integral here is the frequency response (10.13) evaluated at $f = 0$, the relation becomes

$$m_Y = m_X H(0) \qquad (10.14)$$

This relation is not surprising since the mean is the "d.c." component of a random process and $H(0)$ represents the gain of the system at "d.c."

10.3.2 Spectral density functions

Let us now turn to the spectral density functions associated with the linear transformation of a random process. Figure 10.4 shows an LTI system with impulse response

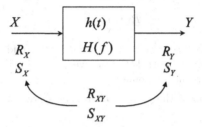

Figure 10.4 Second moment quantities associated with a linear system.

$h(t)$ and frequency response $H(f)$. The input X and output Y are related in the time domain by convolution. The figure also shows the various correlation functions and power spectra associated with the transformation.

The left-hand column of Table 10.1 lists some relations among the correlation functions that are derived in Chapter 9. The right-hand column of the table lists the

time	*frequency*
$R_{YX}(\tau) = h(\tau) \circledast R_X(\tau)$	$S_{YX}(f) = H(f)S_X(f)$
$R_{XY}(\tau) = h^*(-\tau) \circledast R_X(\tau)$	$S_{XY}(f) = H^*(f)S_X(f)$
$R_{YX}(\tau) = h(\tau) \circledast h^*(-\tau) \circledast R_X(\tau)$	$S_Y(f) = H(f)H^*(f)S_X(f)$

Table 10.1 Time and frequency domain relations for a linear transformation.

corresponding relations in the frequency domain derived from well known properties of the Fourier transform [1]. For example, in the entry for R_{XY}, conjugation and reversal of the impulse response corresponds to conjugation of the frequency response while convolution in the time domain results in multiplication in the frequency domain.

Probably the most frequently used relation in this table is the last entry, usually written as

$$S_Y(f) = |H(f)|^2 S_X(f) \qquad (10.15)$$

which results from noticing that $H(f)H^*(f) = |H(f)|^2$.

The following example illustrates computation of a system output PSD using (10.15).

Example 10.6: Example 9.16 of Chapter 9 considers a linear system with impulse response

$$h(t) = e^{-2t}u(t)$$

and an input with autocorrelation function

$$R_X(\tau) = 2 + 3\delta(\tau)$$

corresponding to white noise with non-zero mean. The output autocorrelation function in that example is worked out using convolution in the time domain and the output autocorrelation function is shown to be

$$R_Y(\tau) = \frac{3}{4}e^{-2|\tau|} + \frac{1}{2}$$

Given this result, the output power spectral density could be found directly by taking the Fourier transform of R_Y. It is instructive, however, (and probably easier) to work the problem in the frequency domain.

The transfer function of the system with impulse response $h(t) = e^{-2t}u(t)$ has the well-known form[4]

$$H(f) = \frac{1}{2 + j2\pi f}$$

and the input power spectral density is

$$S_X(f) = \mathcal{F}\{R_X(\tau)\} = \mathcal{F}\{3\delta(\tau) + 2\} = 3 + 2\delta(f)$$

Thus (10.15) becomes

$$
\begin{aligned}
S_Y(f) &= |H(f)|^2 S_X(f) = H(f)H^*(f)S_X(f) \\
&= \left(\frac{1}{2 + j2\pi f}\right)\left(\frac{1}{2 - j2\pi f}\right)(3 + 2\delta(f)) = \frac{3}{4}\left(\frac{1}{1 + \pi^2 f^2}\right) + \frac{1}{2}\delta(f)
\end{aligned}
$$

□

Equation 10.15 is also useful to provide a graphical interpretation of the transformation. The following example illustrates the technique.

Example 10.7: The magnitude response of a bandpass LTI system is given by

$$H(f) = \begin{cases} 1 & \text{for } 200 \text{ Hz} \leq f \leq 300 \text{ Hz} \\ 0 & \text{otherwise} \end{cases}$$

The system response, the power spectral density of a random process applied at the input of the system, and the resulting output PSD are shown opposite. The output PSD is computed by applying (10.15) graphically.

The system response has a magnitude of 1, so squaring it does not change the shape. By multiplying the squared system magnitude by the input power spectral density $S_X(f)$, we obtain the output power spectral density $S_Y(f)$, as shown in the figure.

---

[4] This result is easily derived as an exercise from (10.13).

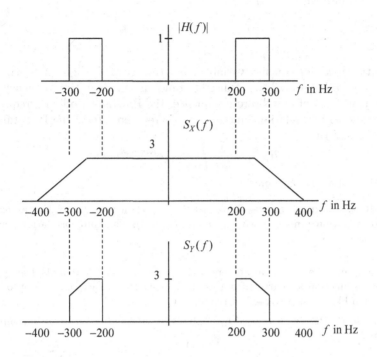

☐

The last example suggests a proof to the important result

$$S_X(f) \geq 0 \qquad \text{for all } f$$

which was stated as (10.5) without proof. The proof now goes like this.

Suppose there were a region of the frequency axis where $S_X < 0$. Then we could build a narrowband filter to cover just that region and apply (10.15):

$$S_Y(f) = |H(f)|^2 S_X(f)$$

The output power density S_Y would be everywhere either zero or negative. If we then were to integrate as in (10.3) we should find

$$\text{Avg power} = \mathcal{E}\left\{|Y(t)|^2\right\} = \int_{-\infty}^{\infty} S_Y(f)df \; < \; 0$$

But the average power of a random process cannot be negative! It's not that we just don't believe in negative power; but mathematically there is no way that the expectation of a squared quantity $\mathcal{E}\left\{|Y(t)|^2\right\}$ can be negative. We conclude that (10.5) must hold.

10.4 Discrete Random Signals

10.4.1 Discrete power spectral density

The power spectral density function for a discrete random signal (sequence) is defined as the discrete-time Fourier transform of the autocorrelation function. The discrete-time Fourier transform is given by[5]

$$S_X(e^{j\omega}) = \sum_{l=-\infty}^{\infty} R_X[l]e^{-j\omega l} \tag{10.16}$$

where ω is the discrete frequency variable (sometimes called "digital frequency"). This function is the basis for analysis of discrete random signals in the frequency domain.

Like the transform of any discrete sequence, the PSD is periodic in frequency with period 2π. The autocorrelation function can be recovered from the PSD function via the inverse transform

$$R_X[l] = \frac{1}{2\pi} \int_{-\pi}^{\pi} S_X(e^{j\omega})e^{j\omega l} d\omega \tag{10.17}$$

Consider the following example.

Example 10.8: A simple but useful model for a random process involves an exponential autocorrelation function (see e.g., Example 9.13). This autocorrelation function has the form

$$R_X[l] = Ca^{|l|}$$

where a and C are real parameters with $0 < a < 1$ and $C > 0$. This discrete autocorrelation function also arises from sampling a continuous exponential autocorrelation function like the one treated in Example 10.2.

The discrete PDF is computed by applying (10.16) and breaking up the sum into two terms.

$$S_X(e^{j\omega}) = C\sum_{l=-\infty}^{\infty} a^{|l|}e^{-j\omega l} = C\sum_{l=-\infty}^{-1} a^{-l}e^{-j\omega l} + C\sum_{l=0}^{\infty} a^l e^{-j\omega l}$$

The term on the far right can be put in closed form by using the formula for a geometric random process:

$$C\sum_{l=0}^{\infty} a^l e^{-j\omega l} = C\frac{1}{1 - ae^{-j\omega}}$$

while the term before it can be written with a little more manipulation as

$$C\sum_{l=-\infty}^{-1} a^{-l}e^{-j\omega l} = C\frac{ae^{j\omega}}{1 - ae^{j\omega}}$$

The PSD is then the sum of these terms

$$S_X(e^{j\omega}) = C\frac{1}{1 - ae^{-j\omega}} + C\frac{ae^{j\omega}}{1 - ae^{j\omega}} = \frac{C(1 - a^2)}{(1 - ae^{-j\omega})(1 - ae^{j\omega})}$$

The equation can be simplified by multiplying terms and using Euler's formula.

$$S_X(e^{j\omega}) = \frac{C(1 - a^2)}{1 + a^2 - 2a\cos\omega}$$

The function is plotted below for for $C = 1$ and $a = 0.8$. The spectrum has a low-pass character as long as the parameter a is positive.

□

Observe that, barring any impulses that may arise from periodic components [2], the PSD of a discrete random process is a *continuous* function of ω. During analysis the spectrum may be sampled for purposes of using an FFT, but the function itself is continuous.

---

[5] The notation $S_X(e^{j\omega})$ is common in the literature and indicates clearly that we are dealing with the PSD of a discrete random process; some authors may shorten this to $S_X(\omega)$, however.

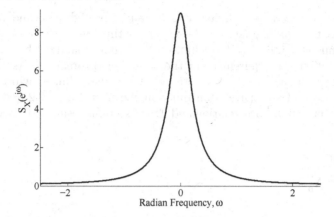

The average power for the random process is computed by setting $l = 0$ in (10.17):

$$R_X[0] = \mathcal{E}\left\{|X[k]|^2\right\} = \frac{1}{2\pi} \int_{-\pi}^{\pi} S_X(e^{j\omega})d\omega \tag{10.18}$$

which indicates that the average power of the discrete random process is spread over one period of the frequency domain, and justifies its name.

The properties for the discrete-time PSD are the same as those for the continuous PSD cited in Section 10.1.2 with the exception that integrals are over the interval $-\pi$ to π instead of $-\infty$ to ∞. When relating the discrete PSD to the spectrum of a sampled continuous random process we must substitute

$$\omega = 2\pi f T$$

where T is the interval between samples in the time domain, and scale the amplitude by the factor T.

10.4.2 White noise in digital systems

The treatment of discrete random processes would be incomplete without a brief discussion of white noise. First, any random sequence for which $R_X[l] = \delta[l]$ where $\delta[l]$ is the discrete time impulse [6] or unit sample function, is referred to as a "white" process or simply "white noise."

In particular, a sequence of IID random variables is white, regardless of its distribution. Let us refer to such a random process as *digital white noise* to distinguish it from the continuous form of white noise. Substituting this delta function in (10.16) shows that the PSD of digital white noise is a constant.

$$R_X[l] = \delta[l] \quad \Longleftrightarrow \quad S_X(e^{j\omega}) = 1 \tag{10.19}$$

Digital white noise could arise from a variety of sources. In earlier chapters it was shown how Gaussian white noise and other Gaussian processes can arise in the real world. If a Gaussian white noise process is sampled, the result is also white and Gaussian. On the other hand, if quantization noise is the primary effect then the added noise is white but uniform (see Appendix D).

Let us comment a bit more on the case of Gaussian noise and how it relates to the bandlimited white noise discussed in Section 10.2.2. While most deterministic

---

[6] Square brackets are used to distinguish the discrete-time impulse from the continuous time impulse $\delta(t)$.

and random signals can be sampled at some suitably high rate and reconstructed from samples, it is impossible to do this for continuous white noise. This follows because continuous white noise by definition has infinite bandwidth, i.e., it does not satisfy the bandlimited requirement of the sampling theorem [2]. As a result, it is necessary to restrict analysis for discrete systems to bandlimited white noise which has the spectrum and autocorrelation function shown in Figs. 10.2 and 10.3. Figure 10.5 shows the discrete autocorrelation and PSD functions resulting from sampling the

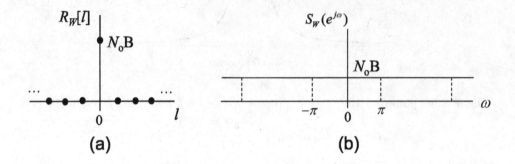

Figure 10.5 Digital white noise. (a) autocorrelation function. (b) Power spectral density.

bandlimited white noise process at time intervals of $T = 1/2B$. Notice that the samples occur exactly at zero crossings in Fig. 10.3 so that the discrete-time autocorrelation function in Fig. 10.5(a) has only a single non-zero value at $l = 0$. The PSD shown in Fig. 10.5(b) is a scaled replica of the bandlimited white noise spectrum of Fig. 10.2; however, like any spectrum resulting from a sampled time function, it is periodically repeated with period 2π.

The average power for the discrete case is given by $R_X[0] = N_oB$ which is the product of the bandwidth and the power density function in Fig. 10.2. The choice to write the noise density parameter as $N_o/2$ is now apparent because the bandwidth of the noise is equal to $2B$ and the product eliminates the factor of 2.

Notice that if the noise bandwidth actually exceeds the bandwidth of the signal that may be present, then sampling too closely (say at $1/4B$) is detrimental to the processing because more noise is folded in. This explains why engineers use a lowpass antialiasing filter before sampling a signal if the bandwidth of the noise exceeds that of the signal.

10.4.3 Transformation by discrete linear systems

Recall that the response Y of a discrete linear system to a random input X is given by the discrete convolution equation

$$Y[k] = h[k] \circledast X[k] = \sum_{i=-\infty}^{\infty} h[i]X[k-i] \qquad (10.20)$$

where $h[k]$ is the system impulse response. Now, refer to Fig. 9.6 of Chapter 9 where the system with input X and output Y is depicted and V is a third random process. It is shown there that the correlation functions R_{XV} and R_{YV} satisfy the convolution relation

$$R_{YV}[l] = h[l] \circledast R_{XV}[l]$$

((9.36) repeated here for convenience), which is analogous to (10.20). Taking the Fourier transform of this equation produces the frequency domain relation

$$S_{YV}(e^{j\omega}) = H(e^{j\omega})S_{XV}(e^{j\omega}) \tag{10.21}$$

where

$$H(e^{j\omega}) = \sum_{k=-\infty}^{\infty} h[k]e^{-j\omega k} \tag{10.22}$$

is the discrete-time *frequency response* of the linear system.

Equation 10.21 forms the basis for a number of relations in the frequency domain. The most important results are listed in Table 10.2 below.

frequency domain
$S_{YX}(e^{j\omega}) = H(e^{j\omega})S_X(e^{j\omega})$
$S_Y(e^{j\omega}) = H(f)S_{XY}(e^{j\omega})$
$S_{XY}(e^{j\omega}) = H^*(e^{j\omega})S_X(e^{j\omega})$
$S_Y(e^{j\omega}) = H(e^{j\omega})H^*(e^{j\omega})S_X(e^{j\omega})$

Table 10.2 Frequency domain transformations by a discrete linear system.

The first two rows of the table result from taking $V = X$ and $V = Y$ respectively as special cases in (10.21). The third row results from applying a Fourier transform identity to the first row. The final, most well-known result comes from combining the second and third rows and can be written as

$$S_Y(e^{j\omega}) = |H(e^{j\omega})|^2 S_X(e^{j\omega}) \tag{10.23}$$

Let us conclude this section with a simple example.

Example 10.9: A first order linear digital filter has an impulse response $h[k] = a^k u[k]$ with parameter $0 < a < 1$. The frequency response is given by

$$H(e^{j\omega}) = \frac{1}{1 - ae^{-j\omega}}$$

Let us find the output PSD and the input/output cross spectrum if the system is driven by a white noise source with parameter $N_o = 1$.

The output PSD is given by

$$S_Y(e^{j\omega}) = |H(e^{j\omega})|^2 S_X(e^{j\omega}) = \left(\frac{1}{1 - ae^{-j\omega}}\right) \cdot \left(\frac{1}{1 - ae^{+j\omega}}\right) \cdot 1 = \frac{1}{1 + a^2 - 2a\cos\omega}$$

The spectrum is real and positive and is plotted below for $a = 0.8$.

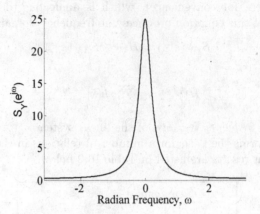

The cross-spectrum is computed from

$$S_{YX}(e^{j\omega}) = H(e^{j\omega})S_X(e^{j\omega}) = \frac{1}{1 - ae^{-j\omega}}$$

This expression is complex and equal to the system frequency response.

The cross spectrum can be written in magnitude and angle form as

$$|S_{YX}(e^{j\omega})| = \frac{1}{\sqrt{1 + a^2 - 2a\cos\omega}} \qquad \angle S_{YX}(e^{j\omega}) = -\tan^{-1}\left(\frac{a\sin\omega}{1 - a\cos\omega}\right)$$

These terms are plotted below (also for $a = 0.8$).

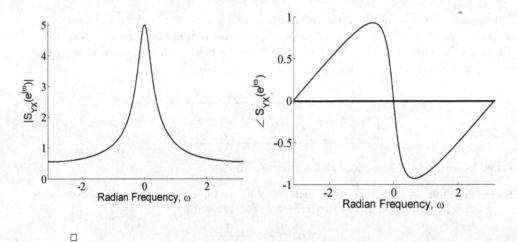

□

Before leaving this example, take a moment to compare the spectrum S_X of the example with the spectrum of the autocorrelation function in Example 10.8. Notice that, except for a scale factor, they are the same! This indicates that the exponential autocorrelation function and the system with exponential impulse response are closely related. In particular, a random process with exponential autocorrelation function can be generated by driving a system having an exponential impulse response with white noise.

10.5 Applications

10.5.1 Digital communication

In baseband transmission of binary data, the binary digits logical 1 and logical 0 are represented using electrical pulses. A variety of pulse shapes are used for this purpose; to name a few: polar non-return to 0, bipolar return to 0, and Manchester. In general these waveforms are referred to as semi-random binary waveforms. Since in a given sequence the binary digits occur randomly, the corresponding pulse waveform is also random, hence the name. The following example develops the autocorrelation and the power spectral density of a semi-random binary waveform for polar signaling.

Example 10.10:   In polar signaling, a logical 1 is represented by a positive voltage of magnitude A and a logical 0 is represented by a negative voltage of magnitude $-A$. The duration of each binary digit is T_b. The figure below shows the waveform for a binary sequence 0 1 1 0 1 0 0 1 0.

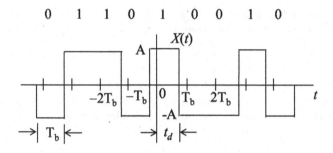

Assuming no pulse synchronization, the starting time of the first complete pulse with respect to a reference time point (say, $t = 0$) is uniformly distributed in the range $[0, \ T_b]$ and is represented as

$$f_{T_d}(t_d) = \begin{cases} \dfrac{1}{T_b}, & 0 \le t_d \le T_b \\ 0, & \text{otherwise} \end{cases}$$

Let us assume that the occurrence of 1s and 0s is equally likely, i.e., the pulse magnitude is $\pm A$ during any bit period with equal probability. Also, the occurrences are independent from bit period to bit period. Consequently, the mean of the process is

$$m_X(t) = \mathcal{E}\{X(t)\} = 0$$

while the autocorrelation function is

$$R_X(t, \tau) = \mathcal{E}\{X(t)X(t - \tau)\}$$

Let us evaluate the autocorrelation function for two cases of the time difference τ, namely $|\tau| > T_b$ and $|\tau| < T_b$.

For $|\tau| > T_b$, the random variables $X(t)$ and $X(t - \tau)$ occur in different bit periods and are independent. As a result

$$R_X(t, \tau) = \mathcal{E}\{X(t)X(t - \tau)\} = \mathcal{E}\{X(t)\}\mathcal{E}\{X(t - \tau)\} = 0$$

For $\tau < T_b$ and $t_d > T_b - |\tau|$, the random variables occur in different bit periods and as above, the autocorrelation function is 0.

For $0 \leq |\tau| < T_b$ and $0 \leq t_d < T_b - |\tau|$, the random variables occur in the same bit period. We thus have a conditional expectation which can be written as

$$R_X(t, t - \tau | t_d) = \mathcal{E}\{X(t)X(t - \tau) \mid t_d < T_b - |\tau|\}$$

$$= \begin{cases} A^2, & 0 < |\tau| < T_b \text{ and } 0 < t_d < T_b - |\tau| \\ 0, & \text{otherwise} \end{cases}$$

The (unconditional) autocorrelation function is then found by taking the expectation of the above term with respect to t_d:

$$R_X(t, t - \tau) = \mathcal{E}\{R_X(t, t - \tau | t_d)\} = \int_{-\infty}^{\infty} \mathcal{E}\{R_X(t, t - \tau | t_d)\} f_{T_d}(t_d) \, dt_d$$

$$= \frac{1}{T_b} \int_0^{T_b - |\tau|} A^2 \, dt_d \text{ for } 0 \leq |\tau| < T_b$$

$$= \frac{A^2}{T_b} (T_b - |\tau|), \text{ for } 0 \leq |\tau| < T_b$$

Since the autocorrelation function depends only on τ, this random process is stationary. We can simplify the expression to

$$R_X(\tau) = \begin{cases} A^2 \left(1 - \dfrac{|\tau|}{T_b}\right) & \text{for } 0 \leq |\tau| < T_b \\ 0 & \text{otherwise} \end{cases}$$

A plot of the autocorrelation function is shown below. Note that the autocorrelation function of this rectangular waveform is triangular in shape.

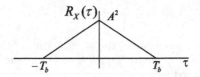

The power spectral density of $X(t)$ is then obtained by taking the Fourier transform of the autocorrelation function and is given by (see Prob. 10.4.)

$$S_X(f) = \mathcal{F}\{R_X(\tau)\} = A^2 T_b \left(\frac{\sin \pi f T_b}{\pi f T_b}\right)^2$$

The $(\sin z / z)^2$ shaped power spectral density is shown below. In baseband commu-

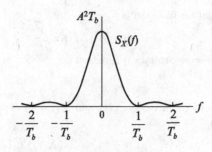

nication applications, the first null width of the power spectral density is used as a figure of merit in evaluating different pulses for transmission. From the figure, the *first null bandwidth* of polar signaling is $1/T_b$.

If the bit period of the polar signal is 1 msec, its first null bandwidth is 1 kHz. In other words, to transmit 1000 bits/sec using polar signaling requires a first null bandwidth of 1 kHz.

□

10.5.2 Optimal filtering in the frequency domain

In Section 9.6.2 we considered the problem of optimal filtering with an FIR linear system described by a finite set of parameters a_i. The situation is depicted in Fig. 9.8 of Chapter 9. We now return to the optimal filtering problem but permit the filter to have an impulse response that extends to infinity in both directions but which we implement in the frequency domain. Such a filter is called IIR, for infinite impulse response, and is noncausal because it acts upon both past and future values of the input sequence.

The estimate for the desired signal can written in the time domain as the convolution

$$\hat{D}[k] = \sum_{i=-\infty}^{\infty} h[i]x[k-i]$$

where $h[i]$ is the impulse response sequence. The error in the estimate is then

$$\mathcal{E}[k] = D[k] - \hat{D}[k] = D[k] - \sum_{i=-\infty}^{\infty} h[i]X[k-i] \qquad (10.24)$$

To minimize the mean squared error, we require

$$\frac{\partial}{\partial h[l]}\left(E\left\{\mathcal{E}^2[k]\right\}\right) = E\left\{2\mathcal{E}[k]\frac{\partial \mathcal{E}[k]}{\partial h[l]}\right\} = -2E\left\{\mathcal{E}[k]X[k-l]\right\} = 0 \qquad (10.25)$$

for each of the terms $h[l]$. The condition on the right is the orthogonality principle. Substituting (10.24) into (10.25) and taking the expectation then yields

$$E\left\{\left(D[k] - \sum_{i=-\infty}^{\infty} h[i]X[k-i]\right)X[k-l]\right\} = R_{DX}[l] - \sum_{i-\infty}^{\infty} h[i]R_X[l-i] = 0$$

Finally, recognizing the convolution and taking the Fourier transform produces the frequency domain condition

$$S_{DX}(e^{j\omega}) - H(e^{j\omega})S_X(e^{j\omega}) = 0$$

and the corresponding solution for the optimal filter

$$\boxed{H(e^{j\omega}) = \frac{S_{DX}(e^{j\omega})}{S_X(e^{j\omega})}} \qquad (10.26)$$

Let us complete this section with an example similar to Example 9.19. The conditions for the signal and noise here are the same as in that example but we use a noncausal filter and apply it in the frequency domain.

Example 10.11: It is desired to estimate the value of a real random signal in white noise using a noncausal filter. The observed sequence is given by

$$X[k] = S[k] + N[k]$$

where

$$R_S[l] = 2(0.8)^{|l|} \quad \text{and} \quad R_N[k] = 2\delta[k]$$

and N is uncorrelated with S.

Since the signal and noise are uncorrelated, it follows that

$$R_{DX}[l] = R_S[l] \quad \text{and} \quad R_X[l] = R_S[l] + R_N[l]$$

(see Example 9.19). Therefore in the frequency domain we have

$$S_{DX}(e^{j\omega}) = S_S(e^{j\omega}) \quad \text{and} \quad S_X(e^{j\omega}) = S_S(e^{j\omega}) + S_N(e^{j\omega})$$

Since the signal has an exponential autocorrelation function, the PSD has the form derived in Example 10.8, while the PSD of the noise is a constant ($S_N = 2$).

The terms needed in (10.26) are then computed as

$$S_{DX}(e^{j\omega}) = S_S(e^{j\omega}) = \frac{2(1 - (0.8)^2)}{(1 - 0.8e^{-j\omega})(1 - 0.8e^{j\omega})} = \frac{0.72}{(1 - 0.8e^{-j\omega})(1 - 0.8e^{j\omega})}$$

and

$$S_X(e^{j\omega}) = \frac{0.72}{(1 - 0.8e^{-j\omega})(1 - 0.8e^{j\omega})} + 2 = \frac{3.2(1 - 0.5e^{-j\omega})(1 - 0.5e^{j\omega})}{(1 - 0.8e^{-j\omega})(1 - 0.8e^{j\omega})}$$

Then from (10.26)

$$
\begin{aligned}
H(e^{j\omega}) &= \frac{0.72}{(1 - 0.8e^{-j\omega})(1 - 0.8e^{j\omega})} \cdot \frac{(1 - 0.8e^{-j\omega})(1 - 0.8e^{j\omega})}{3.2(1 - 0.5e^{-j\omega})(1 - 0.5e^{j\omega})} \\
&= \frac{0.225}{(1 - 0.5e^{-j\omega})(1 - 0.5e^{j\omega})}
\end{aligned}
$$

Simplifying this shows that the optimal filter has zero phase and magnitude given by
$\dfrac{0.225}{1.25 - \cos\omega}$.

□

It is interesting to compare the performance of this filter to that of the FIR filter with three coefficients in Example 9.19 of Chapter 9. Unfortunately the computation of the mean-squared error involves mathematical methods that are beyond the level of this book. The methods are described in [2] where it is shown that the mean-squared error for the optimal noncausal filter is 0.60. In Example 9.19 the mean-squared error for the FIR filter was found to be 0.7647.

10.6 Summary

This chapter focuses on the description and analysis of random signals in the frequency domain. Most of the topics and techniques covered in time domain analysis in Chapter 9 are now reformulated in the frequency domain.

The main tools for analysis are the power spectral density function and the cross power spectral density which are Fourier transforms of the autocorrelation function and cross-correlation function respectively. These functions are introduced early along with the properties that they inherit from their time-domain counterparts.

The next section consists of a discussion of white noise and band-limited white noise as a random process whose power is spread out evenly in frequency, and from whence the process gets its name. The frequency domain description of white noise is rather intuitive and circumvents many of the mathematical difficulties that are encountered when this process is first introduced in the time domain.

The transformation of random processes by linear time-invariant systems is discussed next. Given the time-domain results in Chapter 9, the frequency domain relations follow easily. Where operations on correlation functions are expressed as convolution with the system impulse response, similar operations on the power spectral density functions are accomplished through multiplication by the system frequency response.

The topics in the first three sections of the chapter deal with continuous random signals and continuous linear systems. The section that follows treats the same topics for discrete random signals and systems. The similarities and differences are made clear through discussion and explicit formulas dealing with the two separate cases.

The chapter ends with two applications of frequency domain methods. One of these is formulated in continuous time while the other is carried out in discrete time. A digital communications problem serves as an application of the continuous time concepts, while an optimal filter problem demonstrates application of the discrete time ideas.

References

[1] Alan V. Oppenheim and Alan S. Willsky. *Signals and Systems*. Prentice Hall, Inc., Upper Saddle River, New Jersey, second edition, 1997.

[2] Charles W. Therrien. *Discrete Random Signals and Statistical Signal Processing*. Prentice Hall, Inc., Upper Saddle River, New Jersey, 1992.

Problems

Power spectral density

10.1 Consider the autocorrelation function of the Gaussian random process in Prob. 9.2 of Chapter 9 and determine the power spectral density $S_X(f)$.

10.2 Determine the autocorrelation function of the random process in Prob. 9.4 of Chapter 9 and compute the corresponding power spectral density.

10.3 Consider the the autocorrelation function of the random process $X(t)$ given in Prob. 9.8 of Chapter 9 and determine determine the power spectral density $S_X(f)$.

10.4 Show that the Fourier transform of the triangular-shaped autocorrelation function in Example 10.10 is given by

$$S_X(f) = A^2 T_b \left(\frac{\sin \pi f T_b}{\pi f T_b} \right)^2$$

Hint: Consider the Fourier transform of a square pulse $p(t)$ with width T_b. The convolution $p(t) \circledast p(-t)$ is a triangular waveform similar to $R_X(\tau)$ in the example. Use the properties of the Fourier transform as it relates to convolution to devise $S_X(f)$.

10.5 The autocorrelation function for a certain binary random signal is given by

$$R_X(\tau) = e^{-2|\tau|}$$

(a) Show that the power spectral density function is given by

$$S_X(f) = \frac{1}{1 + \pi^2 f^2}$$

(b) What is the average power of the signal?

(c) The signal is passed through a filter with frequency response

$$H(f) = \begin{cases} 2 & |f| \leq 1 \\ 0 & \text{otherwise} \end{cases}$$

Sketch carefully the output power spectral density function $S_Y(f)$.

(d) Determine the average power of the filter output. Your answer may be left as an integral with the correct limits.

10.6 The cross-correlation function of a random process is given by

$$R_{XY}(\tau) = 5e^{-\tau}u(\tau) + 2$$

Find the cross-power spectral density $S_{XY}(f)$.

Response of linear systems

10.7 Refer to Prob. 9.24 of Chapter 9 and answer these questions: (Assume the input has 0 mean.)

(a) What is the frequency response $H(f)$ of the system?

(b) Take the Fourier transform of $R_Y(\tau)$ as computed in Prob. 9.24 of Chapter 9 to find $S_Y(f)$.

(c) Use the formula $S_Y(f) = |H(f)|^2 S_X(f)$ to check the result of part (b).

(d) Find the average output power $\mathcal{E}\{Y^2(t)\}$ from the autocorrelation function $R_Y(\tau)$ that you computed for the output.

(e) Compute the average output power by integrating the power spectral density function $S_Y(f)$ over frequency from $-\infty$ to ∞. You may use a table of definite integrals in this part.

10.8 The input to an ideal bandpass filter is white Gaussian noise with spectral level $N_o/2 = 3$. The bandpass filter response is shown below:

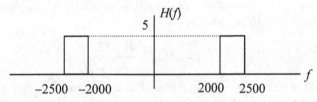

(a) Obtain and plot the input signal power spectral density function.

(b) What is the output power spectral density? (Show it graphically.)

(c) Compute the output power.

10.9 Consider the filter shown in Prob. 10.8. The input to the filter is given by

$$X(t) = 2\cos(4400\pi t + \Phi) + W(t)$$

where $W(t)$ is white Gaussian noise with spectral level $N_o/2 = 4$, Φ is uniform in the range $[-\pi, \pi]$. Assume $W(t)$ and Φ are independent.

(a) Determine and plot the input power spectral density function.

(b) Plot the output power spectral density function.

(c) Determine the output SNR in dB.

10.10 Consider an LTI system having an impulse response function

$$h(t) = 3e^{-2t}u(t).$$

This system is driven by a random process having a spectral density function

$$S_X(f) = 16 \text{ Watts/Hz}.$$

(a) Find the mean of the system output.

(b) Find the variance of the system output.

(c) The input to the system, $X(t)$, is white Gaussian. Now, write the probability density function for the output, Y.

10.11 A linear system is described by the impulse response $h(t) = e^{-at}u(t)$. The input is a white noise process with autocorrelation function

$$R_X(\tau) = \frac{N_o}{2}\delta(\tau).$$

(a) Determine the correlation function of the output random process $R_Y(\tau)$.

(b) What is the power spectral density function of the input?

(c) What is the power spectral density function of the output?

(d) Find the total average power of the output.

10.12 Consider the system shown in Prob. 9.25 of Chapter 9. The input $W(t)$ has 0 mean and $R_W(\tau) = \delta(\tau)$.

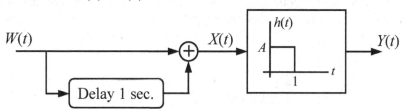

(a) What is the frequency response $H(f)$ corresponding to the filter with impulse response $h(t)$?

(b) What is the power spectral density function $S_X(f)$?

(c) What is the output power spectral density function $S_Y(f)$? Express it in the simplest form possible.

10.13 A certain signal is modeled by a continuous Gaussian random process $X(t)$ with mean zero and autocorrelation function $R(\tau) = e^{-500|\tau|}$.

(a) Find the power spectral density function of the random process $X(t)$.

(b) The signal $X(t)$ is applied to a lowpass filter with the following frequency response:

$$H(f) = \begin{cases} Ae^{-j0.2f} & -500 \le f \le 500 \text{ (Hz)} \\ 0 & \text{otherwise} \end{cases}$$

where A is a constant. Find and sketch the power spectral density function of the output signal.

10.14 Show that the output power spectral density in Example 10.6 given by

$$S_Y(f) = \frac{3}{4}\left(\frac{1}{1+\pi^2 f^2}\right) + \frac{1}{2}\delta(f)$$

can be obtained by directly computing the Fourier transform of the autocorrelation function given by

$$R_Y(\tau) = \frac{3}{4}e^{-2|\tau|} + \frac{1}{2}$$

Discrete random signals

10.15 The autocorrelation function of a discrete random process is given by

$$R_X[l] = 0.3\delta[l+2] + 2\delta[l+1] + 5\delta[l] + 2\delta[l-1] + 0.3\delta[l-2]$$

(a) Determine and sketch the power spectral density $S_X(e^{j\omega})$.

(b) Is this is a valid power spectral density function?

10.16 Refer to Prob. 9.20 of Chapter 9 and answer the following questions: (Assume
the input has 0 mean.)

(a) What is the frequency response of the system $H(e^{j\omega})$?

(b) Compute the power spectral density function $S_Y(e^{j\omega})$ of the output as the
discrete-time Fourier transform of the autocorrelation function $R_Y[l]$.

(c) Confirm the result in part (b) by using the formula $S_Y(e^{j\dot\omega}) = |H(e^{j\omega})|^2 S_X(e^{j\omega})$.

(d) Show that $S_Y(e^{j\omega})$ is real and positive for all values of ω.

10.17 State which of the following represent valid power spectral density functions
for a discrete random process; if not valid tell why.

(a) $S_X(e^{j\omega}) = e^{-\frac{\omega^2}{2}}$

(b) $S_X(e^{j\omega}) = e^{\sin \omega}$

(c) $S_X(e^{j\omega}) = \sqrt{\cos \omega}$

(d) $S_X(e^{j\omega}) = 0.2 + 0.8(1 + \cos \omega)$

(e) $S_X(e^{j\omega}) = \frac{\sin \omega}{\omega^3}$

(f) $S_X(e^{j\omega}) = \delta_c(e^{j\omega} - 1)$

10.18 State which of the following represent valid autocorrelation functions; if not
valid tell why.

(a) $R_X[l] = (\sin l)^2$

(b) $R_X[l] = 1 - u[l-4] - u[-4-l]$

(c) $R_X[l] = \begin{cases} \cos l & -\pi < l < \pi \\ 0 & \text{otherwise} \end{cases}$

(d) $R_X[l] = \frac{1}{|l|!}$

10.19 Find the power spectral density functions corresponding to the following correlation functions and verify that they are real and non-negative.

(a)

$$R_X[l] = \begin{cases} 3 - |l| & -3 \le l \le 3 \\ 0 & \text{otherwise} \end{cases}$$

(b)

$$R_X[l] = 2(-0.6)^{|l|} + \delta[l]$$

10.20 A certain random process is defined by

$$X[k] = A\cos\omega_o k + W[k]$$

where A is a Gaussian random variable with mean zero and variance σ_A^2 and $W[k]$ is a white noise process with variance σ_w^2 independent of A.

(a) What is the autocorrelation function of $X[k]$?

(b) Can the power spectrum of $X[k]$ be defined? If so, what is the power spectral density function?

(c) Repeat parts (a) and (b) in the case when the cosine has a random phase uniformly distributed between $-\pi$ and π.

10.21 A linear time-invariant FIR filter has the impulse response

$$h[k] = \delta[k] - \frac{1}{2}\delta[k-1] + \frac{1}{4}\delta[k-2]$$

The filter is driven by a white noise process $X[k]$ with variance σ_o. Call the output process $Y[k]$.

(a) Determine the autocorrelation function of the output $R_Y[l]$.

(b) Determine the output spectral density function $S_Y(e^{j\omega})$.

10.22 Given a causal linear time-invariant system described by the difference equation

$$Y[k] = 0.2Y[k-1] + X[k]$$

what is:

(a) the correlation function of the output $R_Y[l]$ when the input is white noise with unit variance?

(b) the power spectral density function of the output $S_Y(e^{j\omega})$?

(c) the power spectral density of the output $S_Y(e^{j\omega})$ when the input is a zero-mean random process with autocorrelation function

$$R_X[l] = 2(0.5)^{|l|} \; ?$$

Computer Projects

Project 10.1

In this project, you will compute the autocorrelation and cross-correlation functions for two given sequences as well as their power spectral density and cross-power spectral density functions. You will also compare the experimental results with theoretical results.

1. Generate the following discrete-time sequences:

$$\begin{aligned} y[k] &= 5\cos(0.12\pi k + 0.79) \\ x[k] &= y[k] + 7w[k] \end{aligned}$$

for $0 \le k \le 199$. Here $w[k]$ is a zero-mean white Gaussian noise sequence with variance one. What is the SNR for $x[k]$?

2. Compute the autocorrelation function $R_X[l]$ for $-100 \leq l \leq 99$. Use the *unbiased* form of the estimate.

3. Now compute an estimate for the power density spectrum as the discrete Fourier transform (DFT) of the autocorrelation function. Plot $y[k]$, $x[k]$, $R_X[l]$, and power spectral density function (it should be real). Plot all of these quantities beneath one another on the same page. In your plots, clearly label the axes with appropriate values.

4. Compute the cross-correlation function $R_{XY}[l]$ for $-100 \leq l \leq 99$ and an estimate for the cross-power spectral density function as the DFT of $R_{XY}[l]$. Plot $y[k]$, $x[k]$, $R_{XY}[l]$, and the *magnitude* of the cross-power spectral density function on the same page.

5. Compare all of the computed quantities to theoretical values.

MATLAB programming notes

The MATLAB function "xcorr" can be used to estimate both the autocorrelation function and the cross-correlation function. Use the "unbiased" option. The MATLAB "fft" function can be used for the DFT.

Project 10.2

Computer Project 9.3 of Chapter 9 explores the identification of an unknown system. The procedure is carried out in the time domain by using a white noise input. In this project the same problem is carried out in the frequency domain.

1. Refer to Project 9.3 for the transfer function of the "unknown" system. The frequency response of the system is obtained by substituting $z = e^{-j\omega}$. Compute and plot the magnitude frequency response $|H(e^{j\omega})|$. Use at least 256 frequency samples to plot the frequency response.

2. Generate 10,000 samples of the input $X[k]$ as a unit-variance white Gaussian sequence and plot the autocorrelation function $R_X[l]$ for $-128 \leq l \leq 127$ to verify that the input sequence has the autocorrelation function you expect for white noise.

3. With $X[k]$ as the input, generate the output $Y[k]$ of the unknown system. Compute the cross-correlation function $R_{YX}[l]$ for $0 \leq l \leq 127$. Also compute and plot the magnitude of the cross-spectral density as the Fourier transform of the cross-correlation function and compare it to the magnitude frequency response found in Step 1.

4. Repeat Steps 2 through 5 for input sequences of lengths 1,000 and 100,000.

MATLAB programming notes

You may use the MATLAB function "freqz" to compute the frequency response in Step 1. The MATLAB function "xcorr" can be used to estimate both the autocorrelation function and the cross-correlation function. Use the "unbiased" option. The MATLAB "fft" function can be used for the Fourier transform.

11

Markov, Poisson, and Queueing Processes

This chapter deals with two special types of random processes, namely, the Markov and Poisson processes and their application to Queueing systems. Aspects of both of these processes have been seen in earlier chapters. This final chapter, however, brings knowledge of these processes together in one place and develops it further. In particular, the theory and techniques needed for the analysis of systems characterized by random requests for service and random service times are developed here. Such systems include multi-user and multi-process computer operating systems, and the traffic on computer networks, switches and routers.

The chapter begins by developing the Poisson process from some basic principles. It is seen that many types of random variables encountered earlier are related to this process. The chapter then moves on to develop the discrete form of Markov chain first introduced in Chapter 8. This process is characterized by "states" which determine its future probabilistic behavior, and is a useful model for many physical processes that evolve in time. The Markov and Poisson models are then combined in the *continuous-time Markov chain*. This random process is also characterized by states, but state transitions may occur randomly at *any* time (i.e., not at just discrete epochs of time) according to the Poisson model. This combined model is what is necessary to represent the computer and network problems cited above.

With these tools and concepts firmly established, the chapter moves on with an introduction to queueing theory. This branch of probability and statistics deals with analysis of these combined systems. Among other things, queueing models provide ways to describe message traffic in a system, to estimate service times and predict delays, and to estimate needed resources (or predict catastrophe) under various operating conditions. The study of modern computer networks requires at least a basic understanding of these types of systems.

11.1 The Poisson Model

In earlier chapters of this book we have encountered several types of distributions for random variables dealing with the arrival of "events." Let us just review these ideas.

1. The *Poisson* PMF (Chapter 3) is introduced as a way to characterize the number K of discrete events (telephone calls, print jobs, "hits" on a web page, etc.) arriving within some continuous time interval of length t. The arrival rate of the events was assumed to be λ, and the parameter in the Poisson PMF is then given by $\alpha = \lambda t$.

2. The *exponential* PDF (also Chapter 3) is introduced as characterizing the *waiting time* T to the next event or equivalently, the interarrival time *between* two successive events. Thus T is a continuous random variable.

3. In Chapter 5 the Erlang PDF is derived as the total waiting time $T = T_1 + T_2 + \ldots + T_K$ to the arrival of the Kth event. It was assumed here that the first order interarrival times T_i are exponentially distributed.

All of these distributions are related and are different aspects of of a single probabilistic model for the arrival of events. This model is called the *Poisson model*.

11.1.1 Derivation of the Poisson model

A Poisson model is defined by considering very small increments of time of length Δ_t. Events in these time intervals are considered to be independent. Further, as $\Delta_t \to 0$, the probability of one event occurring in the time interval is proportional to the length of the time interval, i.e., the probability is $\lambda \Delta_t$. The probability that more than one event occurs in the small time interval is essentially zero. These postulates are summarized in Table 11.1.

---

1. Events in non-overlapping time intervals are independent.

2. For $\Delta_t \to 0$:

   (a) $\Pr[1 \text{ event in } \Delta_t] = \lambda \Delta_t$

   (b) $\Pr[> 1 \text{ event in } \Delta_t] \approx 0$

   (c) $\Pr[0 \text{ events in } \Delta_t] = 1 - \lambda \Delta_t$

Table 11.1 Postulates for a Poisson model.

To proceed with developing this model, let us define

$$P[K; T] \overset{\text{def}}{=} \Pr[K \text{ arrivals in time } T] \tag{11.1}$$

The mixed parentheses are used to show that this quantity is a function of both a discrete and a continuous random variable. When considered as a function of K, $P[K; T]$ has a discrete character; when plotted versus T, it is a continuous function. Now consider the time interval depicted in Fig. 11.1. Using the principle of total

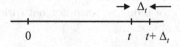

$$0 \qquad\qquad t \quad t + \Delta_t$$

Figure 11.1 Time interval for arrival of events.

probability and the fact that events in non-overlapping intervals are independent, we can write

$$\Pr[k \text{ events in } t + \Delta_t] = \Pr[k-1 \text{ in } t] \cdot \Pr[1 \text{ in } \Delta_t] + \Pr[k \text{ in } t] \cdot \Pr[0 \text{ in } \Delta_t]$$

With the notation of (11.1) and the probabilities from Table 11.1 this equation becomes

$$P[k; t + \Delta_t] = P[k-1; t] \cdot \lambda \Delta_t + P[k; t] \cdot (1 - \lambda \Delta_t)$$

Rearranging, dividing through by Δ_t, and taking the limit as $\Delta_t \to 0$ yields

$$\lim_{\Delta_t \to 0} \frac{P[k; t + \Delta_t] - P[k; t]}{\Delta_t} + \lambda P[k; t] = \lambda P[k-1; t]$$

or

$$\frac{dP[k; t]}{dt} + \lambda P[k; t] = \lambda P[k-1; t] \tag{11.2}$$

This is an interesting equation because it is both a differential equation in the continuous variable t and a difference equation in the discrete variable k. We can verify by direct substitution that the solution to (11.2) is given by

$$P[k;t] = \frac{(\lambda t)^k e^{-\lambda t}}{k!} \qquad t \geq 0, \; k = 0, 1, 2, \ldots \tag{11.3}$$

and that it satisfies the required initial condition[1]

$$P[k;0] = \begin{cases} 1 & k = 0 \\ 0 & k > 0 \end{cases}$$

Equation (11.3) is the essence of the Poisson model. Some of its implications are discussed below.

Distribution for number of arrivals

For any fixed time interval t, the probability of k arrivals in that time interval is given by (11.3). If K is considered to be a random variable, then its PMF is

$$\boxed{f_K[k] = \frac{(\lambda t)^k e^{-\lambda t}}{k!} \qquad k = 0, 1, 2, \ldots} \tag{11.4}$$

This is the same as the Poisson PMF given in Chapter 3 (see (3.8) of Chapter 3) with parameter $\alpha = \lambda t$. Thus the Poisson PMF first introduced in Chapter 3 is based on the Poisson model just derived.

Distribution of interarrival times

The PDF for the kth order interarrival times of the Poisson model can be derived as follows. Let T be the random variable that represents the waiting time to the kth event. The probability that T is within a small interval of time $(t, t + \Delta_t]$ is given by

$$\Pr[t < T \leq t + \Delta_t] = f_T(t)\Delta_t$$

where $f_T(t)$ is the PDF for the interarrival time. Now refer to Fig. 11.1 once again. If the interarrival time is equal to t, that means the kth event falls in the interval of length Δ_t depicted in the figure. In other words, $k-1$ events occur in the interval $(0, T]$ and *one* event occurs in the interval of length Δ_t. The probability of this situation is

$$\Pr[k-1 \text{ in } t] \cdot \Pr[1 \text{ in } \Delta_t] = P[k-1;t) \cdot \lambda\Delta_t = \frac{(\lambda t)^{(k-1)} e^{-\lambda t}}{(k-1)!}\lambda\Delta_t$$

where (11.3) has been used in the last step. Since these last two equations represent the probability of the same event, we can set them equal to each other and cancel the common term Δ_t. This provides the PDF for the waiting time as

$$f_T(t) = \frac{\lambda^k t^{k-1} e^{-\lambda t}}{(k-1)!}; \qquad t \geq 0 \tag{11.5}$$

which is the Erlang density function introduced in Chapter 5 (see (5.57)).

To complete discussion of the Poisson model, notice that for $k = 1$ (11.5) reduces

---

[1] Recall that most differential and difference equations are solved by "guessing" the solution and then verifying that the form is correct. Needed constants are then obtained by applying the initial conditions. In this case, the initial condition states that the probability of 0 arrivals in a time interval of length zero is one, and the probability of more than zero arrivals is zero.

to the exponential density function

$$f_T(t) = \lambda e^{-\lambda t}; \qquad t \geq 0 \tag{11.6}$$

Therefore the first order interarrival times of the Poisson model are in fact exponentially distributed. The mean interarrival time is given by $\mathcal{E}\{T\} = 1/\lambda$ (see Chapter 4, Example 4.7).

11.1.2 The Poisson process

The random process known as a Poisson process counts the number of events occurring from some arbitrary time point designated as $t = 0$.[2] The process is homogeneous in time, so it does not matter where the time origin is taken. If the Poisson process is denoted by $N(t)$, then a typical realization is as shown in Fig. 11.2. The process

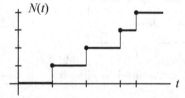

Figure 11.2 Realization of a Poisson process.

is continuous in time but discrete in its values. The random process is completely described by the fact that events in non-overlapping intervals are independent and that for any point in time the probability that $N(t) = k$ is given by the Poisson PMF (11.4).

It is easy to see that the Poisson process is not a stationary random process. First of all, the PMF (11.4) is a function of time (t); the function thus fails the test for stationary in the strict sense. The mean and the variance of this PMF are both equal to the parameter $\alpha = \lambda t$ (see Example 4.8 of Chapter 4) so the mean and variance are time-varying.

11.1.3 An Application of the Poisson process: the random telegraph signal

A classic example of a stationary process related to the Poisson process is known as the *random telegraph signal*. A typical realization of this signal is shown in Fig. 11.3. The signal changes from $+1$ to -1 and back at random times; those change times or "zero crossings" are described by a Poisson process with parameter λ. In these days where digital modulation schemes predominate, this process is less important than it used to be as a model for communications. Nevertheless it arises in many other applications; for example, the digits of a binary register that counts the arrival of random events follow this model. It is traditional however, to assume this process takes on values of ± 1 rather than 0 and 1.

---

[2] In the following, we consider "t" to be the time index of a continuous random process rather than the realization of a random variable T.

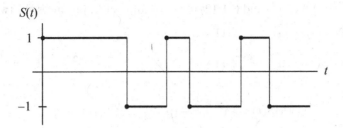

Figure 11.3 Random telegraph signal.

Probabilistic description

Assume that the process starts at time $t = 0$ and that the initial probabilities of a plus one or a minus one at that time are given by

$$\begin{array}{ll} \Pr[S(0)=+1] = p & \text{(a)} \\ \Pr[S(0)=-1] = 1 - p & \text{(b)} \end{array} \qquad (11.7)$$

For any other time we can write

$$\begin{aligned} \Pr[S(t)=+1] \;=\;\; & \Pr[\text{even} \,|\, S(0)=+1] \cdot \Pr[S(0)=+1] \\ & + \Pr[\text{odd} \,|\, S(0)=-1] \cdot \Pr[S(0)=-1] \end{aligned} \qquad (11.8)$$

where the terms "even" and "odd" represent an even and odd number of arrivals (zero-crossings) in the interval $(0, t]$, and likewise,

$$\begin{aligned} \Pr[S(t)=-1] \;=\;\; & \Pr[\text{odd} \,|\, S(0)=+1] \cdot \Pr[S(0)=+1] \\ & + \Pr[\text{even} \,|\, S(0)=-1] \cdot \Pr[S(0)=-1] \end{aligned} \qquad (11.9)$$

Now, the probability of k arrivals in the interval is given by the Poisson PMF (11.4), which does *not* depend on the value of $S(0)$. Therefore

$$\begin{aligned} \Pr[\text{even} \,|\, S(0)=\pm1] \;=\;\; & \Pr[\text{even}] \\ =\;\; & \sum_{k \text{ even}} f_K[k] = e^{-\lambda t} \sum_{k \text{ even}} \frac{(\lambda t)^k}{k!} \end{aligned} \qquad (11.10)$$

To evaluate the infinite sum, consider the Taylor series for the exponentials

$$e^{\lambda t} = \sum_{k=0}^{\infty} \frac{(\lambda t)^k}{k!} \quad \text{and} \quad e^{-\lambda t} = \sum_{k=0}^{\infty} \frac{(-\lambda t)^k}{k!}$$

Observe that the sum of these two expressions produces twice the *even* terms in the series. Equation 11.10 can thus be evaluated as

$$\begin{aligned} \Pr[\text{even} \,|\, S(0)=\pm1] \;=\;\; & \Pr[\text{even}] \\ =\;\; & e^{-\lambda t} \left(\frac{e^{\lambda t} + e^{-\lambda t}}{2} \right) = \tfrac{1}{2} \left(1 + e^{-2\lambda t} \right) \end{aligned} \qquad (11.11)$$

Similarly, the difference of the two Taylor series results in twice the odd terms; therefore

$$\begin{aligned} \Pr[\text{odd} \,|\, S(0)=\pm1] \;=\;\; & \Pr[\text{odd}] \\ =\;\; & e^{-\lambda t} \left(\frac{e^{\lambda t} - e^{-\lambda t}}{2} \right) = \tfrac{1}{2} \left(1 - e^{-2\lambda t} \right) \end{aligned} \qquad (11.12)$$

Substituting (11.7), (11.11) and (11.12) in (11.8) and (11.9) produces the two equations

$$\Pr[S(t)=+1] \;=\; \tfrac{1}{2}\left(1+e^{-2\lambda t}\right)\cdot p + \tfrac{1}{2}\left(1-e^{-2\lambda t}\right)\cdot(1-p)$$

$$\Pr[S(t)=-1] \;=\; \tfrac{1}{2}\left(1-e^{-2\lambda t}\right)\cdot p + \tfrac{1}{2}\left(1+e^{-2\lambda t}\right)\cdot(1-p)$$

or

$$\Pr[S(t)=+1] = \tfrac{1}{2} + \left(p - \tfrac{1}{2}\right)e^{-2\lambda t} \quad \text{(a)}$$
$$\Pr[S(t)=-1] = \tfrac{1}{2} - \left(p - \tfrac{1}{2}\right)e^{-2\lambda t} \quad \text{(b)} \qquad\qquad (11.13)$$

Notice that if $p \neq \tfrac{1}{2}$ the probabilities for $S(t)$ are time-varying and the process is therefore not stationary. This time variation is transient however, and dies out as t gets large. Therefore both $\Pr[S(t) = +1]$ and $\Pr[S(t) = -1]$ approach steady-state values of $\tfrac{1}{2}$. In fact, if $p = \tfrac{1}{2}$ and/or $t \to \infty$ the process become stationary in the strict sense. In this case for any two distinct points t_1 and t_0, the samples are independent with $\Pr[+1] = \Pr[-1] = \tfrac{1}{2}$.

Mean and correlation function

To further describe the random telegraph signal, let us compute its mean and auto-correlation function. Since $S(t)$ can take on only two possible values, computation of the mean is straightforward using (11.13):

$$
\begin{aligned}
m_S(t) \;&=\; \mathcal{E}\{S(t)\} = (+1)\cdot \Pr[S(t)=+1] + (-1)\cdot \Pr[S(t)=-1] \\
&=\; (+1)\cdot \left[\tfrac{1}{2} + \left(p-\tfrac{1}{2}\right)e^{-2\lambda t}\right] + (-1)\cdot \left[\tfrac{1}{2} - \left(p-\tfrac{1}{2}\right)e^{-2\lambda t}\right] \\
&=\; (2p-1)e^{-2\lambda t}
\end{aligned}
$$

The mean is zero if $p = \tfrac{1}{2}$; otherwise it is time-varying but approaches zero as t gets large.

To compute the autocorrelation function consider two points t_1 and t_0. The correlation of the random process at these two points is given by $\mathcal{E}\{S(t_1)S(t_0)\}$. Now, define a random variable X as the product $X = S(t_1)S(t_0)$ and notice that X can take on only two possible values, $+1$ and -1. Therefore we can write

$$\mathcal{E}\{S(t_1)S(t_0)\} = \mathcal{E}\{X\} = (+1)\cdot \Pr[X=1] + (-1)\cdot \Pr[X=-1]$$

Now, assuming $t_1 > t_0$, notice that X is equal to 1 when there is an even number of events in the interval $(t_0, t_1]$ and X equals -1 otherwise. Thus the last equation can be restated as

$$\mathcal{E}\{S(t_1)S(t_0)\} = \mathcal{E}\{X\} = (+1)\cdot \Pr[\text{even}] + (-1)\cdot \Pr[\text{odd}]$$

The probabilities of an even number and odd number of events in the interval $(t_0, t_1]$ are given by the expressions (11.11) and (11.12) with $t = t_1 - t_0$. Substituting these equations yields

$$\mathcal{E}\{S(t_1)S(t_0)\} = (+1)\tfrac{1}{2}\left(1+e^{-2\lambda(t_1-t_0)}\right) + (-1)\tfrac{1}{2}\left(1-e^{-2\lambda(t_1-t_0)}\right) = e^{-2\lambda(t_1-t_0)}$$

If $t_1 < t_0$ a similar expression holds with t_1 and t_0 interchanged. Finally, noting that the autocorrelation function is defined as $R_S(t,\tau) = \mathcal{E}\{S(t)S(t-\tau)\}$ we can make a change of variables in the last equation to obtain

$$R_S(t,\tau) = \mathcal{E}\{S(t)S(t-\tau)\} = e^{-2\lambda|\tau|} \qquad\qquad (11.14)$$

where the absolute value accounts for the ordering of the time points.

Since the autocorrelation function depends only on the time lag τ then under conditions above where the mean is zero the random telegraph signal is wide-sense stationary[3]. A sketch of the autocorrelation function is shown in Fig. 11.4.

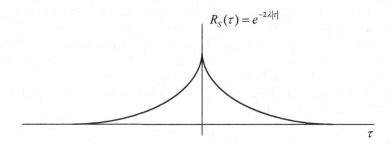

$$R_S(\tau) = e^{-2\lambda|\tau|}$$

Figure 11.4 Autocorrelation function for the random telegraph signal.

11.1.4 Additional remarks about the Poisson process

Sum of independent Poisson processes

One of the remarkable facts about the Poisson model is that the sum of two independent Poisson processes is also a Poisson process. Consider the following example.

Example 11.1: Assume that customers at the website "badticket.com" are from two distinct populations: those wanting to buy rock concert tickets and those wanting to buy classical music tickets. The two groups are described by independent Poisson processes with arrival rates λ_1 and λ_2. We can show that the entire set of customers (independent of their musical tastes) forms a Poisson process with arrival rate $\lambda = \lambda_1 + \lambda_2$.

Since the two processes are independent and each is a Poisson process, events defined on non-ovelapping intervals remain independent. This fulfills the first requirement in Table 11.1. Now consider a small time interval Δ_t. If just one event occurs in the interval, this means there was either one rock customer and no classical customer, or one classical customer and no rock customer. The probability of this event is

$$\begin{aligned}
\Pr[1\ \text{customer}] &= \lambda_1\Delta_t \cdot (1 - \lambda_2\Delta_t) + \lambda_2\Delta_t \cdot (1 - \lambda_1\Delta_t) \\
&= (\lambda_1 + \lambda_2)\Delta_t + \mathcal{O}(\Delta_t^2)
\end{aligned}$$

where $\mathcal{O}(\Delta_t^2)$ represents terms proportional to Δ_t^2, which become negligible as $\Delta_t \to 0$.

The probability of two or more customers in the interval Δ_t is zero because of the properties of Poisson processes, unless a rock *and* a classical customer should both arrive in the interval Δ_t. Since the two Poisson processes are independent, the probability of this event is $(\lambda_1\Delta_t)(\lambda_2\Delta_t)$ which is also of the order of Δ_t^2 and becomes negligible as $\Delta_t \to 0$. Therefore we are left with:

$$\begin{aligned}
\Pr[1\ \text{customer}] &= (\lambda_1 + \lambda_2)\Delta_t \\
\Pr[> 1\ \text{customer}] &\approx 0 \\
\Pr[0\ \text{customers}] &= 1 - (\lambda_1 + \lambda_2)\Delta_t
\end{aligned}$$

This defines a Poisson process with rate $\lambda_1 + \lambda_2$.

$\square$

---

[3] In fact, it has been observed under the same conditions that the process is strict sense stationary. See remarks following (11.13).

Relation to the Bernoulli process

The Poisson process can also be described as a limiting case of a Bernoulli process. Earlier, in Chapter 8, a Bernoulli process was defined as taking on values of 1 with probability p and -1 with probability $1 - p$. For purposes of this development it is more convenient to think of the process as taking on values of 1 and 0 instead of 1 and -1. Time samples in a Bernoulli process of either kind are independent.

Now consider a *continuous* time interval of length t and assume that the interval is divided into N short segments of length $\Delta_t = t/N$. Suppose that events arrive according to the postulates in Table 11.1. If we associate the arrival of an event with a 1 and no arrival with a 0, then we can define a Bernoulli process over the N time segments with probability parameter

$$p = \lambda \Delta_t = \frac{\lambda t}{N}$$

The probability of k events occurring in the N time intervals is given by the Binomial PMF (see Chapter 3 Eq. (3.5)) with p as above:

$$\Pr[k \text{ events}] = \binom{N}{k} \left(\frac{\lambda t}{N}\right)^k \left(1 - \frac{\lambda t}{N}\right)^{N-k}$$

Let us rewrite this expression and take the limit as $N \to \infty$:

$$\lim_{N \to \infty} \binom{N}{k} \cdot \left(\frac{\lambda t/N}{1 - \lambda t/N}\right)^k \cdot \left(1 - \frac{\lambda t}{N}\right)^N \tag{11.15}$$

The first term can be written as

$$\binom{N}{k} = \frac{N!}{k!(N-k)!} \to \frac{1}{k!} N^k$$

where the right-hand side denotes the behavior of the term as N gets very large. The second term becomes

$$\left(\frac{\lambda t/N}{1 - \lambda t/N}\right)^k \to (\lambda t/N)^k$$

The last term in (11.15) has a well known limit[4], namely $e^{-\lambda t}$. Therefore the limit in (11.15) becomes

$$\frac{1}{k!} N^k \cdot (\lambda t/N)^k \cdot e^{-\lambda t} = \frac{(\lambda t)^k e^{-\lambda t}}{k!}$$

which is the same as (11.3).

11.2 Discrete-Time Markov Chains

Chapter 8 introduced the concept of a Markov process and defined a Markov chain as a Markov process with a finite set of states. This section provides some more detailed analysis of Markov chains. We begin in this section with discrete-time Markov chains and move on in the next section to continuous-time Markov chains that are used through the rest of the chapter.

---

[4] The general form is $\lim_{N \to \infty} \left(1 + \frac{a}{N}\right)^N = e^a$. See e.g., [1] p. 471.

11.2.1 Definitions and dynamic equations

A Markov chain can be thought of as a random process that takes on a discrete set of values, which (for simplicity) we denote by the integers $0, 1, 2, \ldots q - 1$ (q may be infinite). A typical realization of a discrete-time Markov chain is shown in Fig. 11.5. At each time point 'k' the Markov chain takes on a value i, which is a non-negative

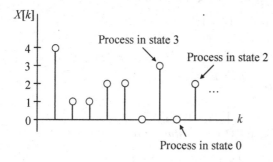

Figure 11.5 Typical realization of a Markov chain.

integer. When $X[k] = i$, the Markov chain is said to be "in state i."

The Markov chain can also be represented by a state diagram such as the one shown in Fig. 11.6. The possible states of the process are denoted by "bubbles" and possible

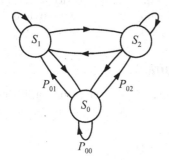

Figure 11.6 State diagram for a Markov chain.

transitions between states are indicated by branches with arrows.

The Markov chain is defined by a set of *state probabilities*

$$p_i[k] \stackrel{\text{def}}{=} \Pr[X[k] = i] = \Pr[\text{state } i \text{ at time } k] \tag{11.16}$$

and a set of state *transition probabilities*

$$P_{ij} \stackrel{\text{def}}{=} \Pr[X[k+1] = j \mid X[k] = i]$$
$$= \Pr[\text{transition from state } i \text{ to state } j] \tag{11.17}$$

The P_{ij} are assumed to be constant in time.

The time behavior of the Markov chain can be described using these two types of probabilities. For example, referring to Fig. 11.6 we can write

$$p_0[k+1] = P_{00} \cdot p_0[k] + P_{10} \cdot p_1[k] + P_{20} \cdot p_2[k]$$

This equation articulates that the probability of being in state 0 at time $k + 1$ is the sum of probabilities of being in state i at time k and transfering from state i to state 0. [Notice that this is just a statement of the *principle of total probability*!] The corresponding equations for all of the states can be written as a matrix equation

$$\begin{bmatrix} p_0[k+1] \\ p_1[k+1] \\ p_2[k+1] \end{bmatrix} = \begin{bmatrix} P_{00} & P_{10} & P_{20} \\ P_{01} & P_{11} & P_{21} \\ P_{02} & P_{12} & P_{22} \end{bmatrix} \begin{bmatrix} p_0[k] \\ p_1[k] \\ p_2[k] \end{bmatrix}$$

By following this procedure, the set of equations for any Markov chain with finite number of states q can be expressed in matrix notation as

$$\mathbf{p}[k+1] = \mathbf{P}^T \mathbf{p}[k] \tag{11.18}$$

where

$$\mathbf{p}[k] = \begin{bmatrix} p_0[k] \\ p_1[k] \\ \vdots \\ p_{q-1}[k] \end{bmatrix} \tag{11.19}$$

is the vector of state probabilities, and $\mathbf{P}$ is the *state transition matrix*

$$\mathbf{P} = \begin{bmatrix} P_{00} & P_{01} & \cdots & P_{0,q-1} \\ P_{10} & P_{11} & \cdots & P_{1,q-1} \\ \vdots & \vdots & \cdots & \vdots \\ P_{q-1,0} & P_{q-1,1} & \cdots & P_{q-1,q-1} \end{bmatrix} \tag{11.20}$$

Notice that the transition matrix appears *transposed* in (11.18)[5]. Notice also that each *row* of the transition matrix contains probabilities conditioned on one particular state (see 11.17); thus *each row of the transition matrix sums to 1*.

11.2.2 Higher-order transition probabilities

Suppose that at time k a discrete-time Markov chain has a set of probabilities $\mathbf{p}[k]$ associated with its states. Then according to (11.18), the vector of state probabilities after ℓ transitions is

$$\mathbf{p}[k+\ell] = \underbrace{\mathbf{P}^T \mathbf{P}^T \cdots \mathbf{P}^T}_{\ell \text{ times}} \mathbf{p}[k] = (\mathbf{P}^\ell)^T \mathbf{p}[k]$$

The matrix $\mathbf{P}^\ell$ is called the ℓth order transition matrix. The elements of this matrix, denoted by $P_{ij}^{(\ell)}$ are interpreted as

$$P_{ij}^{(\ell)} = \Pr\left[X[k+\ell] = j \mid X[k] = i\right] \tag{11.21}$$

In other words, $P_{ij}^{(\ell)}$ is the probability of transition from state i to state j in ℓ transitions. Notice that $P_{ij}^{(\ell)} \neq (P_{ij})^\ell$. That is, the ℓth order transition probabilities are *not* obtained by raising P_{ij} to the ℓth power. Rather, they are the elements of the matrix obtained by multiplying the transition matrix $\mathbf{P}$ by itself ℓ times.

To further understand higher order transition probabilities, let us consider the state diagram of Fig. 11.6 and compute $P_{01}^{(2)}$, the probability of starting in state 0 and arriving in state 1, after two transitions. The possible sequences of states and their probabilities are given by

---

[5] It would seem simpler if we were to define the transition matrix as the transpose of (11.20). The current definition has become standard however, due to the practice in the statistical literature to define the vector of state probabilities as a row vector and write (11.18) as the product of a row vector on the left times a matrix ($\mathbf{P}$) on the right [2, 3, 4, 5].

state sequence	probability
$0 \rightarrow 0 \rightarrow 1$	$P_{00} \cdot P_{01}$
$0 \rightarrow 1 \rightarrow 1$	$P_{01} \cdot P_{11}$
$0 \rightarrow 2 \rightarrow 1$	$P_{02} \cdot P_{21}$

The probability of starting in state 0 and arriving in state 1 after two transitions is therefore the sum

$$P_{01}^{(2)} = P_{00} \cdot P_{01} + P_{01} \cdot P_{11} + P_{02} \cdot P_{21}$$

Note that this is just the element in the first row and second column of the product matrix

$$\mathbf{P}^2 = \mathbf{P} \cdot \mathbf{P} = \begin{bmatrix} P_{00} & P_{01} & P_{02} \\ \times & \times & \times \\ \times & \times & \times \end{bmatrix} \begin{bmatrix} \times & P_{01} & \times \\ \times & P_{11} & \times \\ \times & P_{21} & \times \end{bmatrix}$$

which is obtained by multiplying the first row in the first matrix by the second column in the second matrix. (The $\times$ denotes other entries.) The entire set of second order transition probabilities is given by the terms in the matrix $\mathbf{P}^2$.

The higher-order transition matrix can be used to derive a fundamental result for Markov chains known as the *Chapman-Kolmogorov equation* [6]. Suppose the matrix $\mathbf{P}^\ell$ is factored as

$$\mathbf{P}^\ell = \mathbf{P}^r \, \mathbf{P}^{(\ell-r)} \tag{11.22}$$

where r is any integer between 0 and ℓ; then $\mathbf{P}^r$ and $\mathbf{P}^{(\ell-r)}$ are the rth order and $(\ell - r)^{th}$ order transition matrices respectively. Writing out the matrix product in terms of its elements produces

$$P_{ij}^{(\ell)} = \sum_{m=1}^{q-1} P_{im}^{(r)} P_{mj}^{(\ell-r)} \qquad 0 < r < l \tag{11.23}$$

This equation states that the probability of moving from state "i" to "j" in ℓ transitions is a sum of products of the probabilities of moving from state "i" to *each* of the other states in r transitions and the probability of moving from each of these intermediate states to state "j" in $\ell - r$ transitions. This fundamental principle is elegantly expressed in matrix form by (11.22).

11.2.3 Limiting state probabilities

For most Markov chains of interest, the state probabilities approach limiting values as the time "k" gets large. Figure 11.7 illustrates this effect for a simple two-state Markov chain. The table in the figure lists the state probabilities for $k = 0$ through 6. Starting with the initial state probabilities $p_0[0] = p_1[0] = 0.5$, after just five transitions, the system reaches a steady-state condition with probabilities $\bar{p}_0 = 0.667$ and $\bar{p}_1 = 0.333$. These are the *limiting state probabilities* and their values are *independent of the initial conditions* $p_0[0]$ and $p_1[0]$. Note that

$$\bar{\mathbf{p}} = \lim_{k \to \infty} \mathbf{p}[k] = \lim_{k \to \infty} (\mathbf{P}^k)^T \mathbf{p}[0] \tag{11.24}$$

Thus the existence of a limiting state requires that the ℓth order transition matrix $\mathbf{P}^\ell$ reaches a limit. While it would appear that $\bar{\mathbf{p}}$ depends on the initial probability vector

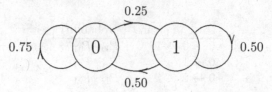

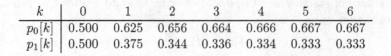

k	0	1	2	3	4	5	6
$p_0[k]$	0.500	0.625	0.656	0.664	0.666	0.667	0.667
$p_1[k]$	0.500	0.375	0.344	0.336	0.334	0.333	0.333

Figure 11.7 Evolution of state probabilities for a two-state Markov chain.

$\mathbf{p}[0]$, it in fact does not, because all of the rows of $\mathbf{P}^\ell$ become identical as $\ell \to \infty$. (It is an interesting exercise to show that this must be the case.) In our example, the sequence of transition matrices is

$$\mathbf{P} = \begin{bmatrix} 0.75 & 0.25 \\ 0.5 & 0.5 \end{bmatrix} ; \quad \mathbf{P}^2 = \begin{bmatrix} 0.688 & 0.312 \\ 0.625 & 0.375 \end{bmatrix} ; \quad \mathbf{P}^3 = \begin{bmatrix} 0.672 & 0.328 \\ 0.656 & 0.344 \end{bmatrix} ;$$

$$\mathbf{P}^4 = \begin{bmatrix} 0.668 & 0.332 \\ 0.664 & 0.336 \end{bmatrix} ; \quad \mathbf{P}^5 = \begin{bmatrix} 0.667 & 0.333 \\ 0.666 & 0.334 \end{bmatrix} ; \quad \mathbf{P}^6 = \begin{bmatrix} 0.667 & 0.333 \\ 0.667 & 0.333 \end{bmatrix} \cdots$$

where it is obvious that the matrix approaches a limit in just a few transitions and that the rows become identical with elements equal to the limiting state probabilities $\bar{p}_0$ and $\bar{p}_1$. Under this condition $\bar{\mathbf{p}}$ computed from (11.24) is the same for any initial probability vector $\mathbf{p}[0]$ that appears on the right.

To find the limiting state probabilities, it is only necessary to note that when the corresponding limits exist,

$$\mathbf{p}[k+1] \to \mathbf{p}[k] \to \bar{\mathbf{p}}$$

as $k \to \infty$. Therefore (11.18) becomes

$$\bar{\mathbf{p}} = \mathbf{P}^T \bar{\mathbf{p}} \quad \text{or} \quad (\mathbf{P}^T - \mathbf{I})\bar{\mathbf{p}} = 0 \tag{11.25}$$

Equation 11.25 determines $\bar{\mathbf{p}}$ only to within a constant; that is, if $\bar{\mathbf{p}}$ is a solution to this equation, then $c\bar{\mathbf{p}}$ is also a solution, where c is any constant. The remaining condition however, that the elements of $\bar{\mathbf{p}}$ sum to 1, insures a unique solution:

$$\sum_{i=0}^{q-1} \bar{p}_i = 1 \tag{11.26}$$

The following example illustrates the procedure of solving for the limiting state probabilities.

Example 11.2: A certain binary sequence, with values of 0 and 1, is modeled by the Markov chain shown in Fig. 11.7. The transition matrix is

$$\mathbf{P} = \begin{bmatrix} 0.75 & 0.25 \\ 0.5 & 0.5 \end{bmatrix}$$

It is desired to find the probability of a run of five 0s and a run of five 1s.

The probabilities of the first binary digit (0 or 1) when the sequence is observed at a random time are the limiting state probabilities. These are found from (11.25)

$$\begin{bmatrix} -0.25 & 0.50 \\ 0.25 & -0.50 \end{bmatrix} \begin{bmatrix} \bar{p}_0 \\ \bar{p}_1 \end{bmatrix} = \begin{bmatrix} 0 \\ 0 \end{bmatrix}$$

and (11.26)

$$\bar{p}_0 + \bar{p}_1 = 1$$

The solution of these two equations is $\bar{p}_0 = 2/3$ and $\bar{p}_1 = 1/3$. The probability of a run of five 0s is then

$$(\tfrac{2}{3})(0.75)^4 = 0.211$$

and the probability of a run of five 1s is

$$(\tfrac{1}{3})(0.5)^4 = 0.021$$

Now compare these results to the probability of runs in a Bernoulli process with the same limiting-state probabilities. The probability of five 0s for the Bernoulli process is

$$(\tfrac{2}{3})^5 = 0.132$$

while the probability of five 1s is

$$(\tfrac{1}{3})^5 = 0.004$$

As you might expect, these probabilities are significantly smaller than the corresponding probabilities for the Markov chain, since successive values of the Bernoulli process are independent, and runs are therefore less likely to occur.

□

11.3 Continuous-Time Markov Chains

A basic model for systems characterized by random traffic and random service times can be obtained by combining the concept of a Markov chain with that of the Poisson process. The result, a continuous-time Markov chain, is the basis for many standard models of traffic in computer networks and other systems.

11.3.1 Simple server system

Consider the simple server system shown in Fig. 11.8. For purposes of providing a

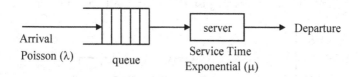

Figure 11.8 A simple queueing system.

familiar application, think of this as a model for customers buying tickets at a box office. Customers line up and are served one at a time. New customers arrive and enter at the back of the line, or queue. Such a system is commonly referred to as a *queueing system* or a *queueing model*, and the mathematics that deals with these systems is known as *queueing theory*. Clearly this model applies to a host of other situations besides the box office scenario. For example, Fig. 11.8 can be a model for print requests sent to a single printer on a computer or computer network. Here, requests for service are placed in an electronic queue similar to the line of customers and are processed one at a time on a first-in first-out basis. In any such system, both the arrivals and the server are frequently described by a Poisson model. The arrivals are Poisson with rate λ; thus the interarrival times are exponential random variables. The service times are likewise assumed to be exponential random variables with parameter μ (mean service time is $1/\mu$); thus departures also comprise a Poisson process with rate μ.

The simple queueing model of Fig. 11.8 can be described by a Markov chain with state diagram shown in Fig. 11.9; here the state represents the number of customers

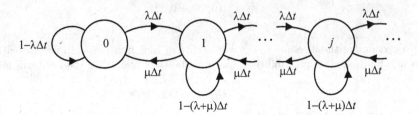

Figure 11.9 State diagram for the simple queueing system.

in the system (those in the queue plus the one being served). Unlike the discrete-time Markov chain however, state transitions may occur *at any time*. According to the Poisson model, in any small interval of time Δ_t, a new customer will arrive with probability $\lambda \Delta_t$; thus the state will change from j to $j + 1$ with probability $\lambda \Delta_t$. Likewise, the service may end within any small interval Δ_t so that the state may change from j to $j - 1$ with probability $\mu \Delta_t$. Since the probabilities of state transitions must sum to one, for all states except the zero state, the probability of remaining in the state is given by $1 - (\lambda + \mu)\Delta_t$.

A Markov chain with state diagram of the form shown in Fig. 11.9 is known as a "birth and death" process. Since state transitions occur only between left or right adjacent states, this model can be used to represent population changes due to births or deaths in a community. The birth and death model also describes the queue of most service systems.

The random process taking on a discrete set of states but where transitions can occur at any point in time is a *continuous-time Markov chain*. The number of customers $N(t)$ in the simple server system is such a process; a typical realization is depicted in Fig. 11.10. The number of customers in just the queue and the number of customers in

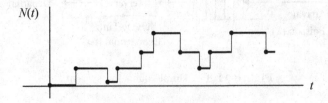

Figure 11.10 Random process representing the number of customers in the system (continuous-time Markov chain).

service (state of the server) are also continuous-time Markov chains. For the single server system discussed here, the server state may be only 0 or 1. When there are no customers in the queue the server process is as shown in Fig. 11.11.

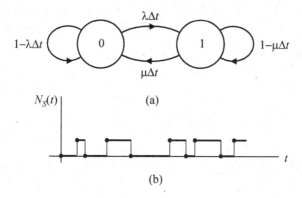

$N_s(t)$

(a)

(b)

Figure 11.11 Random process representing a server. (a) State diagram. (b) Typical Markov chain realization.

11.3.2 Analysis of continuous-time Markov chains

Having introduced the concept of a continuous-time Markov chain, let us move on to describe some of its general characteristics and the tools used for analysis. The process $N(t)$ is characterized by *transition rates* r_{ij} which cause it to change from state i to state j with probability $r_{ij}\Delta_t$. In the case of Fig. 11.9 all of these transition rates are equal to either λ or μ; however, in a more general situation the state diagram and the transition rates could be quite different. It is conventional to represent the continuous time Markov chain $N(t)$ by a modified state diagram known as a *transition diagram*. In such a diagram only the *rates* are represented (not the interval Δ_t) and self loops are eliminated. A transition diagram for the single server process is shown in Fig. 11.12. (Compare this to Fig. 11.11 (a).)

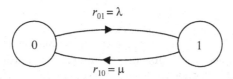

Figure 11.12 Transition diagram for the server process.

Let us now develop the equations that describe the evolution of a general Markov chain $N(t)$ with q states denoted by $0, 1, \ldots, q-1$. A typical state j has multiple transitions to and from other states, as shown in Fig. 11.13. Notice that the rates r_{ij} represent transitions *into* state j while the rates r_{ji} represent transitions *out* of state j. Let us define the time-varying state probabilities $p_i(t)$ for the random process as

$$p_i(t) = \Pr[N(t) = i] \quad i = 0, 1, \ldots, q-1 \tag{11.27}$$

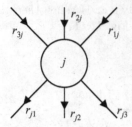

Figure 11.13 Typical state in a general Markov chain.

Then for a particular state j we can write

$$p_j(t + \Delta_t) = \sum_{\substack{i=0 \\ i \neq j}}^{q-1} r_{ij}\Delta_t \cdot p_i(t) + \left(1 - \sum_{\substack{i=0 \\ i \neq j}}^{q-1} r_{ji}\Delta_t\right)p_j(t)$$

Observe again that the terms in the first sum represent transitions into state j while the terms in the second sum represent transitions out of state j. Thus the term in perentheses represents the "self loop" probability, which is not explicitly represented in the transition diagram. Rearranging this equation and dividing through by Δ_t produces

$$\frac{p_j(t + \Delta_t) - p_j(t)}{\Delta_t} = \sum_{\substack{i=0 \\ i \neq j}}^{q-1} r_{ij}p_i(t) - \left(\sum_{\substack{i=0 \\ i \neq j}}^{q-1} r_{ji}\right)p_j(t)$$

Then taking the limit as $\Delta_t \to 0$ yields

$$\boxed{\frac{dp_j(t)}{dt} = \sum_{i=0}^{q-1} r_{ij}p_i(t) \qquad j = 0, 1, \ldots, q-1} \tag{11.28}$$

where the "rate" r_{jj} has been defined as

$$r_{jj} \overset{\text{def}}{=} -\sum_{\substack{i=0 \\ i \neq j}}^{q-1} r_{ji} \tag{11.29}$$

Equations 11.28 are the *Chapman-Kolmogorov* equations for continuous-time Markov chains. These are a coupled set of first order differential equations that can be solved for the probabilities $p_i(t)$. Rather than solve these equations, which requires advanced linear algebra techniques, it is more useful to examine the limiting steady-state condition, where the probabilites approach constant values that do not change with time. As in the case of discrete-time Markov chains, the existence of such limiting state probabilities depends on the topology of the state diagram, or equivalently, the eigenvalues in the matrix form of the Chapman-Kolmogorov equations. In the following, it is assumed that conditions are such that these limiting-state probabilities do in fact exist.

In the limiting condition we have

$$\lim_{t \to \infty} p_j(t) \to \bar{p}_j \quad \text{and} \quad \frac{dp_j(t)}{dt} \to 0$$

Therefore, (11.28) becomes

$$\sum_{i=0}^{q-1} r_{ij}\bar{p}_i = 0 \qquad j = 0, 1, \ldots, q-1$$

Now, bringing the jth term out of the sum and substituting (11.29) for r_{jj} produces

$$\sum_{\substack{i=0 \\ i\neq j}}^{q-1} r_{ij}\,\bar{p}_i - \left(\sum_{\substack{i=0 \\ i\neq j}}^{q-1} r_{ji}\right)\bar{p}_j = 0$$

or

$$\bar{p}_j \sum_{\substack{i=0 \\ i\neq j}}^{q-1} r_{ji} = \sum_{\substack{i=0 \\ i\neq j}}^{q-1} \bar{p}_i\, r_{ij} \qquad j = 0, 1, \ldots, q-1 \tag{11.30}$$

Equations 11.30 are known as the *global balance equations*. These equations show that in the limiting-state condition, the probability of state j times the sum of transition rates *out* of the state must be equal to the sum of probabilities for the other states times the transitions *into* state j. This principle must hold for all states j. The global balance principle is illustrated in Fig. 11.14 where the directions of arrows on the

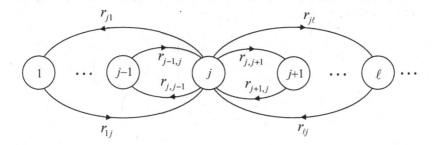

Figure 11.14 Transition rate diagram illustrating the global balance principle.

branches indicate transitions into and out of the states.

The global balance equations (11.30) are not quite enough to solve for the limiting-state probabilities. In fact, as illustrated in the example below, one equation is always redundant. (This set of equations is analogous to (11.25) in the discrete-time case.) One more condition is needed to solve for the $\bar{p}_i$, namely

$$\sum_{i=0}^{q-1} \bar{p}_i = 1 \tag{11.31}$$

Let us now illustrate the procedure with an example.

Example 11.3: Consider the 4-state birth and death process shown below.

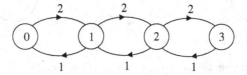

The transition rates in and out of adjacent states are identical.

The global balance equations (11.30) can be written for this process as follows:

$$\text{state 0:} \qquad \bar{p}_0 \cdot 2 \;=\; \bar{p}_1 \cdot 1$$

$$\text{state 1:} \qquad \bar{p}_1(2+1) \;=\; \bar{p}_0 \cdot 2 + \bar{p}_2 \cdot 1$$

$$\text{state 2:} \qquad \bar{p}_2(2+1) \;=\; \bar{p}_1 \cdot 2 + \bar{p}_3 \cdot 1$$

$$\text{state 3:} \qquad \bar{p}_3 \cdot 1 \;=\; \bar{p}_2 \cdot 2$$

These equations can then be written as a single matrix equation

$$\begin{bmatrix} 2 & -1 & 0 & 0 \\ -2 & 3 & -1 & 0 \\ 0 & -2 & 3 & -1 \\ 0 & 0 & -2 & 1 \end{bmatrix} \begin{bmatrix} \bar{p}_0 \\ \bar{p}_1 \\ \bar{p}_2 \\ \bar{p}_3 \end{bmatrix} = \begin{bmatrix} 0 \\ 0 \\ 0 \\ 0 \end{bmatrix}$$

The matrix on the left has determinant zero. (This follows because all of the columns add to zero.) Hence the probabilities $\bar{p}_i$ can only be determined to within a constant, and one of the global balance equations is redundant. The condition

$$\bar{p}_0 + \bar{p}_1 + \bar{p}_2 + \bar{p}_3 = 1$$

corresponding to (11.31) is needed to provide a unique solution. If we replace one of the redundant global balance equations (say the first equation) with this, the foregoing matrix equation becomes

$$\begin{bmatrix} 1 & 1 & 1 & 1 \\ -2 & 3 & -1 & 0 \\ 0 & -2 & 3 & -1 \\ 0 & 0 & -2 & 1 \end{bmatrix} \begin{bmatrix} \bar{p}_0 \\ \bar{p}_1 \\ \bar{p}_2 \\ \bar{p}_3 \end{bmatrix} = \begin{bmatrix} 1 \\ 0 \\ 0 \\ 0 \end{bmatrix}$$

This can now be solved to find

$$\bar{p}_0 = 1/15 \qquad \bar{p}_1 = 2/15 \qquad \bar{p}_2 = 4/15 \qquad \bar{p}_3 = 8/15$$

□

11.3.3 Special condition for birth and death processes

The procedure for finding the limiting-state probabilities discussed in the previous subsection is general for any circumstances where these limiting values exist. In the case of a birth and death process (such as the model for a queue) the global balance equations can be reduced to a simpler set of equations. In this special case, let the variables λ and μ represent transitions into or out of an adjacent state as shown in Fig. 11.15. Note that the rates λ and μ need not be constant but can depend on the state.

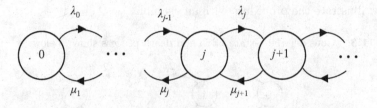

Figure 11.15 Transitions for a birth and death process.

The global balance equations for the birth and death process can then be replaced by the conditions

$$\bar{p}_j \lambda_j = \bar{p}_{j+1} \mu_{j+1} \qquad j = 0, 1, \ldots, q - 2 \qquad (11.32)$$

which say that in steady state, the probabilities of transitioning up and down the chain are equal (see Fig. 11.15). To see how conditions (11.32) arise, let us write the global balance equations (11.30) for the first three states in Fig. 11.15:

$$\begin{aligned}
\bar{p}_0 \lambda_0 &= \bar{p}_1 \mu_1 \\
\bar{p}_1 (\lambda_1 + \mu_1) &= \bar{p}_0 \lambda_0 + \bar{p}_2 \mu_2 \\
\bar{p}_2 (\lambda_2 + \mu_2) &= \bar{p}_1 \lambda_1 + \bar{p}_3 \mu_3 \\
&\vdots
\end{aligned}$$

The first equation is already in the form (11.32). However, if the first equation is transposed (left to right) and subtracted from the second equation, the result is

$$\bar{p}_1 \lambda_1 = \bar{p}_2 \mu_2$$

This is of the form (11.32) and can replace the second equation. Then this second equation can be transposed and subtracted from the third equation above to obtain

$$\bar{p}_2 \lambda_2 = \bar{p}_3 \mu_3$$

This procedure is continued until all of the equations are of the form (11.32).

By solving (11.32) for $\bar{p}_{j+1}$ it can be seen that the limiting-state probabilities for a birth and death process can be computed recursively from

$$\bar{p}_{j+1} = \frac{\lambda_j}{\mu_{j+1}} \bar{p}_j \qquad j = 0, 1, \ldots, q - 2 \qquad (11.33)$$

and the relation (11.31). The procedure is illustrated in the following example.

Example 11.4: Consider the problem of Example 11.3. From the transition rate diagram in that example (11.33) becomes

$$\begin{aligned}
\bar{p}_1 &= 2\bar{p}_0 \\
\bar{p}_2 &= 2\bar{p}_1 = 4\bar{p}_0 \\
\bar{p}_3 &= 2\bar{p}_2 = 8\bar{p}_0
\end{aligned}$$

Substituting these equations into (11.31) then yields

$$\bar{p}_0 + 2\bar{p}_0 + 4\bar{p}_0 + 8\bar{p}_0 = 1$$

or

$$\bar{p}_0 = \frac{1}{1 + 2 + 4 + 8} = \frac{1}{15}$$

The other probabilities then become

$$\bar{p}_1 = 2/15 \qquad \bar{p}_2 = 4/15 \qquad \bar{p}_3 = 8/15$$

as before.

$\square$

11.4 Basic Queueing Theory

This section provides an introduction to the systems known as queueing systems. While these systems occur in many applications, they are of particular interest to electrical and computer engineers to model traffic conditions on computer systems and computer networks.

The most general form of queueing system to be treated in this chapter is shown in Fig. 11.16. This type of system is known in the literature as an $M/M/K/C$ queueing

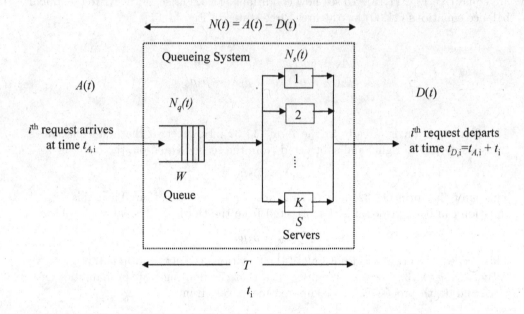

Figure 11.16 An $M/M/K$ queueing system.

system. The M's stand for "Markov," and describe the arrival and service processes; K is the number of servers, and C stands for the *capacity* of the system, i.e., the number of servers plus the maximum allowed length of the queue.

In the figure $A(t)$ and $D(t)$ represent the Poisson arrival and departure processes. We also define the following additional random variables and Poisson random processes:

W	Waiting time (in the queue)	$N_q(t)$	Number of requests in the queue
S	Service time	$N_s(t)$	Number of requests in service
T	Total time spent in the system	$N(t)$	Number of requests in the system

Table 11.2 Definition of random variables and processes for a queueing system.

These variables satisfy the relations (see Fig. 11.16)

$$T = W + S \qquad (11.34)$$

and

$$N(t) = N_q(t) + N_s(t) \qquad (11.35)$$

The following subsections provide an introduction to the topic of queueing systems. Several texts devoted to this topic provide a more extensive and advanced treatment (e.g., [2, 3, 7]).

11.4.1 The single-server system

Consider first a single-server system with no restriction on the length of the queue. It is assumed that the arrival process is a Poisson process with parameter λ and that the service process is a Poisson process with parameter μ. In other words, the service time is a exponential random variable with mean service time of $1/\mu$. Such a system is referred to as M/M/1/∞ or simply an M/M/1 queueing system.

The number of requests in the system can be described by an infinitely long birth and death process as shown in Fig. 11.17.

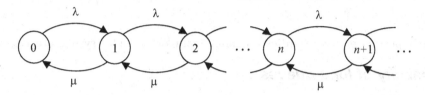

Figure 11.17 Birth and death process model for the number of requests in an M/M/1/∞ system.

Let us develop the PMF for the random variable N which represents the number of requests in the system at some time t. We can define

$$f_N[n] = \Pr[N(t) = n] = \bar{p}_n \tag{11.36}$$

where $\bar{p}_n$ represents the limiting-state probability for state n. To find the limiting-state probabilities, we follow the procedure in Section 11.3.3. Define the new parameter

$$\rho = \frac{\lambda}{\mu} \tag{11.37}$$

This parameter is called the traffic intensity or the *utilization factor*; it relates to how busy the system is. With this definition in (11.33), the limiting-state probabilities satisfy the relation

$$\bar{p}_{j+1} = \rho \, \bar{p}_j \qquad j = 0, 1, 2, \ldots$$

Then applying this equation recursively, starting with $j = 0$ leads to the general expression

$$\bar{p}_n = \rho^n \bar{p}_0 \qquad n = 0, 1, 2, \ldots \tag{11.38}$$

In order to find the probability $\bar{p}_0$ of the zero state, we observe that the probabilities must sum to one and apply the condition (11.31) to obtain

$$\sum_{n=0}^{\infty} \rho^n \bar{p}_0 = \bar{p}_0 \sum_{n=0}^{\infty} \rho^n = \bar{p}_0 \frac{1}{1-\rho} = 1 \tag{11.39}$$

where we have used the formula for an infinite geometric series, assuming that $\rho < 1$. (This last assumption is an important one because if $\rho \geq 1$, the arrival rate λ is equal to or greater than the service rate μ and the system breaks down mathematically as well as physically.) Solving the last equation yields

$$\bar{p}_0 = 1 - \rho$$

Finally, applying this to (11.38) and substituting the result in (11.36) produces the desired expression

$$f_N[n] = \bar{p}_n = (1 - \rho)\rho^n \qquad n = 0, 1, 2, \ldots \tag{11.40}$$

Notice that N is a (type 0) *geometric* random variable with parameter $p = 1 - \rho$. The distribution is sketched in Fig. 11.18.

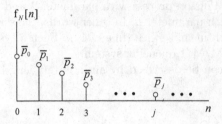

Figure 11.18 PMF for the number of requests in an M/M/1/∞ system.

Two important results regarding the number of requests in the system are as follows.

Probability of long queues

Using the PMF just derived, it is possible to calculate the probability that the number of requests in the system reaches or exceeds some chosen value n. This probability can be written as

$$\Pr[N < n] = \sum_{j=0}^{n-1}(1 - \rho)\rho^j = (1 - \rho)\sum_{j=0}^{n-1}\rho^j$$

$$= (1 - \rho)\frac{1 - \rho^n}{1 - \rho} = 1 - \rho^n$$

where we have used the formula for a finite geometric series. The probability that the number of requests is *greater* than or equal to a chosen value n is therefore

$$\Pr[[N \geq n] = \rho^n \tag{11.41}$$

Average number of requests in the system

The average number of requests in the system is given by the mean of the Geometric PMF. Thus (see Prob. 4.24 of Chapter 4)

$$\mathcal{E}\{N(t)\} = \frac{\rho}{1 - \rho} \tag{11.42}$$

A sketch of this result shown in Fig. 11.19 shows that the average length of the queue

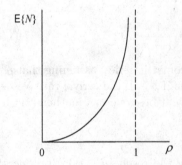

Figure 11.19 Average number of requests in an M/M/1/∞ system as a function of utilization factor ρ.

increases rapidly as the utilization factor ρ approaches one.

Consider some examples to illustrate the use of these formulas.

Example 11.5:   For a certain printer server, jobs arrive at a rate of three jobs per minute while the average printing time for a job is 15 seconds. What is the probability that there are 9 or more requests in the queue? Does this probability change if the arrival rate is doubled but the average service time is cut in half?

First, all information must be expressed in the same units. We have:

$$\lambda = 3 \text{ jobs per minute}$$
$$\mu = 60/15 = 1/0.25 = 4 \text{ jobs per minute}$$
$$\rho = \lambda/\mu = 0.75$$

The probability that there are 9 or more requests in the queue is the probability that there are 10 or more requests in the system. From (11.41)

$$\Pr[[N \geq 10] = \rho^{10} = (0.75)^{10} = 0.056$$

This probability *does not change* if the arrival rate is doubled but the average service time is cut in half.

□

Example 11.6:   What is the *average* number of requests in the system in the previous example? If the arrival rate remains at 3 requests per minute, what is the minimum average service time required to keep the average number of requests in the system less than or equal to 2?

The average number of requests in the system is given by (11.42):

$$\mathcal{E}\{N(t)\} = \frac{\rho}{1-\rho} = \frac{0.75}{1-0.75} = 3$$

To keep the average number of requests in the system less than or equal to 2 we require

$$\mathcal{E}\{N(t)\} = \frac{\rho}{1-\rho} \leq 2$$

or, solving for ρ,

$$\rho \leq 2/3$$

Then since

$$\rho = \lambda/\mu = 3/\mu$$

we find the average service time must satisfy

$$1/\mu \leq 2/9 \text{ (min)}$$

That is, the average service time must be less than or equal to $13\frac{1}{3}$ seconds.

□

11.4.2 Little's formula

Little's formula [8] establishes a relationship between the expected value of the number of requests in the system and the average time that a request spends in the system. Little's result can be stated as follows:

> For systems that reach a steady state condition, the average number of requests in the system is given by
>
> $$\mathcal{E}\{N(t)\} = \lambda\mathcal{E}\{T\} \qquad (11.43)$$
>
> where λ is the arrival rate and $\mathcal{E}\{T\}$ is the average time spent in the system.

Little's result is very powerful since it does not depend on the details of the queueing process and nevertheless applies in a large number of cases. While a totally rigorous proof is well beyond the scope of this discussion, some of the main ideas can be presented here.

A typical set of requests is depicted in Fig. 11.20(a). The time of arrival of the ith

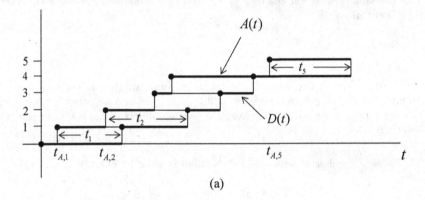

(a)

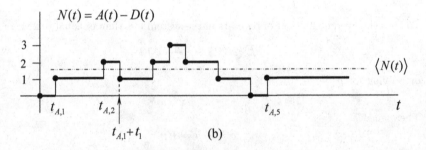

(b)

Figure 11.20 Processes in a queueing system. (a) Arrival and departure processes. (b) Requests in the system.

request $(t_{A,i})$ is indicated by the beginning of the horizontal bar and the departure of the request is represented by the end of the bar. The length of the bar represents the total time t_i that the ith request spends in the system. Thus the departure time of the ith request is given by $t_{D,i} = t_{A,i} + t_i$. The bars are stacked up in the figure so that at any time "t" the upper envelope represents the total number of arrivals $A(t)$ up to that point and the lower envelope represents the total number of departures $D(t)$. The difference between the envelopes

$$N(t) = A(t) - D(t)$$

represents the number of requests in the system at time t. This is plotted in Fig. 11.20(b). Observe that neither $A(t)$ nor $D(t)$ is a stationary random process, since neither is ever decreasing. In most situations the difference $N(t)$ will be stationary, however. In fact, this is a *requirement* for Little's formula, since the value $E\{N(t)\}$ given in (11.43) is independent of time.

To begin the proof, let us consider the time average of the process $\langle N(t) \rangle$ as shown in Fig. 11.20(b). By definition, this time average for any upper limit t_{max} is:

$$\langle N(t) \rangle = \frac{1}{t_{max}} \int_0^{t_{max}} N(t)dt = \frac{1}{t_{max}} \int_0^{t_{max}} (A(t) - D(t))dt = \frac{1}{t_{max}} \sum_{i=1}^{A(t_{max})} t_i \quad (11.44)$$

The last step is critical. It follows from observing that the integral of $A(t) - D(t)$ is the area between the upper and lower envelopes in Fig. 11.20(a), which is just equal to the sum of the times t_i that the requests are in the system. Now let us consider the *average* time spent by requests in the system. Since $A(t_{max})$ is the total number of requests up to time t_{max}, the average time is

$$\bar{t} = \frac{1}{A(t_{max})} \sum_{i=1}^{A(t_{max})} t_i$$

Finally, let us form an estimate for the arrival rate as

$$\hat{\lambda} = \frac{A(t_{max})}{t_{max}}$$

Combining these last two equations with (11.44) then yields

$$\langle N(t) \rangle = \hat{\lambda}\bar{t}$$

If $N(t)$ is ergodic (a step which we are not prepared to show) and t_{max} becomes sufficiently large, then $\langle N(t) \rangle$ and $\bar{t}$ converge to their expected values while $\hat{\lambda} \to \lambda$ and we arrive at (11.43). This is the essence of the proof.

Little's result can also be applied to some of the other system variables. In particular, it is possible to show that

$$E\{N_q(t)\} = \lambda E\{W\} \quad \text{(a)}$$

$$\quad (11.45)$$

$$E\{N_s(t)\} = \lambda E\{S\} \quad \text{(b)}$$

When (11.45) is combined with (11.34) and (11.35), the result is (11.43).

Some explicit expressions for these expectations can now be developed for the M/M/1 system. First, since the service process is Poisson with parameter μ, the expected service time is $E\{S\} = 1/\mu$. Thus, from (11.45)(b),

$$E\{N_s(t)\} = \lambda/\mu = \rho \quad (11.46)$$

To compute the mean total time in the system, Little's formula (11.43) can be combined with (11.42) to obtain

$$E\{T\} = \frac{1}{\lambda}E\{N\} = \frac{1}{\lambda}\frac{\rho}{1-\rho} = \frac{1}{\mu-\lambda} \quad (11.47)$$

where in the last step we substituted $\rho = \mu/\lambda$ and simplified.

Finally, (11.35) together with (11.42) and (11.46) produces

$$E\{N_q(t)\} = E\{N(t)\} - E\{N_s(t)\} = \frac{\rho}{1-\rho} - \rho = \frac{\rho^2}{1-\rho} \quad (11.48)$$

Then applying Little's formula (11.45)(a) and simplifying yields

$$E\{W\} = \frac{1}{\lambda}\frac{\rho^2}{1-\rho} = \frac{\lambda}{\mu(\mu-\lambda)} \tag{11.49}$$

These various results are summarized for convenience in Table 11.3. The average

$$\xrightarrow{\lambda}$$

$$E\{W\} = \frac{\lambda}{\mu(\mu-\lambda)} \qquad E\{N_q(t)\} = \frac{\rho^2}{1-\rho}$$

$$E\{S\} = 1/\mu \qquad E\{N_s(t)\} = \rho$$

$$E\{T\} = \frac{1}{\mu-\lambda} \qquad E\{N(t)\} = \frac{\rho}{1-\rho}$$

Table 11.3 Expected values and relations for quantities involved in an M/M/1/∞ queueing system. Quantities in the right column are related to quantities in the left column according to Little's formula by the factor λ. Alternative forms can be obtained by using $\rho = \lambda/\mu$.

number of requests in the system, $E\{N(t)\}$, and the mean time that a request spends in the system, $E\{T\}$, are especially important quantities for evaluating the condition or performance of the system.

Let us consider an example involving the use of Little's formula and the results in Table 11.3.

Example 11.7: Consider the printer described in Example 11.5. From Table 11.3 we see that the average time spent in the system is given by

$$E\{T\} = \frac{1}{\mu-\lambda} = \frac{1}{4-3} = 1 \text{ (minute)}$$

This result can also be computed using Little's formula, as follows. In Example 11.6 it was determined that $E\{N(t)\} = 3$. Using this result with $\lambda = 3$ in Little's formula (11.43) yields

$$3 = 3 \cdot E\{T\}$$

Therefore $E\{T\} = 1$, as before.

Let us compute some of the other quantities in Table 11.3 for this example. The expected length of the queue is given by

$$E\{N_q(t)\} = \frac{\rho^2}{1-\rho} = \frac{(0.75)^2}{1-0.75} = 2.25 \text{ (requests)}$$

The expected number of jobs in service is $E\{N_s(t)\} = \rho = 0.75$. Therefore the expected number of jobs in the system is $E\{N(t)\} = 2.25 + 0.75 = 3$, which is the same result found in Example 11.6. The average waiting time in the queue is

$$E\{W\} = \frac{\lambda}{\mu(\mu-\lambda)} = \frac{3}{4(4-3)} = 0.75 \text{ min}$$

Thus the total average time of 1 minute spent in the system is comprised of 0.75 minutes waiting in the queue and 0.25 ($1/\mu$) minutes receiving service.

$\square$

The average time spent in the system is sometimes measured in units of mean service time. Using (11.47), this metric for the M/M/1/∞ system would be given by

$$\frac{\mathcal{E}\{T\}}{1/\mu} = \frac{\mu}{\mu - \lambda} = \frac{1}{1 - \rho}$$

For the previous examples with $\rho = 0.75$, the average time that a request spends in the system is equal to $1/(1 - 0.75) = 4$ mean service times.

11.4.3 The single-server system with finite capacity

In some cases it is necessary to take into account that the service system may impose a limit on the number of requests that can be in the system; in other words, the system has *finite capacitiy*. For example, in computer and computer network systems, a finite amount of storage may be allocated to the queue. If this storage is very large so that the probability of exceeding the storage is small, then the M/M/1/∞ model is sufficent to describe the system. If the allocated storage is such that the probability of exceeding the storage is significant, then it is necessary to represent the system by a model with finite capacity, the so-called M/M/1/C queueing model whose state transition diagram is shown in Fig. 11.21. In this model, the queue cannot exceed

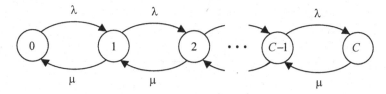

Figure 11.21 Birth and death process model for M/M/1 system with finite capacity C (M/M/1/C).

length $C - 1$; therefore the system can hold at most C requests. When the system is full, requests must be turned away.

Analysis of number of requests in the system

Developing the PMF for the number of requests in the system is identical to the procedure followed in Section 11.4.1 with the exception that the upper limit in the sum in (11.39) that determines $\bar{p}_0$ is changed from ∞ to C. The effect of this change is that

$$\bar{p}_0 = \frac{1 - \rho}{1 - \rho^{C+1}} \tag{11.50}$$

and therefore

$$f_N[n] = \frac{1 - \rho}{1 - \rho^{C+1}} \rho^n \qquad n = 0, 1, 2, \ldots, C \tag{11.51}$$

Notice that for very large traffic intensity, $\rho \to 1$, the ρ^n term has little effect and the distribution approaches a discrete uniform PMF with $f_N[n] = 1/(C+1)$ for $0 \le n \le C$. As with any queueing system, an important metric is the average number of requests in the system. An expression for this average number of requests is most easily obtained using the Probability Generating Function (see Prob. 11.32). The result is

$$\mathcal{E}\{N(t)\} = \frac{\rho}{1 - \rho} - \frac{(C + 1)\rho^{C+1}}{1 - \rho^{C+1}}; \qquad 0 < \rho < 1 \tag{11.52}$$

The first term is the result for an M/M/1/∞ system (let $C \to \infty$ and compare to Table 11.3), while the second term represents the reduction due to the finite-length queue. For $\rho = 1$ the distribution is uniform and we have $\mathcal{E}\{N(t)\} = C/2$.

A plot of $\mathcal{E}\{N\}$ versus C is given in Fig. 11.22 for some selected values of ρ. For ρ

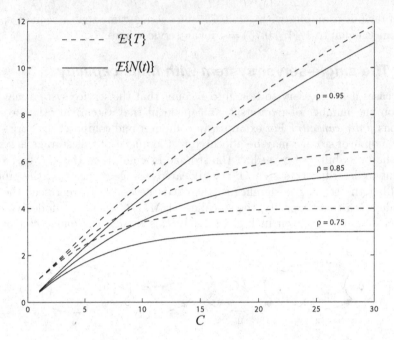

Figure 11.22 Average number of requests in a M/M/1/C system (solid lines) versus capacity C. Average time spent in the system (in units of mean service time) is also shown (dashed lines).

equal to 0.75 and 0.85 the average number of requests in the system is seen to level off at values of 3.0 and 5.67 respectively. These correspond to the values for a system with infinite capacity and can be computed from the formula in Table 11.3. For $\rho = 0.95$, which represents a very high traffic intensity, the average number of requests in the system increases almost linearly as the capacity C ranges from 1 to 30 as shown in the figure. The average number of requests eventually levels off at a value of 19 as the capacity of the system is increased, but this requires the system capacity C to be well over 100.

Mean time spent by requests in the system.

In order to compute the mean time that a request spends in the system, we have to first look at the process of requests actually entering the system. Note that requests cannot enter the system and must "back off" whenever the system is full, i.e., whenever $N(t) = C$. Let us denote the probability of this event by P_b. The probability of back-off is thus found from (11.51) to be

$$P_b = \Pr[N(t) = C] = f_N[C] = \frac{1 - \rho}{1 - \rho^{C+1}} \rho^C \qquad (11.53)$$

In any small increment of time $\Delta_t \to 0$ the probability that a request arrives and must back off is given by

$$(\lambda \Delta_t) \cdot P_b = (\lambda P_b)\Delta_t$$

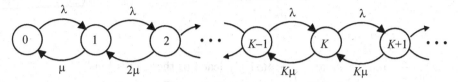

Figure 11.23 Transition diagram for an- $M/M/K/\infty$ multi-server system.

Requests backing off therefore constitute a Poisson process with rate $\lambda_b = \lambda P_b$. Similarly, the probability that a request arrives and does *not* back off is given by

$$(\lambda \Delta_t) P_a = (\lambda P_a) \Delta_t$$

where $P_a = 1 - P_b$ and thus from (11.53)

$$P_a = 1 - \frac{1 - \rho}{1 - \rho^{C+1}} \rho^C = \frac{1 - \rho^C}{1 - \rho^{C+1}} \tag{11.54}$$

Therefore requests actually entering the system are also a Poisson process, with rate $\lambda_a = \lambda P_a$. The two subprocesses constitute the original arrival process with rate $\lambda = \lambda_a + \lambda_b$.

The mean time that a request spends in the system however, depends only on the requests actually entering the system. Thus we refer to λ_a as the *effective arrival rate* and λ_b as the *back-off rate*. The mean time that a request spends in the system is thus computed using Little's formula as

$$\mathcal{E}\{T\} = \frac{\mathcal{E}\{N\}}{\lambda_a} = \frac{\mathcal{E}\{N\}}{\lambda P_a} \tag{11.55}$$

where $\mathcal{E}\{N\}$ is given by (11.52) and P_a is given by (11.54). The mean time (in units of mean service time) is shown as a function of system capacity C by the dashed lines in Fig. 11.22. For $C \to \infty$ the ratio between the average number of requests in the system and the mean time in the system is equal to the traffic intensity ρ.

11.4.4 The multi-server system

In this final section of the chapter, we return to the general case of a queueing system with K servers depicted in Fig. 11.16. To begin the analysis of this system, suppose that k of the K servers are busy. The probability that one of these k servers completes its work within the time interval $(t, t + \Delta_t]$ is $\mu \Delta_t$. Since the servers act independently, the probability that any one of the servers complete is

$$\mu \Delta_t + \mu \Delta_t + \cdots + \mu \Delta_t = (k\mu) \Delta_t \tag{11.56}$$

The transition diagram for the system is depicted in Fig. 11.23. Notice that the backward transition probabilities for the first K states are not the same but are proportional to by $k\mu$ for $k = 1, 2, \ldots, K - 1$. This follows from (11.56) above. For the remaining states, the backward transition probabilities are proportional to $K\mu$.

Following the discussion in Section 11.3.3, the limiting state probabilities for the $M/M/K$ system must satisfy the conditions

$$\bar{p}_0 \lambda = \bar{p}_1 \mu$$
$$\bar{p}_1 \lambda = \bar{p}_2 2\mu$$
$$\vdots$$

$$\bar{p}_{K-1}\lambda = \bar{p}_K K\mu$$

$$\vdots$$

Solving this system of equations recursively leads to the expressions[6]

$$\bar{p}_n = \begin{cases} \dfrac{K^n \rho^n}{n!}\bar{p}_0 & 0 \le n < K \\[3mm] \dfrac{K^K \rho^n}{K!}\bar{p}_0 & K \le n < \infty \end{cases} \qquad (11.57)$$

where it is conventional to *redefine* the parameter ρ for a K-server system as

$$\rho = \frac{\lambda}{K\mu} \qquad (11.58)$$

By requiring that $\sum_{n=0}^{\infty} \bar{p}_n = 1$, the zero-state probability can be found to be

$$\bar{p}_0 = \left[\sum_{n=0}^{K-1} \frac{1}{n!}K^n \rho^n + \frac{K^K \rho^K}{K!(1-\rho)}\right]^{-1} \qquad (11.59)$$

Thus all of the limiting-state probabilities can be computed from (11.57) and (11.59).

The probability that an arriving request finds all servers busy and therefore has to wait in the queue is an important measure of performance. This will be denoted by Pr[queue] and is given by

$$\text{Pr[queue]} = \text{Pr}[N \ge K] = \sum_{n=K}^{\infty} \bar{p}_n \qquad (11.60)$$

Now observe from (11.57) that for $n \ge K$,

$$\bar{p}_n = \rho^{n-K}\bar{p}_K \qquad n \ge K \qquad (11.61)$$

where

$$\bar{p}_K = \frac{K^K \rho^n}{K!}\bar{p}_0 \qquad (11.62)$$

Thus the (11.60) becomes

$$\text{Pr[queue]} = \bar{p}_K \sum_{n=K}^{\infty} \rho^{n-K} = \bar{p}_K \frac{1}{1-\rho} \qquad (11.63)$$

Substituting (11.62) and (11.59) in (11.63) results in

$$\text{Pr[queue]} = C(K,\rho) = \frac{\dfrac{K^K \rho^K}{K!(1-\rho)}}{\left[\displaystyle\sum_{n=0}^{K-1} \frac{1}{n!}K^n \rho^n + \frac{K^K \rho^K}{K!(1-\rho)}\right]} \qquad (11.64)$$

This expression is known as the *Erlang C formula.*

A final measure of performance to be discussed is the mean number of requests in the queue. According to our notation, let N be the number of requests in the system. Then for $N \le K$ the number N_q of requests in the queue is 0; while for $N > K$ it is

---

[6] The details are left as an exercise (see Prob. 11.33).

$N_q = N - K$. Therefore the mean number of requests in the queue is given by

$$E\{N_q\} = \sum_{n=K+1}^{\infty} (n-K)\bar{p}_n = \sum_{n=K+1}^{\infty} (n-K)\rho^{n-K}\bar{p}_K = \bar{p}_K \sum_{k=0}^{\infty} k\rho^k$$

where (11.61) was used to simplify the expression. Now observe that the term on the far right can be written as

$$\bar{p}_K \sum_{k=0}^{\infty} k\rho^k = \left(\frac{\bar{p}_K}{1-\rho}\right) \sum_{k=0}^{\infty} k(1-\rho)\rho^k$$

From (11.63) and (11.64) the term in parentheses can be recognized as $C(K,\rho)$. The rest of the expression is the formula for the mean of geometric random variable, which is given by $\rho/(1-\rho)$. With these observations, the equation for the mean number of requests in the queue becomes

$$E\{N_q\} = \frac{\rho}{1-\rho} C(K,\rho) \qquad (11.65)$$

which is the desired final expression.

Using this expression and the mean service time $E\{S\} = 1/\mu$ in conjuction with Little's formula, a number of other mean values can be computed for the M/M/K system. These are listed in Table 11.4.

$$\xrightarrow{\lambda}$$

$E\{W\} = \dfrac{C(K,\rho)}{K\mu + \lambda}$	$E\{N_q(t)\} = \dfrac{\rho}{1-\rho}C(K,\rho)$
$E\{S\} \;=\; 1/\mu$	$E\{N_s(t)\} = K\rho$
$E\{T\} \;=\; \dfrac{C(K,\rho)}{K\mu+\lambda} + \dfrac{1}{\mu}$	$E\{N(t)\} \;=\; \dfrac{\rho}{1-\rho}C(K,\rho) + K\rho$

Table 11.4 Expected values and relations for quantities involved in an M/M/K/∞ queueing system. Quantities in the right column are related to quantities in the left column according to Little's formula by the factor λ.

The following example illustrates the application of the theory discussed in this section for the multi-server system.

Example 11.8: Technical support for a small company has three technicians on duty. Requests for support are placed in a queue and routed to the next available technician. Assume customers call in for support at a rate of 15 calls per hour and the average time for support is 10 minutes. What is the mean length of the queue and what is the mean time that customers can expect to be waiting in the queue?

The arrival rate is $\lambda = 15/60 = 0.25$ calls/minute. The mean service time is $1/\mu = 10$ minutes. The utilization factor is therefore $\rho = \lambda/K\mu = (0.25/3)10 = 5/6$.

The probability of a queue is given by the Erlang C formula (11.64):

$$C(K,\rho) = \frac{\dfrac{3^3 \left(\frac{5}{6}\right)^3}{3!(1-\frac{5}{6})}}{\left[\displaystyle\sum_{n=0}^{3} \frac{1}{n!}3^n \left(\tfrac{5}{6}\right)^n + \frac{3^3\left(\frac{5}{6}\right)^3}{3!(1-\frac{5}{6})}\right]} \approx 0.7022$$

The expected length of the queue is then given by (11.65):

$$\mathcal{E}\{N_q\} = \frac{\frac{5}{6}}{1 - \frac{5}{6}} 0.7022 \approx 3.5 \text{ (customers)}$$

The mean time spent waiting in the queue can be computed from the expression in Table 11.4.

$$\mathcal{E}\{W\} = \frac{C(K, \rho)}{K\mu + \lambda} = \frac{0.7022}{3/10 + 0.25} = 1.28 \text{ minutes}$$

This is approximately 1 minute and 16 seconds.

□

11.5 Summary

Markov and Poisson models occur in many practical problems in electrical and computer engineering. Various aspects of the Poisson model have already been introduced in earlier chapters. This chapter brings all of these ideas together in the context of a single stochastic process known as the Poisson process. The process is defined by events whose probability of arrival is $\lambda \Delta_t$ and independence of events in non-overlapping time intervals (see Table 11.1). The number of events arriving in a fixed time interval of length t is described by the Poisson PMF with parameter $\alpha = \lambda t$ and the interarrival time T between successive events is described by the exponential PDF.

The Markov random process is a useful model for the data from many physical applications because it limits the conditioning of a random variable in the process to the random variable just preceding. When a Markov process takes on a (finite or infinite) discrete set of values the process is known as a Markov chain. Discrete-time Markov chains can be described algebraically by a state transition matrix or geometrically by a state diagram. Given a set of initial state probabilities, the probabilities of all the states can be computed as a function of time. For most cases of interest, these state probabilities approach limiting values called the "limiting state probabilities."

Continuous-time Markov processes are more challenging to analyze. The continuous-time Markov processes discussed in this chapter are based on the Poisson model. The "events" in this model are state transitions, which can be represented in a transition diagram. The time-dependent state probabilities satisfy a set of coupled first order differential equations. Fortunately, knowledge of this complete time dependence is not essential for many common engineering applications, and we can instead deal with the limiting-state probabilities. These probabilities can be computed from a system of *algebraic* equations called the "global balance equations." These equations can be further simplified when the Markov model takes the form of a "birth and death" process. This is the case for most queueing systems.

A queueing system is characterized by random arrival of "requests" for service, and service times of random length. Requests that arrive and cannot be immediately placed in service are stacked up in a queue. We have introduced a few of these systems known as $M/M/K/C$ queueing systems. In this labeling, the two M's stand for Markov (Poisson) arrival and service models, K is the number of servers in the system, and C is the capacity of the system, determined by the number of servers and the maximum allowed length of the queue. $M/M/1/\infty$, $M/M/1/C$ and $M/M/K/\infty$ were studied explicitly. Little's formula was developed as a powerful tool that relates the average number of requests in the system to the mean time that a request spends in the system. Using the tools developed in this chapter, important engineering systems such as computers, telephone systems, and computer networks can be analyzed for their performance under various traffic conditions. Performance measures such as mean waiting time, mean length of queue and others can be computed and plotted using the techniques developed here.

References

[1] Ross L. Finney, Maurice D. Weir, and Frank R. Giordano. *Thomas' Calculus*. Addison-Wesley, Reading, Massachusetts, tenth edition, 2001.

[2] Leonard Kleinrock. *Queueing Systems - Volume I: Theory*. John Wiley & Sons, New York, 1975.

[3] Leonard Kleinrock. *Queueing Systems - Volume II: Computer Applications*. John Wiley & Sons, New York, 1976.

[4] Howard M. Taylor and Samuel Karlin. *An Introduction to Stochastic Modeling*. Academic Press, New York, third edition, 1998.

[5] Sheldon M. Ross. *Introduction to Probability Models*. Academic Press, New York, eighth edition, 2002.

[6] Alvin W. Drake. *Fundamentals of Applied Probability Theory*. McGraw-Hill, New York, 1967.

[7] Donald Gross and Carl M. Harris. *Fundamentals of Queueing Theory*. John Wiley & Sons, New York, third edition, 1998.

[8] J. D. C. Little. A proof of the queueing formula $L = \lambda W$. *Operations Research*, 9:383–387, 1961.

Problems

The Poisson model

11.1 Email messages arriving at Sgt. Snorkle's computer on a Monday morning can be modeled by a Poisson process with a mean interarrival time of 10 minutes.

    (a) If he turns on his computer at exactly 8:00 and finds no new messages there, what is the probability that there will still be no new messages by 8:10?

    (b) What is the probability that there will be *more than* one new message by 8:10? How about by 8:15?

11.2 The arrival of buses from the campus to student housing in the late afternoon is modeled by a Poisson process with an interarrival time of 30 minutes. Consider the situation where a bus has just arrived at 4:29 pm.

    (a) A student arrives at 4:30 pm and misses the bus. What is the expected value of the time that the student will have to wait for the next bus to student housing?

    (b) At 4:52 the next bus has not yet arrived and another student comes to the bus stop. What is the expected value of the time that these two students will have to wait for the next bus?

11.3 Two overhead projector bulbs are placed in operation at the same time. One has a mean time to failure (MTF) of 3600 hours while the other has an MTF of 5400 hours.

(a) What is the *probability* that the bulb with the shorter MTF fails before the bulb with the longer MTF?

(b) If the bulb with the shorter MTF costs the U.S. government $1.00 while the bulb with the longer MTF costs the government $1.45, which bulb is the "better deal?" That is, which choice would minimize the expected value of the cost of buying bulbs?

(c) What is the most that the U.S. government should pay for the bulb with the longer MTF given that the cost of the bulb with the shorter MTF is fixed at $1.00?

11.4 The mean time to failure for elevators in Spanagel Hall is 1 day. (Assume failure times are exponentially distributed.)

(a) What is the probability of 1 failure in 1 day?

(b) What is the probability of 8 failures in 8 days?

(c) What is the expected value of the waiting time W until the 8th failure?

11.5 A web server receives requests according to a Poisson model with a rate of 10 hits per second.

(a) Suppose that the server goes down for 250 ms. What is the probability that no user accesses are missed by the server?

(b) In a one minute period, there are two outages: 0 to 10 ms and 12,780 to 12,930 ms. What is the probability that the server receives one hit during the first outage and no hits during the second?

(c) What is the probability that the server receives 6 hits during any 0.5-second period?

11.6 An Internet router receives traffic on two separate fiber cables. Arrivals on these feeds are independent and behave according to Poisson models with parameters α packets/sec (pps) and β pps.

(a) Determine the PMF of the total number of packet arrivals at the router during an interval of t seconds.

(b) What is the PDF of the interarrival period on the second feed (the one with β pps)?

11.7 Customers arriving at the 10th Street delicatessen are of two types. Type J are those customers ordering "Jaws" sandwiches and Type T are those ordering low-fat tuna. The shop owner knows that an hour before lunch the two types of customers can be modeled as independent Poisson processes with arrival rates $\lambda_J = 24$ customers/hour and $\lambda_T = 12$ customers/hour. That is, customers liking Jaws will never ask for low-fat tuna and vice-versa.

(a) If a customer arrives at any random time t, what is the probability that the customer will order Jaws (i.e., what is the probability that the customer is of Type J)?

(b) Given that a customer has just come in and ordered Jaws, what is the expected value of the waiting time to the next customer ordering low-fat tuna?

(c) What is the probability that the delicatessen sells exactly 5 Jaws and 3 low-fat tuna sandwiches in a 20 minute interval?

(d) What is the average interarrival time between customers? Two successive customers need not be of the same type.

(e) What is the average waiting time between two customers ordering different sandwiches (i.e., the expected value of the time between either a Type J and a Type T customer or a Type T and a Type J customer)?

(f) If a Jaws sandwich sells for $3.00 and a low-fat tuna sells for $3.50, what is the expected value of the delicatessen's gross earnings in one hour?

11.8 Up in the town of Gold Rush, California public transportation isn't what it used to be, so the arrival of buses headed for Los Angeles, San Francisco, and Durango, Colorado are independent Poisson processes with rates of 3, 2, and 5 buses per day, respectively.

(a) Given that you just missed the bus to Los Angeles, what is the probability that the next bus that comes along is a bus also headed for Los Angeles?

(b) If buses going to Los Angeles stop in San Francisco, what is the expected value of the time you must wait for the next bus that can take you to San Francisco?

Discrete-time Markov chains

11.9 A certain binary random process $X[k]$ that takes on values $+1$ and -1 is represented by a Markov chain with the state diagram shown below:

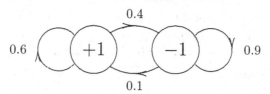

(a) What is the transition matrix for the process?

(b) Given that the process is in state "+1," what is the probability that the process will be in state "−1" three transitions from now?

(c) Given that $X[k] = -1$, what is the probability that $X[k+3] = -1$?

(d) What are the limiting-state probabilities for the process?

(e) If the process is observed beginning at a random time "k", what is the probability of a run of ten $+1$s? What is the probability of a run of ten -1s?

(f) Answer part (e) for a Bernoulli process that has the same probabilities as the limiting-state probabilities for this Markov process.

11.10 The discrete-time random process known as a *random walk* can be defined by the following equation:

$$Y[k] = \begin{cases} Y[k-1] + 1 & \text{with probability } p \\ Y[k-1] - 1 & \text{with probability } 1 - p \end{cases}$$

(a) Draw a state transition diagram for the random walk. Define state "i" by the condition $Y[k] = i$. Be sure to label the transition probabilities.

(b) Do limiting-state probabilities exist for the random walk? If so, what are they?

11.11 The bits in a certain binary message can be modeled by a 2-state Markov process with state diagram shown below:

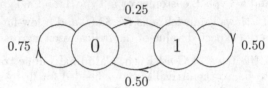

(a) What is the transition matrix?

(b) Given that the first bit in the message is a zero, what is the probability that the third bit is also a 0?

(c) If we wait a long time after the message has started, and observe some bit at random, what is the probability that the observed bit will be a 0?

11.12 In a discrete-time Markov chain, we have $p_{00} = 0.3$ and $p_{11} = 0.6$.

(a) Construct the state transition matrix.

(b) Draw the state transition diagram.

(c) Determine the limiting-state probabilities.

11.13 In a discrete-time Markov chain $p_{01} = 0.6$ and $p_{10} = 0.3$.

(a) Determine the limiting-state probabilities.

(b) Determine the probability of the occurrence of 8 consecutive 0s.

(c) Determine the probability of the occurrence of 8 consecutive 1s.

(d) Now consider a Bernoulli process with the same limiting-state probabilities as in (a). Repeat (b) and (c) for this Bernoulli process.

11.14 Refer to Problem 11.13.

(a) What is the probability of a sequence of 5 0s followed by 3 1s?

(b) Now consider a Bernoulli process with the same limiting-state probability as those used in (a). Repeat (a) for the Bernoulli process.

11.15 A state diagram for a discrete-time Markov chain is shown below.

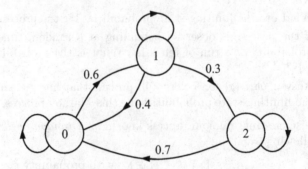

(a) Fill in the transition probabilities of the self-loops in the state diagram.

(b) What is the transition matrix?

(c) Which state sequence has the largest probability: $0, 0, 0$ or $1, 1, 1$ or $2, 2, 2$? What is the value of this largest probability?

Continuous-time Markov chains

11.16 A three-state continuous-time Markov chain is described by rates $r_{01} = 4$, $r_{02} = 5$, $r_{12} = 5$, $r_{10} = 3$, $r_{20} = 2$, and $r_{21} = 2$.

(a) Draw the transition rate diagram for the process.

(b) Find the limiting-state probabilities p_0, p_1, and p_2 for the process.

11.17 In a continuous-time Markov chain with four states, the following transition rates are given: $r_{01} = 2$, $r_{10} = 3$, $r_{02} = 1$, $r_{12} = 1$, $r_{21} = 2$, $r_{23} = 2$, $r_{31} = 2$, and $r_{32} = 3$.

(a) Draw the state transition diagram.

(b) Determine the limiting-state state probabilities.

11.18 In order to study the behavior of communication networks for transmission of speech, speaker activity is modeled as a Markov chain as shown below. The interpretation of the states is as follows: State 0 indicates silence; State 1 indicates unvoiced speech; and State 2 represents voiced speech. The transition rates are given to be: $r_{01} = 1$, $r_{10} = 1.5$, $r_{12} = 2$, and $r_{21} = 3$.

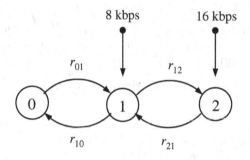

(a) Determine the limiting-state probabilities of this model.

(b) If unvoiced speech is transmitted at 8 kbps and voiced speech is transmitted at 16 kbps, find the average bit rate produced by the transmitter.

11.19 Yoohoo is an Internet traffic monitoring service, which counts the number of unique hits at a given web portal site. The arrival of unique hits is a Poisson process with a rate λ.

(a) Draw a state transition diagram for this process.

(b) In a 4-hour period, what is the probability that the number of hits is 12? Let $\lambda = 2$ hits/hour.

(c) Can you determine the limiting-state state probabilities for this Markov chain? Explain.

Basic queueing theory

11.20 In an M/M/1 queuing system, the arrival rate is 100 packets per second and the service rate is 120 packets per second. Determine the utilization of this system.

11.21 In a certain M/M/1 queueing system the arrival rate is $\lambda = 4$ sec^{-1} and the service rate is $\mu = 5$ sec^{-1}.

(a) What is the utilization ρ?

(b) What is the probability $p_{10} = \Pr[N(t) > 10]$?

(c) What is the average total time in the system $\mathcal{E}\{T\}$? What is the mean waiting time spent in the queue?

(d) To attain $p_{10} \leq 10^{-2}$, what would have to happen to the arrival rate? What are ρ and $\mathcal{E}\{T\}$ at the new arrival rate?

(e) To achieve $p_{10} \leq 10^{-2}$, how should the service rate be changed? What are ρ and $\mathcal{E}\{T\}$ at the new service rate?

11.22 Email messages arrive at a mail server at the average rate of 30 messages per minute. The average time that it takes the server to process and reroute a message is 750 milliseconds.

(a) What is the expected value of the number of messages $N(t)$ waiting in the queue at any time?

(b) What is the average *time* a message spends from the time that it arrives until the time it is rerouted?

(c) How do the answers to parts (a) and (b) change if the arrival rate for messages is doubled, i.e., increased to 60 messages per minute?

11.23 Interrupts on a certain computer occur at an average rate of 10 per second. The average service time for an interrupt is 0.05 sec.

(a) What are the statistics $\mathcal{E}\{N(t)\}$ and $\mathcal{E}\{T\}$ for the queue?

(b) If you wanted to cut $\mathcal{E}\{T\}$ in half, what average service time would you need? What would $\mathcal{E}\{N(t)\}$ be in that case?

11.24 Compare the following two situations with respect to the average time a customer has to spend:

(a) Customers in a small bank with just two tellers have their choice of one of two lines (queues) they may wait in for service. Once they have chosen a line, they may not change to the other one. Both tellers are equally competent; the average service time for each teller is the same and equal to μ.

(b) Customers arrive and wait in a *single* line for two tellers. (Same tellers as before.) The customer at the head of the line goes up to the next available teller.

11.25 Consider an M/M/1 queueing system. The arrivals at the system are described by a Poisson model with $\lambda = 20$ packets/sec (pps). The service rate is 25 pps.

(a) Is the system stable? What is the utilization factor ρ?

(b) The state probabilities are given by $p_k = \rho p_{k-1}$, $k \geq 1$, determine p_0.

(c) What is the average number of packets in the system?

(d) What is the average total delay in the system?

(e) What is the average queueing delay?

(f) Determine the number of packets in the queue.

11.26 Consider a four-state queuing system (M/M/1/3) with an arrival rate of 100 packets per second and a service rate of 120 packets per second.

(a) Determine the limiting state probabilities of the system.

(b) Calculate the average number of packets in the system.

(c) What is the average number of packets in the queue?

11.27 In an M/M/1/3 system, the arrivals are Poisson with $\lambda = 150$ requests per second.

(a) Draw the state transition diagram.

(b) Determine the limiting-state state probabilities.

(c) If p_1 and p_2 are limited to $p_1 = 4/15$ and $p_2 = 2/15$, what average service rate μ is required?

11.28 Traffic arrives at a queueing system in a Poisson manner with parameter $\lambda = 24$ packets/sec (pps).

(a) Consider that the system is M/M/1 with an average service rate of 36 pps. What is the average delay in the system?

(b) Let the queueing system be actually two identical M/M/1 queues in parallel, and the traffic is divided equally among the two queues. The average service rate is 18 pps per queue. Determine the average total delay in the system.

(c) Consider an M/M/2 system for which the average service rate of each server is 18 pps. Find the average total delay in the system.

11.29 Fluorescent light bulbs in Spanagel Hall fail at the rate of 2 per week. (Assume that a week consists of 5 working days and that the probability of failure for light bulbs on weekends is negligible. Therefore weekends may be disregarded entirely.) It takes an average of 2 working days for Public Works to respond and replace a light bulb. Only one man works to replace the bulbs. If additional requests for service come in while he is out replacing a bulb those requests are placed in a queue and will be responded to on a "first come/first served" basis.

(a) What is the mean number of light bulbs that will be out of working condition in Spanagel at any given time?

(b) Assuming that an employee reports a bulb as soon as it fails, what is the average length of time an employee must put up with a burned out or blinking fluorescent bulb until it is replaced?

11.30 Requests on a network printer server are modeled by a Poisson process and arrive at a nominal rate of 3 requests/minute. The printer is very fast, and the average (mean) time required to print one of these jobs is 10 seconds. The printing time is an exponential random variable independent of the number of requests.

(a) What is the expected value of the waiting time until the *third* request?

(b) Assuming that the system can accommodate an infinitely long queue, what is the probability that when a job arrives there are no requests in the queue? (Be careful about answering this!)

(c) All real systems must have some finite capacity C. If the system (which includes the printer and its queue) is completely full, requests will be sent back to the user with the message "PRINTER BUSY: TRY AGAIN LATER." What minimum value of K is needed if the probability that a request arrives and is sent back is to be less than 0.001?

11.31 The color printer in the main ECE computing lab has a mean time to failure (MTF) of 11 days. The IT department (Code 05) has a mean time of response (MTR) of 2 days. Assume that both are described by a Poisson model; i.e., the failure time for the printer is an exponential random variable with parameter $\lambda = 1/11$ and the service time is also an exponential random variable with parameter $\mu = 1/2$. Assume that as soon as the IT technician arrives, the printer starts working again, so there is no extra waiting time while the technician services the machine. We can model the situation by a continuous-time Markov chain with two states

$$\text{State 0:} \quad \text{printer is working}$$
$$\text{State 1:} \quad \text{printer is down}$$

What are the limiting-state probabilities $\bar{p}_0$ and $\bar{p}_1$ that the printer is or is not working?

11.32 The probability generating function (PGF) corresponding to a discrete random variable is defined by (4.21) of Chapter 4.

(a) Show that the PGF for the distribution (11.51) for a finite-capacity queueing system is given by

$$G_N(z) = \frac{1 - (\rho z)^{C+1}}{1 - \rho z} \bar{p}_0$$

where $\bar{p}_0$ is given by (11.50).

(b) Use this to show that the mean of the distribution is given by (11.52).

11.33 Show how the equations (11.57) follow from the conditions for the birth and death process preceding those equations. Also, by summing the limiting state probabilities and setting the sum equal to 1, derive the expression (11.59) for $\bar{p}_0$ for the K-server case.

11.34 In Example 11.8, what is the probability that at least one of the technicians is not busy?

11.35 One of the technicians in Example 11.8 goes on a coffee break, but does not come back for a long time. How does this change the mean waiting time that a customer calling in will experience?

11.36 Consider a queueing system with K servers and no queue. In other words, if requests arrive and all servers are busy, those requests are turned away. Show that the probability that an arriving request is turned away is given by the expression

$$\bar{p}_K = \mathcal{B}(K, \rho) = \frac{(K\rho)^K / K!}{\sum_{n=0}^{K} (K\rho)^n / n!}$$

This expression is known as the *Erlang B formula*.
Hint: Begin by drawing the transition diagram for the system.

Computer Projects

Project 11.1

In this project you will generate random sequences of 0s and 1s and compute their relative frequencies. You will compare the cases where the 0s and 1s occur independently to the cases where dependency is introduced using a simple Markov chain model.

1. Simulate and plot a random sequence of 0s and 1s where each successive digit is chosen independently from a Bernoulli PMF with parameter $p = 0.5$. Call this Case A. Plot a sequence of $N = 50$ successive digits and then repeat the experiment; plot another set of 50 digits and demonstrate that it is different from the first.

2. Repeat Step 1 but now choose the parameter p of the Bernoulli distribution to be 0.8. Call this Case B. Again plot two realizations to be sure they are different. Cases A and B each represent a *Bernoulli process* but with different values for the parameter p.

3. Now assume that the digits are not independent; in particular, the sequence of 0s and 1s is generated by a Markov chain with state diagram shown below.

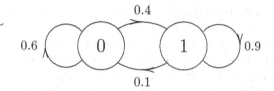

Plot a sequence of $N = 50$ digits for this Markov model of binary data assuming the initial value for the state probabilities is $\mathbf{p}[0] = \begin{bmatrix} 0.5 & 0.5 \end{bmatrix}^T$. Call this Case C. Repeat this step, but with different initial conditions; in particular, plot another realization of 50 binary digits assuming that $\mathbf{p}[0] = \begin{bmatrix} 0.2 & 0.8 \end{bmatrix}^T$. Call this Case D.

4. Now generate sequences of N points for each of Cases A, B, C and D and for the values of N specified below. Do *not* try to plot these sequences on paper. For each sequence of Cases A through D, estimate the probability of a 0 or 1 in the sequence by computing the relative frequency. That is, for a given sequence of length N the relative frequency of a 1 is the number of 1s appearing in the sequence divided by N. Summarize your results in a table for N=50, 100, 1000, and 10,000.

5. Compare the relative frequencies of 0s and 1s for Cases A through D. Are there any differences you can observe in the characteristics of the sequences that have similar relative frequencies? Look at plots on your display monitor and zoom in on some of the longer sequences to better answer this question.

6. Compute the relative frequency of 0s and 1s for a Markov chain with initial condition $\mathbf{p}[0] = \begin{bmatrix} 0.8 & 0.2 \end{bmatrix}^T$ and compare it to Cases C and D above. What does this tell you about the properties of a Markov chain?

7. Solve for the limiting-state probabilities of the Markov chain and compare them to the relative frequency results.

MATLAB programming notes

See programming notes for Project 2.1 for generating the random binary sequence.

Project 11.2

In this project, you will simulate and study an M/M/1 queueing system.

1. Consider an M/M/1 packet queueing system with an arrival rate λ and a service rate μ. For $\lambda = 0.1$ packets per millisecond and $\mu = 0.125$ packets per millisecond, calculate the mean (expected value) number of packets in the system and in the queue, and the mean time spent by packets in the system and in the queue.

2. Simulate this M/M/1 queue by following these steps:

   (a) The arrival process $A(t)$ can be realized as a sequence of 0s and 1s where a 1 indicates the arrival of a packet. Assume that in a given 1 ms interval only one arrival can occur. The interarrival time (time between successive arrivals) is characterized by an exponential random variable with parameter λ.

   (b) Generate the departure process $D(t)$. Each arriving packet is served by the queue; the service time is characterized by an exponential random variable with parameter μ. At a given time, the system can only serve one packet on a first-come-first-served basis. Additional packets must wait in the queue for service. If there are no packets waiting to be served, the server stays idle until the next packet arrives.

   (c) Measure the following quantities:

      i. Number of packets in the system $N(t)$ as the difference between the arrival and departure processes: $N(t) = A(t) - D(t)$. Use a time resolution of 1 ms.

      ii. The time spent by the ith packet in the system $t_i = t_{Ai} - t_{Di}$, where t_{Ai} is the arrival time of the ith packet and t_{Di} is the departure time of the same packet.

   (d) Plot the following for a simulation time period of 200 ms (200 points):

      i. The arrival process $A(t)$ and the departure process $D(t)$ vs. time on the same graph.

      ii. The number of packets in the system $N(t)$ vs. time.

      iii. A graph of the total time t_i spent by the ith packet vs. its arrival time t_{Ai}.

3. From your simulation and the measurements in Step 2(c), obtain the time average value $\overline{N} = \langle N(t) \rangle$ and the average time $\overline{t} = 1/K \sum_{i=1}^{K} t_i$ spent by K packets in the system. Compare these with the corresponding theoretical values $\mathcal{E}\{N(t)\}$ and $\mathcal{E}\{T\}$ found in Step 1. Repeat these calculations for simulation time periods of 200 ms, 1000 ms, 10,000 ms, 100,000 ms, and 1,000,000 ms.

4. Repeat Steps 2(a), 2(b), and 2(d) for a simulation time period of 10,000 ms, a fixed $\mu = 0.10$, and a range of values of λ from 0.005 to 0.095 in increments of 0.01. In each case, calculate $\overline{N}$ and $\overline{t}$ as in Step 3. Plot $\overline{N}$ and $\overline{t}$ as a function of $\rho = \lambda/\mu$. Do these experimental results obey Little's formula?

MATLAB programming notes

The function "expon" from the software package can be used to generate exponential random variables.

Appendices

 **Basic Combinatorics**

Elements of combinatorics (counting arrangements of objects) appear in several problems on probability. While advanced knowledge of this topic is not necessary for the study of probability, some rudimentary knowledge is essential. A summary of the minimum requirements for the main part of the text is presented in this appendix.

A.1 The Rule of Product

To begin this discussion, consider the problem of computing the number of 4-digit hexadecimal numbers that are possible, using the the sixteen characters { 0,1, 2, 3, 4, 5, 6, 7, 8, 9, A, B, C, D, E, F }. A typical such hexadecimal number is

$$06FF$$

In forming these numbers, there are 16 choices for the first digit, sixteen choices for the second digit, and so on. Therefore the number of possible 4-digit hexadecimal numbers is

$$16 \cdot 16 \cdot 16 \cdot 16 = 16^4 = 65,536$$

The principle behind this computation is known as the Rule of Product and applies in many similar situations. The most general form of the rule is as follows.

> In an experiment to form a sequence of k elements (k-tuple), where there are N_i possible choices for the ith element, the number of unique k-tuples that can be formed is $\prod_{i=1}^{k} N_i$.

In many applications, such as in the foregoing example, the number of choices for all elements is the same (i.e., $N_i = N$). In this case the the number of unique k-tuples is simple N^k.

Let us consider a more complicated example.

Example A.1: A digital logic gate has four inputs. How many different Boolean functions can a four-input logic gate realize?

The question can be answered more generally. Let us denote the number of inputs by n. Then the input-output relation of the gate can be represented by a so-called "truth table" which is illustrated below for the case of $n = 4$.

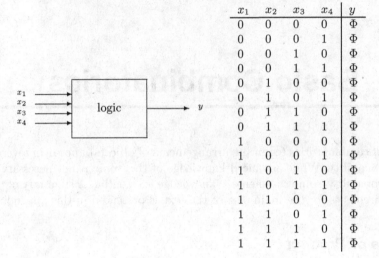

x_1	x_2	x_3	x_4	y
0	0	0	0	Φ
0	0	0	1	Φ
0	0	1	0	Φ
0	0	1	1	Φ
0	1	0	0	Φ
0	1	0	1	Φ
0	1	1	0	Φ
0	1	1	1	Φ
1	0	0	0	Φ
1	0	0	1	Φ
1	0	1	0	Φ
1	0	1	1	Φ
1	1	0	0	Φ
1	1	0	1	Φ
1	1	1	0	Φ
1	1	1	1	Φ

The 1s and 0s represent combinations of inputs and the Φ represents either a 0 or a 1 depending on the function that is being realized.

Notice that there are 16 rows in this table. That is because there are four inputs and two possible choices for each input (0 or 1). Thus, applying our principle, there are $2^4 = 16$ combinations of inputs or rows in the table. In general, for a logical function with n inputs, there will be 2^n rows in the truth table. This is the first application of the Rule of Product in this example.

Now consider the main question: how many possible functions are there? In defining a function, the output y for each row of the truth table can be chosen to be either a 0 or a 1. If $k = 2^n$ is the number of rows in the truth table, the number of possible functions is 2^k or 2^{2^n}. This is a second use of the Rule of Product. Therefore a logic gate with four inputs could implement $2^{2^4} = 2^{16} = 65,536$ possible different functions.

$\square$

A.2 Permutations

A related combinatorial problem is the following. Eight people are to ride in a vehicle with eight seats. How many different ways can the people be seated?

To answer this question, first note that there are eight choices for the driver. Once the driver is chosen, there are seven choices for who will ride "shotgun," i.e., in the front passenger seat. After that, there are six choices for the passenger who will sit directly behind the driver, and so on. Thus, according to the Rule of Product, the number of ways of arranging the riders is

$$8 \cdot 7 \cdot 6 \cdot 5 \cdot 4 \cdot 3 \cdot 2 \cdot 1 = 8! = 40,320$$

(a surprisingly large number!)

A particular arrangement of the riders is called a *permutation*. If there are N riders (and N seats) then the number of permutations is given by[1]

$$P^N = N! \tag{A.1}$$

---

[1] The notation here is just P with a superscript N to indicate the order of the permutation. It does *not* mean that some number P is raised to the Nth power.

As a variation on this problem, suppose there are ten potential riders, but still only eight seats. Then the number of ways of arranging the people (with two left out) is

$$10 \cdot 9 \cdot 8 \cdot 7 \cdot 6 \cdot 5 \cdot 4 \cdot 3 = 1,814,400$$

Notice that this computation can be expressed as

$$\frac{10!}{2!} = 1,814,400$$

This kind of arrangement is referred to as a permutation of N *objects taken k at a time*. The number of such permutations is given by

$$P_k^N = \frac{N!}{(N-k)!} \tag{A.2}$$

It is interesting to compare the essential features of this problem to the one described in the previous section. In many probability studies it is traditional to develop the problems described here as problems of drawing colored balls from an urn (jug). The balls in the urn all have different colors so they can be distinguished. In the type of problem described in the previous section a ball is selected but then is replaced in the urn. Thus the total number of choices remains fixed at N for each drawing. In permutation problems, described here, the ball is not replaced in the urn, so the number of choices is reduced by one after each drawing. Because of this analogy, problems of this sort are referred to by combinatorial mathematicians as "sampling with replacement" and "sampling without replacement" respectively.

The number of permutations of a moderate number of objects can become very large. Consider the following example taken from the arts.

Example A.2: Abstract-expressionist painter Johnson Polkadot begins his daily work by selecting a can of paint from his supply of 50 different colors. This paint is applied randomly to the canvas and allowed to dry. He then selects another color and applies it over the first and continues in this way until he is finished. Assume he can apply only one color per hour. If he is now 35 years old and works for 12 hours a day every day of the year, can he live long enough to be forced to repeat a color sequence?

The number of sequences of 12 colors is given by

$$P_{12}^{50} = \frac{50!}{(50-12)!} = 50 \cdot 49 \cdot \ldots \cdot 39 \approx 1.01 \times 10^{19}$$

If Johnson is now 35 and lives to be 105, the number of paintings he will produce is $70 \cdot 365 = 25,550$ (assuming he rests one day a year on leap years). Since this number is nowhere close to the number of possible color sequences, he will never be forced to repeat the sequence of colors used in a painting.

□

A.3 Combinations

The last topic to be described is that of *combinations*. This is an important combinatorial result probably used more than the others, but is also a bit more complicated to explain.

Let us consider the transmission of N binary data bits of information and ask, "In how many different ways can k errors occur in N bits?" To answer this question, let us pretend that errors are caused by a little demon who shoots colored arrows into the bits (see illustration below) which cause the errors. This demon is an excellent shot, so if it decides to attack a particular bit, it never misses. Assume the demon has k arrows in its quiver and that each arrow is a different color so they can be distinguished. (This mode of operation is known as NTSC for Never Twice the Same Color, which refers to the former analog color television standard in the USA.)

If the demon decides to cause k errors, the number of ways to place k arrows in N bits is given by (A.2). To be specific, pretend that there are $N = 5$ bits and the demon causes $k = 3$ errors; then this number would be

$$P_3^5 = \frac{5!}{(5-3)!} = 60$$

We can show that this number is *much* too large! To see this, assume that the arrows are colored red, green and blue. If all the sequences with $k = 3$ errors are listed, one of the sequences would be (for example)

$$\Phi \quad R \quad \Phi \quad G \quad B$$

where Φ indicates that no error has occurred and R, G, or B indicates that the bit has been destroyed by the corresponding colored arrow. There are other configurations with the colored arrows in the same set of positions, however. These other configurations are:

$$
\begin{array}{ccccc}
\Phi & R & \Phi & B & G \\
\Phi & G & \Phi & R & B \\
\Phi & G & \Phi & B & R \\
\Phi & B & \Phi & R & G \\
\Phi & B & \Phi & G & R \\
\end{array}
$$

and these are included in the count given by (A.2).

If we are concerned with just the errors, all of these configurations are equivalent. In other words, the color of the arrow is irrelevant; an arrow is an error regardless of its color. So it is clear that using (A.2) results in overcounting. The factor of overcounting is equal to the number of permutations of the colors; in this case where $k = 3$ the factor is $3! = 6$. (Check this out, above!) Thus, the number of errors when $N = 5$ and $k = 3$ is

$$P_3^5/3! = \frac{5!}{2!\,3!} = 10$$

The general result is given by

$$C_k^N = \binom{N}{k} = \frac{N!}{k!(N-k)!} \tag{A.3}$$

and represents the number of *combinations* of N objects taken k at a time. The expression $\binom{N}{k}$ is known as the *binomial coefficient* and is typically read as "N *choose* k." It can be thought of as the number of ways that k errors may occur in N bits, although it applies in a multitude of other problems.

Example A.3: A technician is testing a cable with six conductors for short circuits. If the technician checks continuity between every possible pair of wires, how many tests need to be made on the cable?

The number of tests needed is the number of combinations of 6 wires taken 2 at a time or the number of ways to choose 2 wires from a set of 6. This number is

$$C_2^6 = \binom{6}{2} = \frac{6!}{2!\,4!} = \frac{6 \cdot 5}{2} = 15$$

□

B

The Unit Impulse

The unit impulse is one of the most frequently used, but frequently misunderstood concepts in engineering. While well-founded in physical applications, this seemingly innocuous idea led to consternation among mathematicians of the last century until given a proper foundation through the theory of Distributions and Generalized Functions (see e.g., [1, 2]). Although the unit impulse has its greatest importance in system theory, it also has significant applications in the study of random variables from an engineering perspective. This appendix provides an introduction to the unit impulse for students who may not yet have encountered this "function" in their other studies. References [3] and [4] provide some further discussion from an engineering perspective.

B.1 The Impulse

In the study of engineering and physics, situations are sometimes encountered where a finite amount of energy is delivered over a very brief interval of time. Consider electronic flash photography as an example. The light produced by the flash tube is extremely bright but lasts only a very small fraction of a second. The light intensity arriving at the camera focal plane is integrated to produce the exposure. The light intensity can be modeled as shown in Fig. B.1. Although the flash output may not

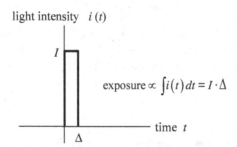

Figure B.1 Model for a pulse of light used in electronic flash photography.

be precisely a square pulse, this is not important, since the camera responds to the integral or total amount of light reaching the focal plane.

As another example, consider the crash-testing of an automobile. The car is driven into a solid stationary wall at some velocity v. When the car hits the wall there is an immediate change of momentum, since the velocity goes to 0. Newton's Law requires that this change of momentum be produced by a force at the instant of time that the car hits the wall. This force, which exists for only a brief moment, must be very large so that the integral of the force over time produces the finite change of momentum dictated by the situation. The effect on the car is significant!

In these situations, as well as in others, there is a need to deal with a pulse-like function $g_\Delta(x)$ which is extremely narrow, but also sufficiently large at the origin so that the integral is a finite positive value, say unity. Such a function, while having clear application in these physical scenarios, turns out to very useful in other areas of electrical engineering, including the study of probability. Figure B.2 shows a few

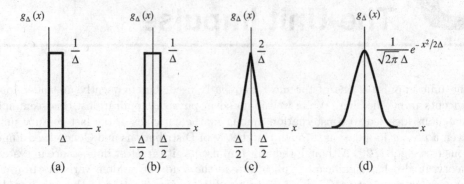

Figure B.2 Various approximations to a unit impulse. (a) Square pulse. (b) Centered square pulse. (c) Triangular pulse. (d) Gaussian pulse.

possible ways that this pulse-like function could be defined. In all cases the pulses have an area of 1, and we can write

$$\lim_{\Delta \to 0} \int_{-\infty}^{\infty} g_\Delta(x)dx = 1 \tag{B.1}$$

As Δ approaches 0, the details of the pulse shape ultimately do not matter. In engineering, we think of the *unit impulse* $\delta(x)$ as the idealized limiting form of the function $g_\Delta(x)$ as $\Delta \to 0$. We could write

$$\delta(x) = \lim_{\Delta \to 0} g_\Delta(x)$$

but this is at best misleading since this "function" assigns a nonzero value to only a single point ($x = 0$) where it assigns infinity.[1] A more correct statement is given by (B.1), using any of the limiting forms above and the statement

$$\int_{-\infty}^{\infty} \delta(x)dx = 1 \tag{B.2}$$

to define the unit impulse.

Graphically, the impulse is represented as shown in Fig. B.3, i.e., as a bold thick

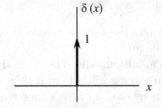

Figure B.3 Graphical representation of the unit impulse.

upward-pointing arrow at the origin. The number "1" which is optionally placed next to the arrow indicates that when integrated, the impulse has unit area.

---

[1] The late professor Sam Mason was said to have quoted an M.I.T. engineering student as remarking, "You mean the impulse is so small everywhere that you can't see it except at the origin where it is so large that you can't see it either? In other words, you can't see it at all! At least I can't."

B.2 *Properties*

Although the impulse is not a true function according to the mathematical definition, it can be manipulated as if it were a proper function. This makes the impulse extremely convenient to use in engineering applications. For example, a scaled impulse, i.e., an impulse with area A is simply represented by

$$A\delta(x)$$

In a graphical representation like Fig. B.3, we may write the area A next to the arrow instead of the value 1 to indicate a scaled impulse with area A.

Impulses may occur shifted along the x axis as well. For example, we could plot the expression

$$2.5\delta(x) + 1.5\delta(x - 2)$$

as shown in Fig. B.4. This is a set of two scaled impulses with the second impulse shifted

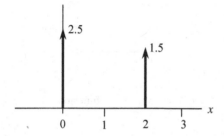

Figure B.4 Set of two impulses with different areas.

to the point $x = 2$. In the graphical representation it is common (but not universal) to represent the difference in areas by different heights of the arrows as illustrated in this figure. In the case of a negative scale factor such as $-2.5\delta(x)$, downward-pointing arrows are sometimes used in the representation. While all of this may be convenient, it is important to remember that the actual height of the impulse is *infinite* and we should not be mislead by the size of the arrows when they are drawn in this manner.

One of the most important properties of the impulse is known as the "sifting property." For any function $f(x)$ we may write

$$f(x) = \int_{-\infty}^{\infty} f(\xi)\delta(\xi - x)d\xi \tag{B.3}$$

which is valid at least for values of x where there is no discontinuity.[2] This equation, which may seem puzzling when encountered for the first time, is easily explained by the picture of Fig. B.5 where the integrands are plotted as a function of the dummy variable

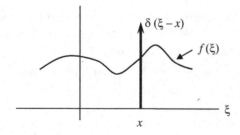

Figure B.5 Illustration of the "sifting" property of an impulse.

---

[2] Discontinuities can lead to inconsistency in interpretation of the integral. Although this is often ignored in casual use of the sifting property, it is good to be aware of the problem.

ξ. The product of the function $f(\xi)$ and the impulse $\delta(\xi - x)$ is 0 everywhere except at the point $\xi = x$. At this point the impulse is scaled by the value of the function, which is $f(x)$. Integration then produces the *area* of the scaled impulse, which is equal to $f(x)$. While the usefulness of (B.3) may not seem immediately apparent, the relation is extremely important in simplifying integrals involving impulses.

A final property to be mentioned is that the unit impulse can be considered as the formal derivative of the unit step function. The unit step function, conventionally denoted by $u(x)$, is defined by

$$u(x) = \begin{cases} 1 & x \geq 0 \\ 0 & x < 0 \end{cases} \tag{B.4}$$

and sketched in Fig. B.6. From Fig. B.6, it can be seen that $u(x)$ satisfies the relation

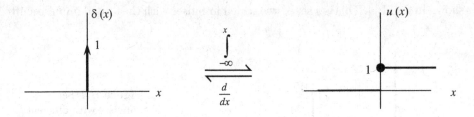

Figure B.6 Impulse as the derivative of the unit step function.

$$u(x) = \int_{-\infty}^{x} \delta(\xi)\,d\xi \tag{B.5}$$

In particular, for any negative value of the upper limit x, the integral in (B.5) is 0, while for any positive value of x the integral is 1. An ambiguity occurs for $x = 0$ which is generally resolved by defining the step to be continuous from the right. Since the integral of the unit impulse is found to be the unit step, the unit impulse can be regarded as the *derivative* of the unit step.

This relation can be generalized to the cases where functions have simple discontinuities. Chapter 3 discusses how cumulative distribution functions which have such discontinuities result in probability density functions that have impulses at the corresponding locations.

References

[1] M. J. Lighthill. *An Introduction to Fourier Analysis and generalized Functions*. Cambridge University Press, New York, 1959.

[2] A. H. Zemanian. *Distribution Theory and Transform Analysis*. McGraw-Hill Book Company, New York, 1955.

[3] Athanasios Papoulis. *Signal Analysis*. McGraw-Hill, New York, 1977.

[4] William McC. Siebert. *Circuits, Signals, and Systems*. McGraw-Hill, New York, 1986.

C The Error Function

The *"error" function*, defined by

$$\text{erf}(x) = \frac{2}{\sqrt{\pi}} \int_0^x e^{-z^2} dz \;\; ; \;\; 0 \le x < \infty$$

can be used as an alternative to the Q function in evaluating integrals of the Gaussian density. This function is more common in studies involving basic statistics and is included in many software libraries for programming languages as well as in tables such as that appearing on the next page. The error function is provided in MATLAB as a standard mathematical function; while the Q function is provided only in the communications toolbox.[1] The error function is compared to the Q function and the normalized CDF function in the table below.

function	symmetry relation
$\text{erf}(x) = \dfrac{2}{\sqrt{\pi}} \displaystyle\int_0^x e^{-z^2} dz$	$\text{erf}(-x) = -\text{erf}(x)$
$Q(x) = \dfrac{1}{\sqrt{2\pi}} \displaystyle\int_x^{\infty} e^{-\frac{z^2}{2}} dz$	$Q(-x) = 1 - Q(x)$
$\Phi(x) = \dfrac{1}{\sqrt{2\pi}} \displaystyle\int_{-\infty}^x e^{-\frac{z^2}{2}} dz$	$\Phi(-x) = 1 - \Phi(x)$

Table C.1 Functions for evaluation of Gaussian density integrals.

Two common probabilities are given below for a Gaussian random variable with mean μ and variance σ^2.

$$\Pr[X > x] = \frac{1}{\sqrt{2\pi\sigma^2}} \int_x^{\infty} e^{-(z-\mu)^2/2\sigma^2} dz \;\;\Longleftrightarrow\;\; \frac{1}{2}\left[1 - \text{erf}\left(\frac{x-\mu}{\sqrt{2}\sigma}\right)\right] \quad \text{(C.1)}$$

$$\Pr[X \le x] = \frac{1}{\sqrt{2\pi\sigma^2}} \int_{-\infty}^x e^{-(z-\mu)^2/2\sigma^2} dz \;\;\Longleftrightarrow\;\; \frac{1}{2}\left[1 + \text{erf}\left(\frac{x-\mu}{\sqrt{2}\sigma}\right)\right] \quad \text{(C.2)}$$

The last equation represents the CDF for X.

---

[1] MATLAB also provides the *complementary error function*

$$\text{erfc}(x) = \frac{2}{\sqrt{\pi}} \int_x^{\infty} e^{-z^2} dz$$

but this function is less prominent in the literature.

THE ERROR FUNCTION

Table of error function values

x	erf(x)	x	erf(x)	x	erf(x)	x	erf
0.00	0.00000	0.40	0.42839	0.80	0.74210	1.20	0.91
0.01	0.01128	0.41	0.43797	0.81	0.74800	1.21	0.91
0.02	0.02257	0.42	0.44747	0.82	0.75381	1.22	0.91
0.03	0.03384	0.43	0.45689	0.83	0.75952	1.23	0.91
0.04	0.04511	0.44	0.46623	0.84	0.76514	1.24	0.92
0.05	0.05637	0.45	0.47548	0.85	0.77067	1.25	0.92
0.06	0.06762	0.46	0.48466	0.86	0.77610	1.30	0.93
0.07	0.07886	0.47	0.49375	0.87	0.78144	1.35	0.94
0.08	0.09008	0.48	0.50275	0.88	0.78669	1.40	0.95
0.09	0.10128	0.49	0.51167	0.89	0.79184	1.45	0.95
0.10	0.11246	0.50	0.52050	0.90	0.79691	1.50	0.96
0.11	0.12362	0.51	0.52924	0.91	0.80188	1.55	0.97
0.12	0.13476	0.52	0.53790	0.92	0.80677	1.60	0.97
0.13	0.14587	0.53	0.54646	0.93	0.81156	1.65	0.98
0.14	0.15695	0.54	0.55494	0.94	0.81627	1.70	0.98
0.15	0.16800	0.55	0.56332	0.95	0.82089	1.75	0.98
0.16	0.17901	0.56	0.57162	0.96	0.82542	1.80	0.98
0.17	0.18999	0.57	0.57982	0.97	0.82987	1.85	0.99
0.18	0.20094	0.58	0.58792	0.98	0.83423	1.90	0.99
0.19	0.21184	0.59	0.59594	0.99	0.83851	1.95	0.99
0.20	0.22270	0.60	0.60386	1.00	0.84270	2.00	0.99
0.21	0.23352	0.61	0.61168	1.01	0.84681	2.05	0.99
0.22	0.24430	0.62	0.61941	1.02	0.85084	2.10	0.99
0.23	0.25502	0.63	0.62705	1.03	0.85478	2.15	0.99
0.24	0.26570	0.64	0.63459	1.04	0.85865	2.20	0.998
0.25	0.27633	0.65	0.64203	1.05	0.86244	2.25	0.998
0.26	0.28690	0.66	0.64938	1.06	0.86614	2.30	0.998
0.27	0.29742	0.67	0.65663	1.07	0.86977	2.35	0.999
0.28	0.30788	0.68	0.66378	1.08	0.87333	2.40	0.999
0.29	0.31828	0.69	0.67084	1.09	0.87680	2.45	0.999
0.30	0.32863	0.70	0.67780	1.10	0.88021	2.50	0.999
0.31	0.33891	0.71	0.68467	1.11	0.88353	2.55	0.999
0.32	0.34913	0.72	0.69143	1.12	0.88679	2.60	0.999
0.33	0.35928	0.73	0.69810	1.13	0.88997	2.65	0.999
0.34	0.36936	0.74	0.70468	1.14	0.89308	2.70	0.999
0.35	0.37938	0.75	0.71116	1.15	0.89612	2.75	0.999
0.36	0.38933	0.76	0.71754	1.16	0.89910	2.80	0.999
0.37	0.39921	0.77	0.72382	1.17	0.90200	2.85	0.999
0.38	0.40901	0.78	0.73001	1.18	0.90484	2.90	0.999
0.39	0.41874	0.79	0.73610	1.19	0.90761	2.95	0.999

As an illustration, Example 3.12 of Chapter 3 is repeated here and solved using the error function.

Example C.1: A radar receiver observes a voltage V which is equal to a signal (due to reflection from a target) plus noise:

$$V = s + N$$

Assume that the signal s is a constant and the noise N is a Gaussian random variable with mean $\mu = 0$ and variance $\sigma^2 = 4$ (mv^2). Assume further that the radar will detect the target if the voltage V is greater than 2 volts.

When the target is a certain distance from the radar, the signal s is equal to 3 volts. What is the probability that the radar detects the target?

The voltage V in this example is a Gaussian random variable with mean $\mu = s = 3$ and variance $\sigma^2 = 4$. The probability that the target is detected is given by

$$
\begin{aligned}
\Pr[V > 2] &= \frac{1}{2}\left[1 - \operatorname{erf}\left(\frac{2-3}{\sqrt{2}\cdot 2}\right)\right] \\
&= \frac{1}{2}\left[1 + \operatorname{erf}\left(\frac{1}{2\sqrt{2}}\right)\right] = \frac{1}{2}(1 + 0.3829) = 0.6915
\end{aligned}
$$

(here MATLAB has been used to compute the value of the error function).

□

D Noise Sources

Typically noise is an unwanted random process (or interfering signal). Noise that interferes with the operation of electronic systems originates from a variety of sources; most sources produce noise in a limited spectral range. As a result, the noise of interest depends on the frequency range of operation of the electronic system. Analog as well as digital systems are hampered by noise in different ways. In the following, we first detail the sources of noise for analog systems and follow that by a discussion of quantization noise, a source of noise in digital systems.

Whether or not a given noise waveform is correlated depends on the nature of its power spectral density. If the power spectral density is relatively flat over the frequency band of interest, then for all practical purposes, we call the noise uncorrelated. Recall that we have *white* noise if in addition to being uncorrelated it also has 0 mean. There are essentially two types of uncorrelated noise: external and internal. Noise generated outside of the system under study is termed external noise. We have no control over the generation or the power spectral density of external noise. We typically attempt to shield the system from external noise. Atmospheric noise, extraterrestrial noise, and man-made noise are some of the externally produced noise. Depending on the frequency band of operation of the system, each of these can become an issue in the system design and performance analysis.

Noise generated by devices and circuits within the system under study is termed internal noise. Internal noise cannot be completely eliminated, but through proven design techniques and by choosing appropriate devices its effects can be somewhat mitigated. Shot noise, excess noise, transit-time noise, and thermal noise are predominant internal noise. All four types of noise are a result of the mobility of carriers of electric current in an electronic device. An electronic device, such as a transistor, depends for its operation on the movement of subatomic carriers known as electrons and holes. Excess noise and transit-time noise are little understood while shot noise and thermal noise have been studied extensively. These internal noise sources are a hindrance in analog systems.

In digital systems, quantization noise is the predominent source of system performance degradation. Quantization noise occurs due to rounding errors during the analog to digital conversion process. In this appendix, we examine the thermal noise and quantization noise as representative sources of noise in analog and digital systems, respectively.

D.1 Thermal noise

Thermal noise is due to random motion of electrons in a conductor caused by thermal agitation. Thermal noise is also referred to as Brownian noise, Johnson noise, or white Gaussian noise.

Electric current in a conductor is a result of electron movement. The flow of electrons is, however, not deterministic but random. As the electrons move through the

conductor, they collide with molecules, thus generating short pulses of electric distur-
bances. The root mean square (rms) velocity of electrons is known to be proportional
to the absolute temperature; further, the mean square voltage *density* that appears
across the terminals of a resistor due to thermal noise is given by

$$\mathcal{E}\{V_T^2\} = 2\kappa TR \text{ volts}^2/\text{hz} \qquad (D.1)$$

where T is temperature in degrees Kelvin, R is the resistance in ohms, and κ is is
Boltzmann's constant, 1.38×10^{-23} joules per degrees Kelvin [1].

Example D.1: Consider an electronic device operating at a temperature of 300° K over a
bandwidth of 5 MHz with an internal resistance of 1000 Ω. We can determine the rms
noise voltage across the terminals of this device as follows.

From (D.1), the mean square voltage density across a 1000-Ω resisitor is given by

$$\mathcal{E}\{V_T^2\} = 2\kappa TR = 2 \times 1.38^{-23} \times 300 \times 1000 = 8.28 \times 10^{-18} \text{ volts}^2/\text{Hz}$$

The rms voltage across the terminals of this resistor is then given by

$$\sqrt{\mathcal{E}\{V_T^2\}(2B)} = \sqrt{8.28 \times 10^{-18} \times 2 \times 5 \times 10^6} = 9.0995 \times 10^{-6} \text{ volts}$$

where B is the bandwidth in hertz.

Commonly, the definition of bandwidth indicates only the positive side of the spectral
range. Thus, the factor of $2B$ above accounts for the positive and the negative side of
the spectrum.

□

To see how the above formula relates to the noise power spectral density $S_W(f)$
studied in this book consider the circuit diagram of Fig. D.1 in which the thermal noise

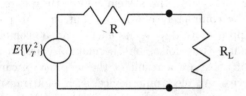

Figure   D.1 Thermal   noise
source with a mean square volt-
age spectral density $\mathcal{E}\{V_T^2\}$
and an internal resistance R.

source is the resistance R. The mean squared voltage spectral density so generated is
transferred to a load resistor R_L. The power spectral density of the source is given by

$$S_S(f) = \frac{\mathcal{E}\{V_T^2\}}{R + R_L} = \frac{2\kappa TR}{R + R_L}$$

It is well known that the maximum power spectral density is delivered to a load
resistor when the load resistance is matched to the source resistance, i.e., $R = R_L$.
As a consequence, the source resistor R and the load resistor R_L share the power
spectral density equally. Under these conditions, the power spectral density delivered
to the load resistor (also known as available power spectral density) is one half of that
generated by the source

$$S_W(f) = \frac{1}{2}S_S(f) = \frac{1}{2}\frac{2\kappa TR}{2R} = \frac{\kappa T}{2} = \frac{N_o}{2}\text{watts/Hz} \qquad (D.2)$$

Since it is customary to represent the power spectral density of white noise in terms
of N_o, we have $N_o = \kappa T$.

Example D.2: Let us determine the available power spectral density and the available average noise power generated for the electronic device in Example D.1.

The power spectral density of thermal noise generated in the device, from (D.2) is

$$S_W(f) = \frac{N_o}{2} = \frac{\kappa T}{2} = \frac{1.38 \times 10^{-23} \times 300}{2} = 2.07 \times 10^{-21} \text{ watts/Hz}$$

From this, we have $N_o = 4.14 \times 10^{-21}$.

The available average noise power is

$$P_N = \int_{-B}^{B} S_W(f)df = \frac{N_o}{2}(2B) = N_o B$$
$$= 2.07 \times 10^{-21} \times 10^7 = 2.07 \times 10^{-14} \text{ watts}$$

Notice that the available average power is not dependent on the value of resistance; however, the temperature and the bandwidth are relevant parameters.

□

Now suppose that a perfectly clean sinusoid is sent into an electronic device with an internal thermal noise source as depicted in Fig. D.2. At the output of the device, the

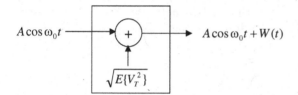

$A \cos \omega_o t$ → + → $A \cos \omega_o t + W(t)$

$\sqrt{E\{V_T^2\}}$

Figure D.2 Depiction of noise added by a thermal noise source in an electronic device.

signal now has additive thermal noise. Let us define the *signal-to-noise ratio* (SNR) in deciBels as the log ratio of signal power to average noise power:

$$\text{SNR}_{\text{dB}} = 10 \log_{10} \frac{P_S}{P_N} \tag{D.3}$$

where P_S is the signal power of the sinusoid

$$P_S = \frac{A^2}{2}$$

and P_N is the average noise power (above). Substituting these into (D.3) yields

$$\text{SNR}_{\text{dB}} = 10 \log_{10} \frac{A^2}{2} \frac{1}{N_o B} = 10 \log_{10} \frac{A^2}{2N_o B}$$

The input SNR is infinite while the output SNR is finite, indicating that there is signal quality degradation.

Example D.3: Let us continue Example D.2 by assuming that the input signal to the device is $3 \cos \omega_o t$. the SNR at the output of the device from (D.3) is

$$\text{SNR}_{\text{dB}} = 10 \log_{10} \frac{A^2}{N_o \cdot 2B} = 10 \log_{10} \frac{9}{4.14 \times 10^{-21} \times 10^7} = 17.4 \text{dB}$$

This signal quality degradation is significant but tolerable. Since the signal in this case is a sinusoid, the SNR can be improved by narrowing the bandwidth since noise power increases linearly with the bandwidth.

□

D.2 *Quantization noise*

Most signals of interest originate in analog form and are converted to digital form due to many advantages of digital processing, storage, and transmission. Digitization requires rounding or truncation of the analog signal samples. Rounding (truncation) causes loss of precision, which is equivalent to adding noise to the signal. This approximation error is termed *quantization noise* or digitization noise.

Figure D.3 is a block diagram of the the digitization process showing the quantizer block that introduces quantization noise.

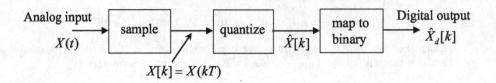

Figure D.3 Digitization of an analog signal.

Figure D.4 shows the transfer characteristic of a typical quantizer. The continuous magnitude input to the quantizer at any time k is denoted by $X[k]$ while the corresponding output value is $\hat{X}[k]$. Given an input sample $X[k]$, we can represent the

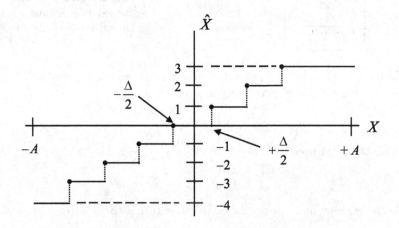

Figure D.4 Quantizer with an input range of $-A$ to $+A$.

quantized output sample $\hat{X}[k]$ as

$$\hat{X}[k] = X[k] + W_q[k]$$

where $W_q[k]$ is a random variable representing the quantization (rounding) error. For the input range $-\frac{\Delta}{2} \leq X < \frac{\Delta}{2}$, the corresponding output value is 0, where Δ is called the *step* size. If the input dynamic range is $2A$, i.e., it extends from $-A$ to A, then the step size is given by

$$\Delta = \frac{2A}{N}$$

where N is the number of steps (or levels) in the quantizer which is invariably a power of 2. In particular, let us assume

$$N = 2^b$$

where b is the number of bits used by the encoder.

The quantization errors $W_q[k]$ are assumed to be IID and uniformly distributed

$$f_{W_q}(w_q) = \begin{cases} \dfrac{1}{\Delta} & \text{for } -\frac{\Delta}{2} \leq X_c < \frac{\Delta}{2} \\ 0 & \text{otherwise} \end{cases}$$

The mean of $W_q[k]$ is assumed to be zero while the variance is given by (see Section 4.2.1 of Chapter 4).

$$\sigma^2_{W_q} = \frac{\Delta^2}{12}$$

Substituting $\Delta = 2A/2^b$ into the above equation leads to

$$\sigma^2_{W_q} = \frac{A^2}{3 \times 2^{2b}} \tag{D.4}$$

Since the quantization noise is assumed to be IID, the process is white and the power spectral density is

$$S_{W_q}(e^{j\omega}) = \frac{A^2}{3 \times 2^{2b}}$$

with the corresponding autocorrelation function

$$R_{W_q}[l] = \frac{A^2}{3 \times 2^{2b}} \delta[l]$$

The quantization noise power is given by

$$P_Q = R_{W_q}[0] = \frac{A^2}{3 \times 2^{2b}}$$

Let us assume that the signal being digitized is a pure sinusoid (i.e., no additive noise) with amplitude A and power $P_S = A^2/2$. After digitizing, the signal is now equivalent to a sinusoid in additive white noise. The corresponding signal to quantization noise ratio (SQNR) is given by

$$\begin{aligned} \text{SQNR}_{\text{dB}} &= 10\log_{10}\frac{P_S}{P_Q} = 10\log_{10}\left(\frac{A^2}{2} \cdot \frac{3 \times 2^{2b}}{A^2}\right) \\ &= 10\log_{10} 1.5 + 2\,b\log_{10} 2 = 1.76 + 6.02\,b \tag{D.5} \end{aligned}$$

This is commonly approximated by

$$\text{SQNR}_{\text{dB}} \approx 6b \text{ dB}$$

and used as the rule of thumb: "*6 dB per bit.*"

Example D.4: A given analog to digital converter has a resolution of 8 bits. The upper bound on the signal to quantization noise ratio in dB is then given by $\text{SQNR}_{\text{dB}} = 48$ dB, i.e., in the absence of any other noise, the SNR degrades from infinity to 48 dB due to digitization.

Suppose that we require about 72 dB of SQNR in a certain application. Then we need to use at least a 12-bit analog to digital converter.

□

References

[1] Wilbur B. Davenport, Jr. and William L. Root. *An Introduction to the Theory of Random Signals and Noise*. McGraw-Hill, New York, 1958.

Index

Printed in the United States
by Baker & Taylor Publisher Services